T0371983

BIOFILMS IN INFECTION PREVENTION AND CONTROL

BIOFILMS IN INFECTION PREVENTION AND CONTROL

A Healthcare Handbook

Edited by

STEVEN L. PERCIVAL
DAVID W. WILLIAMS
JACQUELINE RANDLE
TRACEY COOPER

Amsterdam • Boston • Heidelberg • London • New York • Oxford
Paris • San Diego • San Francisco • Singapore • Sydney • Tokyo
Academic Press is an imprint of Elsevier

Academic Press is an imprint of Elsevier
525 B Street, Suite 1800, San Diego, CA 92101-4495, USA
32 Jamestown Road, London NW1 7BY, UK
225 Wyman Street, Waltham, MA 02451, USA

British Library Cataloguing-in-Publication Data
A catalogue record for this book is available from the British Library

Library of Congress Cataloging-in-Publication Data
A catalog record for this book is available from the Library of Congress

ISBN: 978-0-12-397043-5

For information on all Academic Press publications
visit our website at http://elsevierdirect.com

Printed and bound in the United States
14 15 16 17 18 10 9 8 7 6 5 4 3 2 1

Steven L. Percival would like to dedicate this book
to Carol, Alex, Tom, Mum and Dad. Thank you!

David W. Williams would like to dedicate this book
to Lorna, Daniel, Ailish Calum, Sioned and Anne.
In memory of Eirwyn.

Thank you again for all your support!

CONTENTS

PREFACE

A biofilm is considered to be a complex community of micro-organisms attached to each other, or associated with a surface or interface, and encased in extracellular polymeric substance (EPS). The composition of the EPS is complex and may contain polysaccharides, proteins, nucleic acid, lipids and metals. The EPS provides the 'house' of the biofilm, giving the residing micro-organisms a safe haven from the effects of host immunity or administered antimicrobials.

Micro-organisms within the biofilm can be responsible for causing and prolonging infection and/or disease. Biofilm infections are, in general, healthcare-related, including those associated with use of medical devices such as urinary and central venous catheters, endotracheal (ET) tubes and orthopaedic prostheses. Other biofilm-related infections are prostatitis and those of chronic wounds. Many of the infections are of growing significance, because they are related to the ever-increasing ageing population. It is important to note that the Centers for Disease Control and Prevention (CDC) reports that more than 65% of all healthcare-associated infections (HCAI) originate from biofilms.

Of great concern is that the micro-organisms within the biofilm are significantly more tolerant to antimicrobials compared to their planktonic counterparts. This antimicrobial tolerance by biofilms can be 1000-fold higher than the susceptibility of free-floating or planktonic micro-organisms. Consequently, biofilms pose a significant challenge to patients in both hospital and community healthcare settings. Addressing the prevention and control of biofilms will dramatically help in decreasing infection rates, patient morbidity and mortality. This in turn will reduce the escalating costs of biofilm-related infections faced by the healthcare profession.

Biofilms in Infection Prevention and Control: A Healthcare Handbook is the first of its kind that deals specifically with the fundamentals of infection control and biofilms in healthcare. The book is divided into two parts and 18 chapters; it begins by describing the rudiments of infection control in Chapter 1. This chapter introduces the reader to infection control and prevention and discusses the basic principles of safe practice. Chapters 2 through 5 then address the challenges facing healthcare providers—infection prevention, hand hygiene, decontamination, and the significance of changing practices in healthcare. The rest of the chapters in Part 1 introduce readers to infections associated with invasive devices and wounds.

Part 2 of the book focusses explicitly on the area of biofilms and the problems they pose to healthcare professionals. Chapter 9 introduces the reader to the area of 'biofilmology'—that is, 'the study of biofilms' and the fundamentals of biofilm development. Subsequent chapters offer extensive reviews on biofilms and HCAI, biofilms and ventilator-associated pneumonia (VAP), antimicrobials, microbial resistance and superbugs and biofilms and their association with urethral and central venous catheters. Chapter 17 discusses the importance of the supply of water to hospitals and healthcare facilities and the role that systems for it play in the spread and dissemination of clinically significant micro-organisms to patients. The final chapter covers biofilms and their role in wound infections.

Biofilms in Infection Prevention and Control provides biologists, medical personnel, healthcare workers, infection control professionals, microbiologists, as well as students and academics, with a practical text to support clinical practice. It will help healthcare workers understand the evidence-base and rationale for infection prevention in an easy-to-follow format, including exhibits that contain lists of common skills and procedures. This handbook is multi-formatted, with some chapters providing guidelines for healthcare practice to combat biofilms and others presenting information on specific biofilm-related infections.

Overall, this contemporary handbook will provide its readers with a comprehensive, concise and informative text that highlights the significance of biofilms in infection control and the urgent need to prevent their formation. This is an area that is frequently overlooked and neglected in modern medical and healthcare education.

Professor Steven L. Percival
Professor David W. Williams

LIST OF CONTRIBUTORS

Debra Archer
Veterinary Clinical Science, Equine Hospital, Neston

Mitch Clarke
University Hospitals Trust, Nottingham

Christine A. Cochrane
Research Fellow, Institute of Ageing and Chronic Disease, Faculty of Health and Life Sciences, University of Liverpool, Liverpool

Tracey Cooper
Consultant Nurse, University Hospitals NHS Trust, Infection Control, Southampton

Simon Finnegan
Department of Chemistry, University of Sheffield, Sheffield

Robin Howe
Consultant Microbiologist, Public Health Wales Microbiology, University Hospital of Wales, Cardiff

Gavin J. Humphreys
Lecturer, School of Pharmacy and Pharmaceutical Sciences, University of Manchester, Manchester

Eleri M. Jones
Institute of Ageing and Chronic Disease, University of Liverpool, Liverpool

Lim Jones
Clinical Lecturer, Cardiff Institute of Infection and Immunity, Cardiff

Rachael P.C. Jordan
School of Dentistry, College of Biomedical and Life Sciences, Cardiff University, Cardiff

Tadahiro Karasawa
Director of Clinical Pathology and Head of Infectious Disease Group, Department of Internal Medicine, Fujimi Kogen Medical Centre, Nagano

Claire Kilpatrick
Nurse Consultant, Infection Control, Health Protection, HAI and IC Group 1, Glasgow

Tomoari Kuriyama
Head, Kuriyama Dental Practice, Honorary Research Fellow, Graduate School of Medical Science, Kanazawa University, Toyama

Andrew McBain
Senior Lecturer, School of Pharmacy and Pharmaceutical Sciences, University of Manchester, Manchester

Sara McCarty
Institute of Ageing and Chronic Disease, Faculty of Health and Life Sciences, University of Liverpool, Liverpool

Ginny Moore
Biosafety Unit, Research Department, Health Protection Agency, Salisbury

Lindsay Nicolle
Health Sciences Centre, Winnipeg, MB

Steven L Percival
Vice President R&D, Scapa Healthcare, Manchester and Professor of Microbiology and Anti-infectives, Surface Science Research Centre and Institute of Ageing and Chronic Disease, University of Liverpool, Liverpool

Jacqui Prieto
Faculty of Health Sciences, University of Southampton, Southampton

Jacqueline Randle
Senior Lecturer, School of Health Sciences, University of Nottingham, Nottingham

Greoff Sjogren
Decontamination Manager. Royal West Sussex NHS Trust, St. Richards Hospital, Chichester

Louise Suleman
Institute of Ageing and Chronic Disease, Faculty of Health and Life Sciences, University of Liverpool, Liverpool

Lauren Tew
Infection Control, NHS Bath and North East Somerset Trust, St. Martin's Hospital, Bath

Natalie Vaughan
Lead in IPC, Nottingham University Hospitals Trust, Nottingham

Jimmy Walker
Principal Investigator, Decontamination, Biosafety Unit, Research Department, Health Protection Agency, Salisbury

David Williams
Professor of Oral Microbiology, Tissue Engineering and Reparative Dentistry, School of Dentistry, Cardiff University, Cardiff

Emma Woods
Innovation Manager, Scapa Healthcare, Research and Development, Manchester

LIST OF FIGURES

LIST OF TABLES

LIST OF EXHIBITS

PART 1

Fundamentals of Infection Control

CHAPTER ONE

Introduction to Infection and Infection Prevention

Tracey Cooper and Steven L. Percival

INTRODUCTION

Globally millions of people receive some form of healthcare every year in every country of the world. This ranges from basic non-invasive care to highly complex and technically advanced interventions and treatments. One unintended consequence of healthcare interventions can be the development of healthcare-associated infections (HCAI).

For many years the term *hospital-acquired infection* (HAI) was used, denoting that an infection was acquired during a hospital admission. In recent years it has been recognised that relatively complex care is increasingly being delivered in the community, including the patient's own home, as well as in acute care hospital settings. The definition of HCAI takes this fully into account and includes all infections that develop as a result of healthcare, no matter where the care is delivered.

Rates of infection are measured in many countries using a variety of methods. This measurement is described as infection surveillance. Infection rates are usually reported as either prevalence rates or incidence rates. Prevalence rates are identified from surveys which collect information on the number of infections present at a given point in time and include those newly identified as well as those that are resolving and being treated. Incidence rates are identified from surveys that collect information over a period of time and include all new cases of infection as they occur. Prevalence rates are therefore higher than incidence rates.

Understanding of the importance of infection prevention and HCAI has developed over the past few decades, and there has been a corresponding increase in the amount of political and media attention given to HCAI in the United Kingdom and globally. It is clear that not all HCAI are avoidable, but much work has been done clearly demonstrating that most countries can reduce avoidable HCAI significantly if clinical practice is improved.

In 2006 the World Health Organisation (WHO) launched a Global Patient Safety Challenge in which it identified five major challenges for all countries. Included in this

Biofilms in Infection Prevention and Control
Doi: http://dx.doi.org/10.1016/B978-0-12-397043-5.00001-3

was the reduction of avoidable infection, including HCAI, by the promotion of hand hygiene. In the United Kingdom focussed action has been led by the government, the Scottish Executive and the Welsh Assembly. In England and Wales, the National Patient Safety Agency launched the 'cleanyourhands' campaign in 2004, and in Scotland the 'Germs. Wash Your Hands of Them' campaign was launched in 2007. These campaigns demonstrate that the prevention of HCAI is recognised as a critical factor for safe healthcare. The media across the United Kingdom and the United States continue to publish stories of patient infection and avoidable morbidity and mortality, and the public increasingly expects that all healthcare workers will act to reduce infection risk.

To reduce the risk of infection and to minimise adverse effects from infection when it does occur, healthcare workers need to:

- Implement good practice.
- Utilise evidence-based practice.
- Detect infection promptly through diagnostic tests and patient observation.
- Initiate effective treatment, including correct prescribing of systemic and topical antimicrobial agents.
- Document actions and interventions correctly and clearly.
- Communicate effectively with colleagues, patients, relatives and caregivers.
- Review infections that occur and learn from them so that we can improve care for other patients.

THE IMPACT OF INFECTION

When we consider the international perspective in more detail, it is evident that infection risks change radically across the globe. In many developing countries communicable disease and blood-borne viruses, such as HIV, spread readily. This is the result of financial and resource constraints that cause a lack of vaccination programmes and an inability to implement basic control measures during healthcare interventions. In countries with the resources to implement effective control measures, vaccination programmes and more complex healthcare procedures, it is rare that these infections spread. Instead, other micro-organisms present risks to those who receive healthcare and necessitate the need to implement effective control measures.

Healthcare-associated infections not only contribute to an increase in morbidity and mortality rates in hospitalised patients, but they also are associated with a substantial increase in healthcare costs.[1] Research has been conducted to identify the burden and economic cost of HCAI. Studies have estimated that in 1996, in England alone, the cost of HCAI was approximately £1 billion annually.[2,3]

In England and Wales there have been four national prevalence surveys of infections in hospitals. The most recent, in 2011, identified an infection prevalence rate in England of 6.4%.[4] The most common infection types were respiratory tract, urinary tract, and surgical site. Prior to this latest national point prevalence survey, the National Audit Office estimated that 5000 patients died every year in England because of HCAI[3]; specifically, meticillin-resistant *Staphylococcus aureus* (MRSA) was responsible for many cases.[5–7] The Scottish National HAI Prevalence Survey[8] found similar trends.

The World Health Organisation conducted a worldwide study and concluded that more than 1.4 million people are affected by HCAI. This study was carried out in 55 hospitals across 14 countries and indicated that the rates of HCAI have increased. Globally it is estimated that 5 to 10% of hospitalised patients suffer from HCAI[9] and that HCAI is responsible for one of the five top causes of death in the United States.[10]

Although the England and Wales survey[11] and the 2011 point prevalence survey[4] provide a useful picture of infections in hospitals, they take no account of HCAI occurring either after discharge or as a result of healthcare interventions delivered outside the acute care setting. It is therefore important to remember that the real rate of infection is likely to be higher than the rate identified in these national surveys.

In addition, the Centers for Disease Control and Prevention (CDC) has proposed that 65% of all hospital-associated infections are a result of the presence of microorganisms growing in a biofilm. This is discussed in more detail in Part 2 of this book.

Key Points

- On any day in an 'average' ward of 30 patients, 3 patients are suffering from an healthcare-associated infection.
- These 3 patients will be suffering avoidable pain and anxiety; they require extra medication, extra interventions (e.g., wound dressings and extra care) from a range of healthcare workers as well as longer hospital stays.
- When these patients return home, they are likely to require extra care from community healthcare workers and experience a delay in returning to normal activities.
- Biofilms cause 65% of HCAI.

ICUS AND HEALTHCARE-ASSOCIATED INFECTIONS

Patients in intensive care units (ICUs) and high-dependency units, paediatric intensive care units (PICUs) and neonatal intensive care units (NICUs) are at greater risk of HCAI. The incidence of HCAI in ICUs has been reported to be 5 to 30%.[12,4] The reasons for this include the following facts about intensive care patients:

- They are critically ill and at the same time may be exposed to the most resistant and pathogenic microbes.[13–20]

- They are at an increased risk of infection because of their serious underlying diseases and the high number of invasive interventions they receive.[21]
- Typically they receive antibiotic therapy and so are at added risk of infection due to antibiotic-resistant micro-organisms.[22]
- They require high rates of direct contact with healthcare workers and so are at high risk of cross-infection via healthcare workers' hands.[23–25]
- They are cared for by healthcare workers who may carry antibiotic resistant micro-organisms on their hands.[26] Karabey and colleagues[27] found that approximately 30% of healthcare workers' hands were contaminated with MRSA.
- They are recipients of new technologies and invasive procedures which provide an access point for bacteria to spread into a patient's tissues as they bypass the body's normal defence mechanisms.[28]
- They are cared for in units that have the highest levels of environmental contamination with resistant micro-organisms such as MRSA.[26] Weist and coworkers[29] found a significant increase in cross-transmission rates of HCAI (approximately 37.5%) associated with ICU settings. Similarly Randle and Flemming[30] found that healthcare workers in PICUs can easily contaminate their hands with micro-organisms from inanimate objects.[30]
- They are cared for in units which are under great pressure in terms of bed occupancy, which can result in staff 'cutting corners' because of workload pressure.[30]

Worldwide, the most common HCAI in ICU settings are respiratory tract infections,[31–33,4] urinary tract infections,[34,35] gastrointestinal infections[36] and surgical site infections.[37]

Key Points

- Units caring for critically ill patients have a greater risk of healthcare-associated infection, with incidences ranging from 5 to 30%.
- Contributing factors include increased need for invasive interventions with these patients, greater need for direct contact with healthcare workers and increased susceptibility and risk of exposure to resistant organisms.
- The most common HCAIs are respiratory tract, urinary tract, gastrointestinal and surgical site.

THE LAW AND HEALTHCARE-ASSOCIATED INFECTION

In the European Union, the United States and the United Kingdom many existing laws and regulations incorporate aspects of infection prevention and control. Healthcare organisations and healthcare workers therefore have a range of legal duties, with compliance on a daily basis usually through policies and procedures to be followed. Compliance by all members of staff and temporary healthcare workers helps to ensure that the organisation fulfils its legal duties and protects patients from infection risk.

Professional Responsibility and Public Expectation

Public awareness of HCAI and subsequent media attention have grown over the past decade. The public increasingly and rightfully expect that they will receive safe, clean healthcare wherever it is delivered and that they will not be exposed to the risk of avoidable infection. Infection with meticillin-resistant *Staphylococcus aureus* and *Clostridium difficile* has grabbed the attention of the public, resulting in frequent media headlines criticising 'dirty hospitals'. National infection reduction targets in England were introduced in 2005 in response to public concern. Complaints about standards of cleanliness and poor hand hygiene have increased globally, as has litigation against healthcare providers and healthcare workers alleging avoidable infection.

The media have been a very powerful influence on people's perceptions, especially on patients who have never been admitted to hospital or who have limited knowledge of HCAI.[38,39] This has influenced where patients wish to be treated, and findings from the London Patient Choice Project of 2000 showed that two-thirds of patients reported they would opt to be treated in a hospital with lower MRSA and healthcare-associated infection levels than in one that had been reported to have a higher incidence rate.[40]

It also seems that patients feel they should be involved in helping reduce HCAI[41]; this marks a shift in the role patients play in healthcare. A pilot study carried out by Madeo et al.[42] found that patients have a high level of awareness of the risk of HCAI but have little knowledge about how infections spread or about their prevention.

All healthcare workers have a responsibility to protect the health and welfare of their patients, and for registered healthcare workers these duties are enshrined in codes of conduct for each of the professions. Duties under these codes do not refer specifically to infection prevention, but clearly contracting a healthcare-associated infection causes additional pain and suffering for the patient. If a healthcare-associated infection occurs because of a failure to follow policy and good practice, or as a result of a deliberate decision not to follow policy and good practice, then the healthcare worker will be in breach of her or his code of conduct. In a landmark ruling, failure to follow infection prevention policy by not washing hands was cited by the Nursing and Midwifery Council as one of the reasons for removal of a registered nurse from the professional register in 2007.

There are a multitude of national and international documents—Winning Ways,[43] A Matron's Charter,[44] Saving Lives,[45] Guidelines on Hand Hygiene in Healthcare[46] and the Health and Social Care Act 2008: Code of Practice on the Prevention and Control of Infections (hereafter Code of Practice),[47] among others—that provide

Table 1.1 Examples of UK Legislation

Legislation	**Relevance**
Health and Safety at Work Act (1974)	Requirement to take measures that are reasonably practicable to protect patients, healthcare workers and the public from infection risk.
Control of Substances Hazardous to Health Regulations (2002)	Pathogenic micro-organisms are substances hazardous to health. Some of the chemical agents used to control them may be hazardous to health as well.
Environmental Protection Act (1990) and Hazardous Waste Regulations (2006)	Duties to protect the environment from waste that may pose a risk of infection to the public; includes clinical wastes and liquid wastes discharged to sewers.
Food Safety Act (1990) and associated regulations	Duty to ensure that food and drink are safe to consume. Applies to food and beverages that are served to patients, which also includes ice and sip feeds.
Health and Social Care Act 2008: Code of Practice on the Prevention and Control of Infections and Related Guidance (2010)	Responsibility for all registered providers of healthcare and adult social care in England to comply with specific requirements on infection prevention. This forms part of the Care Quality Commission Regulatory Framework in England.

guidance on how healthcare workers can practice safely and reduce the patients' risk of infection (see Table 1.1). In 2009 the Council of the European Union adopted a recommendation on patient safety which includes a request that member states adopt and implement a strategy to prevent and control HCAI. The majority of countries in the Union have responded by reinforcing the responsibility of professionals to reduce HCAI through national strategy.[48]

It is clear that every person involved in the delivery of care can make a difference in the infection risk that patients are exposed to. Understanding the individual responsibility we each have and always protecting patients from infection are at the core of delivering good care.

Key Points

- Increased public awareness of HCAI has resulted in media criticism and increased patient complaints and has affected where patients choose to be treated.
- It is a healthcare worker's responsibility to ensure that standards of good practice are followed in order to reduce the risk of healthcare-associated infections and to deliver a high standard of patient care in all settings.

WORKING WITH PATIENTS AND THEIR FAMILIES AND/OR CAREGIVERS

Not all patients are capable of taking an active part in their care, but most understand information if it is given in a clear and simple manner. Information that should be provided to patients and their families or caregivers when appropriate includes the following:

- The potential infection risks associated with a procedure. This may be part of a formal process, such as seeking written consent for surgery, or part of gaining informal consent, such as prior to insertion of a peripheral cannula.
- Measures that healthcare workers should take routinely to prevent infection so that patients know the standard of care they can expect. This can include hand hygiene.
- Measures that patients themselves can take to reduce their risk of infection. This can include washing their hands after using the toilet and not picking at wounds and dressings.
- Information about local rates of infection should also be provided if available.

Surveillance

Surveillance is now commonly viewed as an important quality indicator in the clinical audit process[49] and is frequently an integral part of risk management and clinical governance systems. There are some similarities in the definitions and aims of surveillance and audit, and the two terms are sometimes used interchangeably, even though they do represent different methodologies and purposes.[50]

Hadden and O'Brien[51] identify the following as the main purposes of the surveillance process:

- Early warning of changes in incidence of infection.
- Detection of outbreaks.
- Evaluation of the efficacy of infection prevention measures.
- Identification of groups at risk of infection.
- Prioritisation of resource allocation.

Results from surveillance studies indicate that active surveillance can reduce infection rates.

The seminal study on the efficacy of nosocomial infection control (SENIC) showed a reduction of approximately 30% in HCAI when surveillance was combined with active feedback of results to clinical teams.[52,53] Ayliffe and colleagues[54] concluded that in order to keep infection at a minimum level and to prevent and control outbreaks, some surveillance is necessary for identifying problems rapidly.[55]

Numerous other studies have indicated the potential success of surveillance in reducing HCAI. Guebbels et al.[56] report five 'success stories' in reducing surgical site infections (SSIs) in The Netherlands. Kelleghan and colleagues[57] report a surveillance programme in the United States that achieved a 57% reduction in the incidence of ventilator-associated pneumonia (VAP).

The surveillance programme that was done by Sykes and colleagues[58] reported having achieved a fall in the prevalence of surgical site infection by 38% over a 12-year period. The programme included feedback and follow-up to healthcare workers. Feedback was then relaxed for a period of 15 months, creating a multiple crossover study of its effect. Rates of SSIs rose during this period, but when the surveillance programme was restarted they showed a second reduction.

Despite the apparent correlation between the surveillance process and a reduction in HCAI, specific criticisms of surveillance exist; they include

- *Inconsistency in benchmarking, case definitions and datasets.* This makes comparisons across geographical sites and between countries difficult or even meaningless.[59,60]
- *Financial cost.* Edmond et al.[59] estimate that in one US state surveillance would cost an additional £5.5 million. However, this does not take into account the cost reductions achieved as a result of lower infection rates.

These criticisms can be countered by participation in national surveillance programmes, such as the schemes operated in England, Wales and Scotland which have standard definitions of infection and allow comparison between different organisations and benchmarking.

Criticisms of the cost can be easily countered when the cost-effectiveness of surveillance is taken into account. The investment in staff time and systems for surveillance can be quickly offset by the cost reductions achieved with a very small drop in the rates of infection.

Key Point

- Since 2003 it has become mandatory in England to report cases of *Clostridium difficile* in patients who are 65 years of age and older. This partly explains the rapid rise in cases since 2003; before this time reporting was patchy because a robust surveillance system was not in place. In recent years mandatory reporting has been extended to all patients aged two years and older.

Communicating with Colleagues to Reduce the Risk of Infection

Communication between healthcare workers and the documentation of care is essential if infection is to be prevented. For example, some invasive devices can lead to a significant risk of infection if left *in situ* for too long. Communication and documentation ensure that staff know when the device must be removed.

Effective communication is also vital to identify signs of potential infection and ensure action is taken. Early detection of possible infection can mean earlier treatment and hopefully less overall impact on the patient. Pittet and Donaldson[9] emphasise that just recognising and explaining the cause of trends in infections is not sufficient in hospitals, and that effective multidisciplinary communication regarding infection issues is essential. This has also been highlighted in the National Audit Office executive summary,[3] which recommends that healthcare workers accept greater ownership of HCAI control. Central to this is feedback to healthcare workers, which can help reinforce good practice and identify areas where improvements can be made.[61]

TRAINING, EDUCATION AND ROLES

Infection prevention and control directives in many countries stress the need for education in order to emphasise that a comprehensive approach is required to combat HCAI. As an example, in 2003 the Department of Health in England published the *Winning Ways* report, which called for healthcare workers to work together to reduce HCAI.[43] Within this was the recommendation that infection prevention and control be part of induction programmes for all healthcare workers and for it to form part of healthcare workers' personal development plans.

Additionally, the Saving Lives programme[45] was aimed at reducing HCAI, including MRSA. It involved a requirement for hospitals to undertake a self-assessment to ensure that all employees had a programme of education and training in order to understand their responsibility for infection prevention and control and of the actions they need to take. The National Health Service (NHS) in England set up an online training programme which healthcare workers could complete at their own pace at any time.

The recommendations on education probably arise from studies indicating that a lack of education and training increases non-compliance with the fundamentals of infection prevention and control practice, thus increasing the incidence of HCAI.[62–64] The House of Lords Select Committee report, *Fighting Infection*,[65] found that training in infection prevention and control for all healthcare workers was inadequate. For instance, in 2003 few infection control teams (ICTs) maintained training records. Those that did reported that 10 to 20% of staff did not receive induction training in most non-medical groups and that the proportion of staff receiving annual update training was on average 60%.[3]

In 2000, the National Audit Office (NAO)[66] recommended that NHS trusts review their policies on education and training in infection prevention and control procedures to ensure that that all healthcare workers are targeted by induction training and that key healthcare workers who have day-to-day contact with patients are kept up to date on good infection prevention and control practice. By the time the NAO published its follow-up study in 2009, however, it reported that there was sufficient priority for infection prevention training in induction and for ongoing training for staff.[67]

It is essential that all staff participate in the training provided by their employer in infection prevention and control, not to 'tick a box' but to ensure that each individual has the knowledge required to deliver safe care to patients.

Infection Prevention and Control Nurses' Role

In the United Kingdom the first infection control nurse was appointed in 1959 in Torquay, Devon. This was a temporary appointment in response to high levels of infection by meticillin-sensitive *Staphylococcus aureus*. The following year the first full-time infection control nurse was appointed in Exeter, Devon, UK, because the Torquay appointment was very successful in reducing infection rates. The nurse's job was to visit hospitals' wards and laboratories and advise on the management of infection cases to prevent spreading, as well as to take swabs and specimens for analysis.[68]

Since then, the role of infection prevention and control nurses in the United Kingdom has developed and evolved significantly, with involvement in a diverse range of issues, including:

- Management of individual patients.
- Outbreak management.
- Policy and clinical practice based on international evidence.
- Audit and monitoring to ensure that practice reflects policy.
- Education and training programmes for all staff.
- Monitoring of cleanliness and the environment.
- Advice on planning and refurbishment of buildings, including new builds.
- Advice on management and maintenance of buildings.
- *Legionella* and *Pseudomonas* control measures in water systems.
- Waste and linen management.
- Advice on purchase of new equipment.
- Decontamination standards.

In addition to these issues, infection prevention and control teams have to apply their knowledge over a wide range of clinical practice settings in which they may not have experience working in. For instance, within a Specialist Teaching Trust the team may have to advise on areas as diverse as the care of older people and long-term respite care, cardiac surgery and cardiac theatre practice, neonatal and paediatric intensive care and specialist nutritional interventions.

Within community teams the range of advice can extend from acute and community hospital wards and theatres to small group homes for those with mental illness or learning impairment, home births, community nursing practice and GP surgeries and prison healthcare. This requires that infection prevention and control teams apply the evidence base by blending key principles, situational factors and common sense to ensure that infection prevention measures are appropriate and suitable for the area of practice.

Since the 1990s in the United Kingdom, infection prevention and control teams have become increasingly involved in organisational assurance and governance in order to ensure that organisations meet national requirements. In England in the 1990s, this originally included the Controls Assurance Standards, which were then replaced by the document *Standards for Better Health*, as well as NHS Litigation Authority standards and most recently the *Code of Practice*.[47]

The knowledge base and leadership skills required are substantial and are built over a number of years through a mixture of experiential learning, usually supported by a senior, more experienced colleague, plus specialist study. Increasingly infection prevention and control nurses are qualified to a master's degree level in response to the need for highly developed skills in critical appraisal of research and other evidence, as well as highly developed leadership and influencing skills. Infection control doctors are also increasingly undertaking additional study in specific areas (e.g., environmental ventilation systems) through master's degree programmes or through postgraduate diplomas.

In addition to formal study, the Infection Prevention Society of the United Kingdom and Ireland has produced a document containing core competencies for infection prevention nurses. This document has been endorsed by health departments in England, Wales, Scotland and Northern Ireland to help members recognise and document their specialist skills and to identify development needs. It can be used by many healthcare professionals working in infection prevention and control, including full-time infection prevention and control nurses and link staff.

Infection Prevention and Control Link Professional's Role

Infection Prevention and Control Link Professional programmes have been developed in many hospitals in an attempt to provide effective infection prevention and control role models that are based in clinical practice.[69,70] These programmes have also developed over time, with many link groups now including a wide range of staff rather than just nurses.

Link staff are based in a clinical area or field of practice and, in addition to their clinical role, take on responsibility for specific infection prevention and control activities in that area. Educational preparation, programme content and delivery and role responsibilities vary according to local need.[71]

Activities usually include acting as a role model for good practice and challenging poor practice. In many organisations the role variously includes acting as a resource for information and knowledge; maintaining a resource folder of information for staff; delivering informal or formal education at a clinical level; and auditing or monitoring practice, linking to the infection prevention and control team. Link staff usually meet with the infection prevention and control team to discuss key issues and problems and to receive specific education.

A number of studies have been performed that demonstrate the value of link staff in focusing on practice at the local level. Teare and Peacock[70] emphasise the value of increasing the profile of infection prevention, and Ching and Seto[72] describe measureable improvement in standards for urinary catheter care following intervention by link nurses in Hong Kong. Cooper[73] describes measurable improvement in facilities for hand hygiene following targeted work by link staff in a district general hospital in England. The Healthcare Commission reviewed arrangements for link staff and levels of MRSA *bacteraemia* and *Clostridium difficile* infection in NHS trusts as part of a large survey performed in 2006. It concluded that link staff in all clinical areas, with dedicated time for their responsibilities, results in lower rates of infection.[74]

Infection prevention and control is increasingly being viewed as everyone's responsibility, and the link professional can reinforce this and work locally to support the infection prevention and control team, especially when supported by managers and an effective educational programme that meets the team's needs.[75]

CONCLUSION

This chapter provides an introduction to some important issues regarding healthcare workers, patients and HCAI. Although rates of HCAI and figures for financial costs and morbidity and mortality rates can be provided, it is difficult to fully describe the severe negative impact a healthcare-associated infection can have on an individual patient and his or her family. Clearly it is recognised that some HCAI can be avoided, and the actions of healthcare workers are critical in making this happen.

The remainder of this book focusses on the key principles of infection prevention and the specific issues associated with biofilms. It guides the reader in common practical procedures; the source of infections; areas and devices where micro-organisms reside, specifically focussing on biofilms; and the measures required to prevent infection. By better understanding the factors responsible for infections, we can apply those measures, which will help to achieve the safe care patients expect and deserve, to our practice.

REFERENCES

1. Kim KM, Kim MA, Chung YS, Kim NC. Knowledge and performance of the universal precautions by nursing and medical students in Korea. *Am J Infect Control* 2001;**29**(5):295–300.
2. Plowman R, Graves N, Griffin M, Roberts JA, Swan AV, Cookson B, Taylor L. *The socio-economic burden of hospital acquired infection*, vols. I, II, III and Executive Summary. London: Public Health Laboratory Service; 1999.
3. National Audit Office. *Improving patient care by reducing the risk of hospital acquired infection: a progress report*. London: The Stationery Office; 2004.
4. Health Protection Agency. *English National Point Prevalence Survey on healthcare associated infections and antimicrobial use, 2011*. London; 2012.
5. Morgan M, Evans-Williams D, Salmon R, Hosein I, Looker DN, Howard A. The population impact of MRSA in a country: the national survey of MRSA in Wales, 1997. *J Hosp Infect* 2000;**44**:227–39.
6. National Nosocomial Infections Surveillance (NNIS) System Report, data summary from January 1992–June 2001, issued in August. *Am J Infect Control* 2001;**29**:404–21.

7. Waters V, Larson E, Wu F, San Gabriel P, Haas J, Cimiotti J, et al. Molecular epidemiology of Gram-negative bacilli from infected neonates and health care workers' hands in neonatal intensive care units. *Clin Infect Dis* 2004;**38**:1682–7.
8. Reilly J, Stewart S, Allardice GA, et al. Results from the Scottish National HAI Prevalence Survey. *J Hosp Infect* 2008;**69**(1):62–8.
9. Pittet D, Donaldson L. Clean care is safer care: a worldwide priority. *Lancet* 2005;**366**:1246–7.
10. Sickbert-Bennet EE, Weber DJ, Gender-Teague MF, Sobsey MD, Samsa GP, Rutal WA. Comparative efficacy of hand hygiene agents in the reduction of bacteria and viruses. *Am J Infect Control* 2005; **33**(2):66–77.
11. Hospital Infection Society *Third prevalence survey of healthcare-associated infections in acute hospitals manual. Protocol version 1.2*. London: HIS; 2006.
12. Maki D. Nosocomial infection in the intensive care unit. In: Parrillo J, Bone R, editors. *Critical care medicine: principles of diagnosis and management*. Mosby; 1995.
13. Haddadin AS, Fappiano SA, Lipsett PA. Methicillin-resistant *Staphylococcus aureus* (MRSA) in the intensive care unit. *Postgrad Med J* 2002;**78**(921):385–92.
14. Rayner D. MRSA: an infection control overview. *Nurs Stand* 2003;**17**(45):47–53.
15. Vincent JL. Nosocomial infections in adult intensive-care units. *Lancet* 2003;**361**:2068–77.
16. Starnes MJ, Brown CV, Morales IR, Hadjizacharia P, Salim A, Inaba K, et al. Evolving pathogens in the surgical intensive care unit: a 6-year experience. *J Crit Care* 2008;**23**(4):507–12.
17. Blatnik J, Lesnicar G. Propagation of methicillin-resistant *Staphylococcus aureus* due to the overloading of medical nurses in intensive care units. *J Hosp Infect* 2006;**63**:162–6.
18. Creedon S. Health care workers' hand decontamination practices. An Irish study. *Clin Nurs Res* 2006; **15**(1):6–26.
19. Eckmanns T, Schwab F, Bessert J, Wettstein R, Behnke M, Grundmann H, et al. Hand rub consumption and hand hygiene compliance are not indicators of pathogen transmission in intensive care unit. *J Hosp Infect* 2006;**63**:406–11.
20. Inan D, Saba R, Yalcin AN, Yilmaz M, Ongut G, Ramazanoglu A, et al. Device-associated nosocomial infection rates in Turkish medical-surgical intensive care units. *Infect Control Hosp Epidemiol* 2006;**27**:343–8.
21. McKinley L, Moriarty HJ, Short T, Johnson C. Effect of comparative data feedback on intensive care unit infection rates in a Veterans Administration Hospital Network System. *Am J Infect Control* 2003;**31**(7):397–404.
22. Gastmeier P, Sohr D, Geffers C, Nassauer A, Dettenkofer M, Röden H. Occurrence of methicillin-resistant *Staphylococcus aureus* infections in German intensive care units. *Infection* 2002;**30**: 198–202.
23. Fridkin SK, Gaynes RP. Antimicrobial resistance in intensive care unit. *Clin Chest Med* 1999; **20**:303–16.
24. Pessoa-Silva CL, Hugonnet S, Pfister R, Touveneau S, Dharan S, Posfay-Barbe K, et al. Reduction of health care associated infection risk in neonates by successful hand hygiene promotion. *Pediatrics* 2007;**120**(2):e382–90.
25. Allegranzi B, Pittet D. Role of hand hygiene in healthcare-associated infection prevention. *J Hosp Infect* 2009;**73**(4):305–15.
26. Hardy K, Oppenheim B, Gossain S, Gao F, Hawkey P. A study of the relationship between environmental contamination with methicillin-resistant *Staphylococcus aureus* (MRSA) and patients' acquisition of MRSA. *Infect Control Hosp Epidemiol* 2006;**27**:127–32.
27. Karabey S, Ay P, Derbentli S, Nakipoglu Y, Esen F. Handwashing frequencies in an intensive care unit. *J Hosp Infect* 2002;**50**(1):36–41.
28. Parliamentary Office of Science and Technology. *Infection control in healthcare settings, postnote*. London; 2005.
29. Weist K, Pollege K, Schulz I, Rüden H, Gastmeier P. How many nosocomial infections are associated with cross-transmission? A prospective cohort study in a surgical intensive care unit. *Infect Control Hosp Epidemiol* 2002;**23**(3):127–32.
30. Randle J, Fleming K. The risk of infection from toys in the intensive care setting. *Nurs Stand* 2006; **20**(40):50–4.

31. Del Giudice P, Blanc V, Durupt F, Bes M, Martinez J-P, Counillon E, et al. Emergence of two populations of methicillin-resistant *Staphylococcus aureus* with distinct epidemiological, clinical and biological features, isolated from patients with community-acquired skin infections. *Br J Dermatol* 2006; **154**:118–24.
32. Muto C. Methicillin-resistant *Staphylococcus aureus* control: we didn't start the fire, but it's time to put it out. *Infect Control Hosp Epidemiol* 2006;**27**(2):111–6.
33. Rosenthal V, Maki D, Salomao R, Alvarez-Moreno C, Mehta Y, Higuera F, et al. Device-associated nosocomial infections in 55 intensive care units of 8 developing countries. *Ann Intern Med* 2006; **145**(8):582–92.
34. Marena C, Lodola L, Zecca M, Bulgheroni A, Carretto E, Maserati R, et al. Assessment of handwashing practices with chemical and microbiologic methods: preliminary results from a prospective crossover study. *Am J Infect Control* 2002;**30**:334–40.
35. Bassim H, El Maghraby M. Methicillin-resistant *Staphylococcus aureus* (MRSA). A challenge for infection control. *Ain Shams J Obstet Gynecol* 2005;**2**:277–9.
36. Starakis I, Marangos M, Gikas A, Pediaditis I, Bassaris H. Repeated point prevalence survey of nosocomial infections in a Greek university hospital. *J Chemother* 2002;**14**(3):272–8.
37. Soleto L, Pirard M, Boelaert M, Peredo R, Vargas R, Gianella A, et al. Incidence of surgical-site infections and the validity of the National Nosocomial Infections Surveillance System risk index in a general surgical ward in Santa Cruz, Bolivia. *Infect Control Hosp Epidemiol* 2003;**24**:26–30.
38. Meikle J. Strategy to shame super bug hospitals. *Guardian* 2003 6 December.
39. Storr J, Topley K, Privett S. The ward nurse's role in infection control. *Nurs Stand* 2005;19(41):56-64.
40. South West London Health Authority. *London studies put patient choice under the microscope.* <www.pickereurope.org/>; 2005.
41. Randle J, Clarke M, Storr J. Hand hygiene compliance in healthcare workers. *J Hosp Infect* 2006;**64**: 205–9.
42. Madeo M, Shields L, Owen E. A pilot study to investigate patients' reported knowledge, awareness, and beliefs on health care-associated infection. *Am J Infect Control* 2008;**36**(1):63–9.
43. Department of Health. *Winning Ways: Working together to reduce healthcare-associated infection in England.* London; 2003.
44. Department of Health. *A Matron's Charter: an action plan for cleaner hospitals.* London; 2004.
45. Department of Health. *Saving Lives: a delivery programme to reduce healthcare associated infection including MRSA—skills for implementation.* London; 2005.
46. WHO. *Guidelines on hand hygiene in healthcare.* Geneva; 2009.
47. Department of Health. The Health and Social Care Act 2008. *Code of practice on the prevention of and control of infections and related Guidance*; 2010 <https://www.gov.uk/government/uploads/system/uploads/attachment_data/file/216227/dh_123923.pdf>.
48. European Commission. *Report from the Commission to the Council on the Basis of Member States' Reports on the Implementation of the Council Recommendation* (2009/C 151/01) on Patient Safety, Including the Prevention and Control of Healthcare Associated Infections. <http://ec.europa.eu/health/patient_safety/docs/council_2009_report_en.pdf>; 2012.
49. Gaynes RP, Solomon S. Improving health acquired infection rates: the CDC experience. *Jt Comm J Qual Improv* 1996;**7**:457–67.
50. Hay A. Audit in infection control. *J Hosp Infect* 2006;**62**:270–7.
51. Hadden F, O'Brien S. Assessing acute health trends: surveillance. In: Pencheon D, Guest C, Melzer D, Gray J, editors. *Oxford handbook of public health practice.* Oxford University Press; 2001. p. 14–19.
52. Hayley RW, Quade D, Freeman HE, Bennett JV, The CDC SENIC Planning Committee Study on the Efficiency of Nosocomial Infection Control (SENIC PROJECT): summary of the design. *Am J Epidemiol* 1980;**111**:472–8.
53. Hayley RW. Surveillance by objective: a new priority-directed approach to the control of nosocomial infection. *Am J Infect Control* 1985;**13**:78–89.
54. Ayliffe GAJ, Babb JR, Taylor LJ. *Hospital acquired infection: principles and prevention*, 3rd ed. Butterworth-Heinmann; 2000.
55. Emmerson AM, Ayliffe GA. *Surveillance of nosocomial infections. Ballieres clinical infectious diseases, vol.* **3**. Balliere Tindall; 1996. p. 2.

56. Guebbels E, Bakker H, Houtman P, van Noort-Klaassen M, Pelk M, Sassen T, et al. Promoting quality through surveillance of surgical site infections: five prevention success stories. *Am J Infect Control* 2004;**32**:424–30.
57. Kelleghan SI, Salemi C, Padillo S, et al. An effective continuous quality improvement approach to the prevention of ventilator-associated pneumonia. *Am J Infect Control* 1993;**21**:322–30.
58. Sykes PK, Brodribb RK, McLaws ML, McGregor A. When continuous surgical site infection surveillance is interrupted: the Royal Hobart Hospital Experience. *Am J Infect Control* 2005;**33**(7):422–7.
59. Edmond M, White-Russell M, Ober J, Woolard C, Bearman G. A statewide survey of nosocomial infection surveillance in acute care hospitals. *Am J Infect Control* 2005;**33**:480–2.
60. El-Masri M, McLeskey S, Korniewicz D. Nosocomial bloodstream infection surveillance in trauma centers: the lack of uniform standards. *Am J Infect Control* 2004;**32**:370–1.
61. Gaynes R, Richards C, Edwards J, Emori T, Horan T, Alonso-Echanove J, et al. The National Nosocomial Infections Surveillance (NNIS) system hospitals. Feeding back surveillance data to prevent hospital acquired infections. *Emerg Infect Dis* 2001;**7**:295–8.
62. Larson E, Oram L, Hedrick E. Nosocomial infection rate as an indicator of quality. *Med Care* 1988; **26**:676–84.
63. Gould D, Chamberlain A. The use of a ward based educational teaching package to enhance nurses' compliance with infection control procedures. *J Clin Nurs* 1997;**6**:55–67.
64. Safdar N, Abad C. Educational interventions for prevention of healthcare-associated infection: a systematic review. *Crit Care Med* 2008;**36**(3):933–40.
65. House of Lords Select Committee. *Fighting infection. Fourth report.* <www.publications.parliament.uk/pa/ld200203/ldselect/ldsctech/138/138.pdf>; 2002.
66. National Audit Office. *Report by the comptroller and auditor general: the management and control of hospital acquired infection in acute NHS Trusts in England.* HC 230 session 1999-2000. London: The Stationary Office; 2000.
67. National Audit Office. *Reducing healthcare associated infections in hospitals in England.* <www.nao.org.uk/report/reducing-healthcare-associated-infections-in-hospitals-in-england/>; 2009.
68. Ayliffe GAJ. The emergence of the ICNA and progression to the IPS. *Br J Infect Control* 2008; **9**(4):6–9.
69. Horton R. Down to earth infection control. *Nurs Stand* 1992;**6**(18):22–3.
70. Teare L, Peacock A. The development of an infection control link-nurse programme in a district general hospital. *J Hosp Infect* 1996;**34**:267–78.
71. Cooper T. Delivering an infection control link nurse programme: implementation and evaluation of a flexible teaching approach. *Br J Infect Control* 2004;**5**(5):24–6.
72. Ching TY, Seto WH. Evaluating the efficacy of the infection control liaison nurse in the hospital. *J Adv Nurs* 1990;**15**:1128–31.
73. Cooper T. Delivering an infection control link nurse programme: improving practice. *Br J Infect Control* 2004;**5**(6):24–7.
74. Healthcare Commission. *Healthcare associated infection. What else can the NHS do?* London; 2007.
75. Cooper T. Delivering an infection control link nurse programme: an exploration of the experiences of the link nurses. *Br J Infect Control* 2005;**6**(1):20–3.

CHAPTER TWO

Infection Prevention

Principles of Safe Practice in Healthcare

Jacqui Prieto, Claire Kilpatrick and Jacqueline Randle

INTRODUCTION

The prevention of healthcare-associated infections (HCAI) in the United Kingdom and globally is key to the provision of high-quality, safe healthcare.[1] There are many opportunities for infection to spread from one person to another in the healthcare environment. These are more easily recognised when a person has a known communicable infection and the measures to prevent spread have been clearly defined. However, opportunities for the spread of infection frequently remain unrecognised, particularly when infection is undiagnosed or when a person is colonised rather than infected with potentially pathogenic micro-organisms.

In view of this, it is essential to apply infection prevention measures at all times to minimise the risk of spread from known and unknown sources of infection. Patients receiving healthcare are often more vulnerable to infection because of their condition or the need for treatment, particularly when this involves an invasive procedure. At times, healthcare workers and visitors may also be at risk of exposure to infection. It is therefore important to recognise the range of measures required to protect patients, healthcare workers and visitors.

A safe, clean, fit-for-purpose care environment is also essential to minimise risks of infection transmission. At times, standards of infection prevention and control can be difficult to achieve because of the facilities available. For example, in the hospital setting the requirement to isolate a patient with a known or suspected communicable infection may not always be met in a timely manner because there are no single rooms.[2] However, it is crucial that all possible prevention and control measures be adopted and that incidents where this has not been possible, and cross-infection has occurred, be reported and acted on to make healthcare environments as safe as possible for patients, healthcare workers and others.

This chapter introduces the reader to Standard Precautions. One standard precaution is hand hygiene and, because of its importance to and influence on healthcare policy.

Biofilms in Infection Prevention and Control
Doi: http://dx.doi.org/10.1016/B978-0-12-397043-5.00002-5

The following aspects of Standard Precautions are addressed here:

- Use of personal protective equipment (including glove and apron selection and face, eye, nose and mouth protection)
- Respiratory hygiene/cough etiquette
- Safe management of used linen
- Safe disposal of clinical waste
- Safe handling and disposal of sharps, including safe injection practice
- Occupational exposure management, including sharps injury
- Management of blood and other body fluid spillages

Also addressed are the broad principles of:

- Maintaining asepsis using an aseptic technique
- Safe collection of specimens
- Transmission-based precautions

STANDARD PRECAUTIONS

Standard Precautions are the minimum measures applied routinely when performing healthcare and social care activities in order to minimise the risk of spreading micro-organisms, from both known and unknown sources of infection, to patients, healthcare workers and visitors. This includes preventing micro-organisms from colonising a patient at a vulnerable site on that person, such as through a mucous membrane or non-intact skin, or via an invasive medical device.

The key principal underpinning Standard Precautions is that *any* body fluid or moist body site may contain infectious micro-organisms that pose a risk of infection. The measures included in Standard Precautions are known under other names such as 'Standard Infection Control Precautions', 'Standard Principles', 'Standard Practice' and 'Safe Working Practice'. This reflects the ongoing evolution of the concept of Standard Precautions, the evidence base for which has recently undergone review in both the United Kingdom[3] and the United States.[4]

Use of Personal Protective Equipment

Personal protective equipment (PPE) is the specialised disposable clothing or equipment used by healthcare workers or visitors to protect themselves from exposure to infectious substances. This includes disposable gloves, aprons/gowns, goggles/visors and masks/respirators.

For healthcare workers in the National Health Service (NHS), it is important to remember that the NHS has a responsibility to provide appropriate PPE. All healthcare workers should be educated in the its use to prevent injury or harm.[5] The importance of PPE as an essential component for reducing cross-infection is emphasised by Slyne et al.[6] However, they summarise studies which have shown that there is no consistent approach towards the use of PPE and there is confusion as to which item to wear and when.

Table 2.1 Risk Assessment for the Selection of PPE

Task/situation	Disposable apron/gown	Disposable gloves	Face, eye/mouth protection
Contact with body fluid not expected	✗	✗	✗
Contact with body fluid expected	✓	✓	✗
Contact with body fluid expected with high risk of splashing	✓	✓	✓

Therefore the decision to use PPE should be based on an assessment of the risk of a task or situation (Table 2.1). PPE should be donned at the start of a procedure before exposure/contamination might occur. When gloves and an apron are worn primarily to protect the wearer, the importance of their prompt removal between tasks on the same patient or between patients may be overlooked, giving rise to the possibility of cross-contamination/infection.

The need to remove or change contaminated PPE during, as well as after patient care, must be considered to avoid cross-contamination/infection. Likewise, hand hygiene should always be performed following removal of PPE, given the potential to contaminate the hands. Personal protective equipment should be changed when damaged and/or torn or soiled and, in the case of surgical masks, if wet with moisture, including from breath. A wide variety of PPE is available, and it is essential that the type used be the most appropriate for the care activity to be undertaken.

Key Point

- Uniforms are not classified as personal protective equipment.

Glove Selection

Gloves suitable for healthcare purposes are made from various materials, including chlorethene polymer, neoprene and vinyl. Polythene gloves are not suitable for healthcare use.[7] Gloves made from this material possess different properties; it is therefore important to select the appropriate glove for any given task,[8] remembering that not all gloves can protect the wearer and/or the patient during all healthcare activities.[7]

Figure 2.1 summarises the properties and recommended uses of different types of disposable glove based on the approach adopted by Health Protection Scotland.[9] The problem of allergy, both to latex proteins and the chemical accelerators used in latex glove manufacture, has increased substantially in recent years; it influences glove choice for those with an allergy, including the wearer and the patient receiving care. Specifically, Social Care and Social Work Improvement Scotland[8] identifies the best

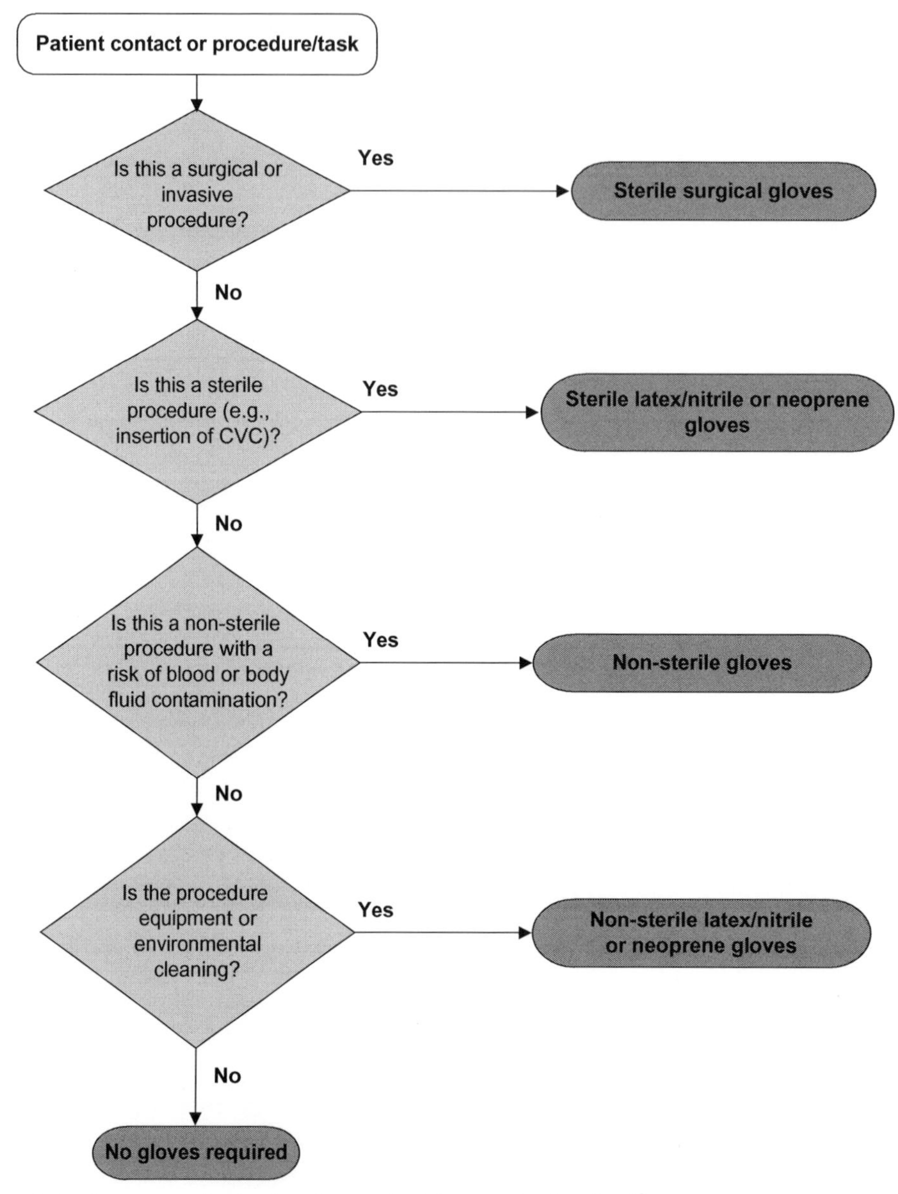

Figure 2.1 Glove use and selection. *Source: Health Protection Scotland.*[9]

evidence-based practice as being when healthcare workers decontaminate their hands and wear a pair of clean, non-sterile gloves before certain patient activities, and decontaminate their hands after removing them.

Gloves can only protect the wearer and do not remove the contaminant from the environment, so they should not replace hand hygiene.[7,10] The limited evidence base in this area is contradictory, with some research suggesting that glove use decreases hand hygiene compliance and other research suggesting that it improves it.[10] Importantly, Hakko et al.[11] find that the belief that glove use prevents the spread of infection is widespread; however, gloves are not a substitute for hand hygiene.

Apron/Gown Selection

A disposable plastic apron should be worn when there is a risk that clothing may be exposed to blood or body fluids, secretions or excretions.[12] When there is a risk of significant splashing (e.g., in the operating theatre or obstetric department), a long-sleeved, disposable, fluid-repellent gown should be worn to give added protection to the legs and body but often, more importantly, to the arms, which might be at the highest risk from contamination during procedures.[12] To coincide with this, appropriate footwear may be required to provide full coverage to the bottom half of the legs and the feet.

Selection of Face, Eye and Nose/Mouth Protection

Face or eye/nose/mouth protection should be worn when there is a risk of blood or other body fluid splashing into the face. Occasionally, respiratory protection may be required for exposure to pathogenic micro-organisms spread by the airborne route (see Transmission-Based Precautions section).

There are various forms of facial PPE, including surgical masks, particulate respirator masks (often referred to as FFP3 respirator masks and most commonly used to protect from pathogenic micro-organisms spread via the airborne route), goggles and visors. It is important to select the most appropriate PPE for each procedure/situation, ensuring it fits closely and securely so as to provide adequate protection. For example, a visor or a well-fitting surgical mask worn in conjunction with goggles provides protection from splashing of blood or body fluids. Full protective equipment should be worn when undertaking any type of aerosol-generating procedure and regular corrective spectacles are inadequate.[9]

Respiratory Hygiene/Cough Etiquette

The latest US guideline for isolation precautions[5] includes recommendations for respiratory hygiene/cough etiquette as part of Standard Precautions in order to minimise the spread of respiratory infections. This applies to all persons with signs and symptoms of respiratory tract infection (cough, congestion, runny nose, increased production of respiratory secretions), including patients, healthcare workers and visitors.

Suggested preventative measures include covering the mouth/nose with a tissue when coughing/sneezing, promptly disposing of used tissues, using surgical masks on the coughing person when tolerated and appropriate, hand hygiene after contact with respiratory secretions (given this counts as exposure to a body fluid) and spatial separation of persons with respiratory infections. Health Protection Scotland[9] does not recommend the use of antimicrobial hand wipes in a hospital setting but concurs that these can be used for hand hygiene if running water is unavailable in community settings. Their use should be followed by the use of an alcohol-based handrub. Contaminated hands should also be kept away from the mucous membranes of the eyes and nose. Healthcare workers with a respiratory infection should avoid direct patient contact, especially with patients deemed to be at high risk for susceptibility to infection. Ideally, staff should remain off work altogether until symptoms have resolved.

Safe Management of Used Linen

Within healthcare settings, used linen can harbour large numbers of potentially pathogenic micro-organisms. It is therefore important to handle it carefully to avoid contamination of the surrounding environment and to minimise the risk of transmission of infection. Exhibit 2.1 details the principles of safe practice for the management of used linen. It should be read in conjunction with the Department of Health's *Choice Framework for Local Policy and Procedures 01-04: Decontamination of linen for health and social care—Management and provision*[19] and local policy, since guidelines may vary.

EXHIBIT 2.1 Management of Used Linen

Principles of safe practice

- A disposable plastic apron should always be worn when handling used linen (and disposable gloves if linen is soiled/fouled).
- Linen should be removed from the bed smoothly, without shaking to minimise the dispersal of micro-organisms and skin scales into the environment.
- Used linen should be placed immediately in a designated bag/receptacle close to the point of use.
- Soiled/fouled linen should be placed in an impermeable bag and sealed and the bag placed in an outer clear bag to alert laundry staff not to sort it by hand before it is washed, thus avoiding risk of exposure to infection.
- Used linen receptacles should never be overfilled and should be appropriately tagged for identification.
- Linen should be stored in a designated, safe and lockable area while awaiting transfer.
- Hand hygiene should be performed after handling of linen and removal of PPE.
- Relatives/friends who take soiled clothing home to launder should be given it in a bag, which should then be placed in a sealed plastic bag or impermeable purpose-made patient laundry bag; they should also be offered advice about safe laundering (see next section).
- Laundering of personal items should first be agreed to with patients, relatives/friends or local procedures and with those who wash personal items (e.g., in the community).

Advice for laundering soiled personal items at home

- If possible, launder the soiled clothing separately from other clothes in a domestic washing machine at the highest temperature it will tolerate.
- For clothing which is heavily soiled, it may help to use a pre-wash cycle prior to the main wash cycle.
- Soaking in disinfectant before washing to reduce contamination is not necessary and may bleach coloured fabrics. Normal washing powder/solution should be used.
- Tumble-drying and hot ironing, if possible, should be carried out after washing.
- The outside of the washing machine should be wiped down with hot water and detergent after the soiled clothing is loaded.
- Thorough handwashing is required after handling soiled clothing as part of ensuring clean hands following exposure to blood or body fluids.
- Clean clothing/linen should not be placed in the same bag used to transport soiled linen.

Source: Adapted from Health Protection Scotland.[9]

Safe Disposal of Clinical Waste

The safe disposal of all waste by those involved in handling, transporting or processing it is an essential part of health and safety and is covered by legislation and national guidance. Blenkharn[13] identifies the risks of clinical waste as being the transmission of blood-borne virus infection; respiratory, enteric and soft tissue cross-transmission; physical injury; and adverse local or systemic effects through contact with potentially harmful pharmaceuticals. The safe disposal of clinical or hazardous healthcare waste, particularly when it might be contaminated with blood, other body fluids, secretions or excretions, is part of Standard Precautions. Exhibit 2.2 details the principles of safe practice in relation to the disposal of clinical waste. This should be read in conjunction with local policy, since guidelines may vary.

EXHIBIT 2.2 Disposal of Clinical Waste

Principles of safe practice

- Consider wearing gloves and, where necessary, an apron before handling waste.
- Waste should be segregated and disposed of immediately after use by the person generating it and as close to the point of use as possible.
- To hold waste, use UN-approved waste bags/containers which are of an appropriate strength to ensure that they are capable of containing the waste without spillage or puncture, of the correct colour to denote the type of waste and labelled with a 'hazardous healthcare waste for treatment/incineration' sign as appropriate.

- Use identified bag holders which have hands-free/pedal-operated lids so that hands do not become contaminated during waste disposal.
- Use UN-approved sharps containers/boxes and assemble correctly, following the manufacturer's instructions, before use. Dispose of sharps containers/boxes after three months if not before.
- Where patients can dispose of their own waste (e.g., used tissues), they should be encouraged to do so and be provided with appropriate waste receptacles for this, preferably a small, leak-proof bag.
- In the community, the responsibility for waste disposal is the householder's, but clinical waste can be collected, on request, by a local authority.
- If the householder is treated by a healthcare worker, the clinical waste produced as a result of the treatment is the responsibility of the healthcare worker.
- Bags/sharps containers should not be allowed to become more than two-thirds full.
- When two-thirds full, seal the bag/container appropriately, in accordance with local policy, before it is transported for processing.
- In healthcare settings, the tagging of waste to identify its point of origin is essential.
- The storage area for healthcare waste awaiting collection must be clearly labelled and secured to prevent unauthorised access.
- Domestic (black bag) waste must be stored separately from clinical waste.
- Wherever possible, waste storage areas should be sited away from clinical, food preparation and general storage areas.
- Spillages that occur while waste is being handled must be dealt with immediately.
- Hand hygiene should be performed after disposing of waste and PPE as part of ensuring clean hands after exposure to blood and body fluids.

Source: Adapted from Health Protection Scotland.[9]

Safe Handling and Disposal of Sharps, Including Safe Injection Practice

Sharps are any items that have the potential to cause penetration injury; they include

- Needles
- Lancets
- Scalpel blades
- Syringes
- Glass vials
- Broken glass
- Slides
- Biopsy needles
- Disposable razors
- Intravascular guide wires
- Intravenous-giving sets
- Cannulas

- Arterial blood sample packs
- Other disposable sharps
- Spicules of bone and teeth

The safe handling and disposal of sharps is essential to avoid injury, and all sharps' injuries are considered to be potentially preventable.[12]

Wherever possible, the use of equipment that avoids the need for sharps (e.g., needle-free devices) should be used. Used sharps pose a significant risk of infection, and it is the responsibility of those using them to use them correctly and dispose of them immediately after use in an approved puncture-resistant sharps container. Those using sharps should be trained to use and dispose of them safely.

Occupational Exposure Management, Including Sharps Injury

Occupational exposure to blood, body fluids, secretions and excretions through spillages or sharps injury/splashing poses a potential risk of infection to healthcare and social care workers. The routine use of preventative measures to avoid such exposures is part of Standard Precautions. Familiarity with local policy and support guidelines, particularly out-of-hours arrangements, is essential to ensure that exposure incidents are managed immediately in a safe and appropriate manner. Figure 2.2 details the actions to be taken in the event of an exposure incident; these should be followed in conjunction with local policy.

Management of Blood and Other Body Fluid Spillages

The spillage of blood or other body fluids represents a potential risk of infection to those who may come into contact with it and, in particular, the person required to deal with a spill. For this reason, it must be dealt with immediately by a person who has participated in both a hepatitis B vaccination programme and training in safe and effective spillage management. The requirement to disinfect the spillage before clearing it depends on the body fluid in question. Figure 2.3 details the risk-assessment and procedural steps involved in managing blood and body fluid spillages, including the use of a chlorine-releasing agent. It is important for staff to refer to their local guidelines for the correct protocol and strength of chlorine-releasing solution to use when there has been a blood spillage.

MAINTAINING ASEPSIS USING AN ASEPTIC TECHNIQUE

In healthcare, many treatments and procedures place the patient at increased risk of infection because they are invasive. This includes surgery and the use of invasive medical devices such as intravascular and urinary catheters. To manage such devices, care bundles[14] designed to optimise patient outcomes have been developed—for example, the Department of Health Saving Lives programme and Health Protection Scotland/Scottish Patient Safety Programme Infection Control Bundles.[15]

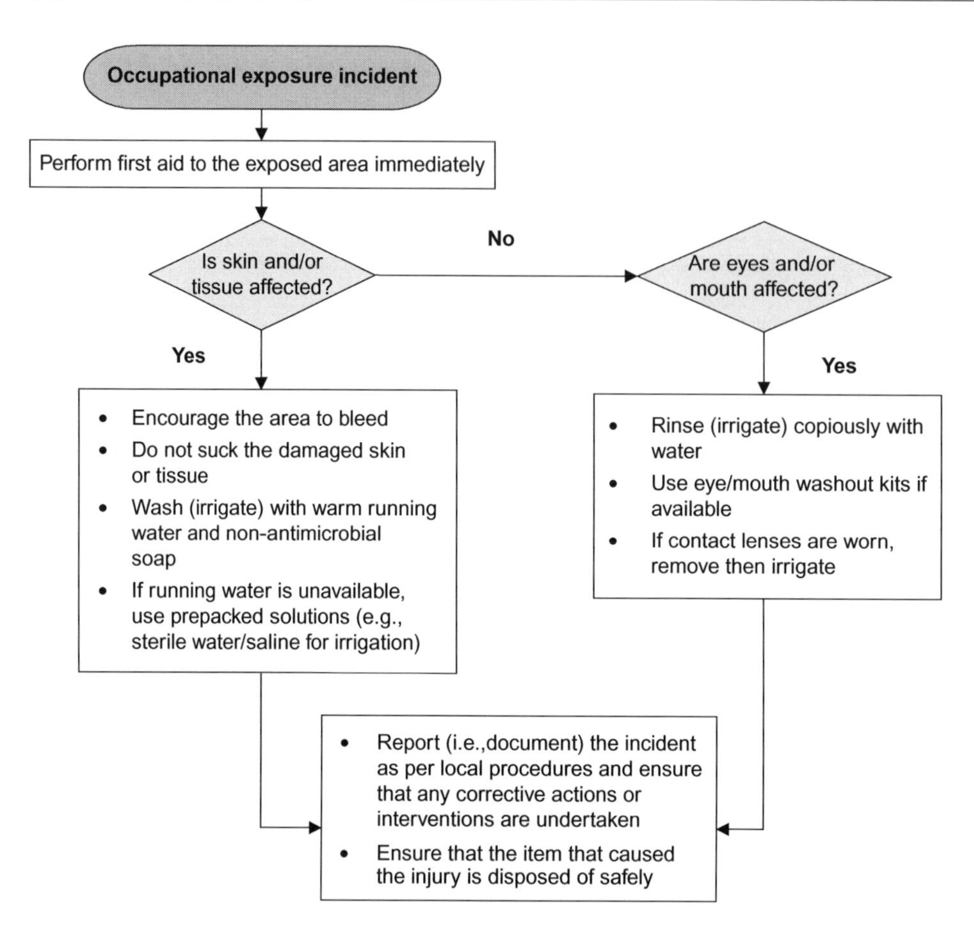

Figure 2.2 Action to be taken in the event of an exposure incident. *Source: Health Protection Scotland.*[9]

Maintaining asepsis is considered one of the key elements in ensuring safe patient care. The term *asepsis* means being free from living pathogenic micro-organisms. The aseptic technique is used to prevent micro-organisms from coming into contact with hands, surfaces or equipment being introduced to sterile equipment and susceptible body sites such as surgical wounds and device insertion sites. Various terms have traditionally been used to denote a procedure that is aseptic, including *sterile technique*, *aseptic technique* and *aseptic non-touch technique*.[16]

Rowley and Clare,[17] however, suggest that a variation in terms and practice has resulted in sub-optimal care in some cases. To improve quality, reduce infections and standardise care, they suggest that the *aseptic non-touch technique* be used as an umbrella term. This means that all clinical activities that have a goal of asepsis, such as IV therapy or the insertion of a peripheral cannula, should be undertaken using a non-touch aseptic technique. The only thing that changes, depending on the clinical activity to be undertaken, is the level of precaution and the size and management of the aseptic field.

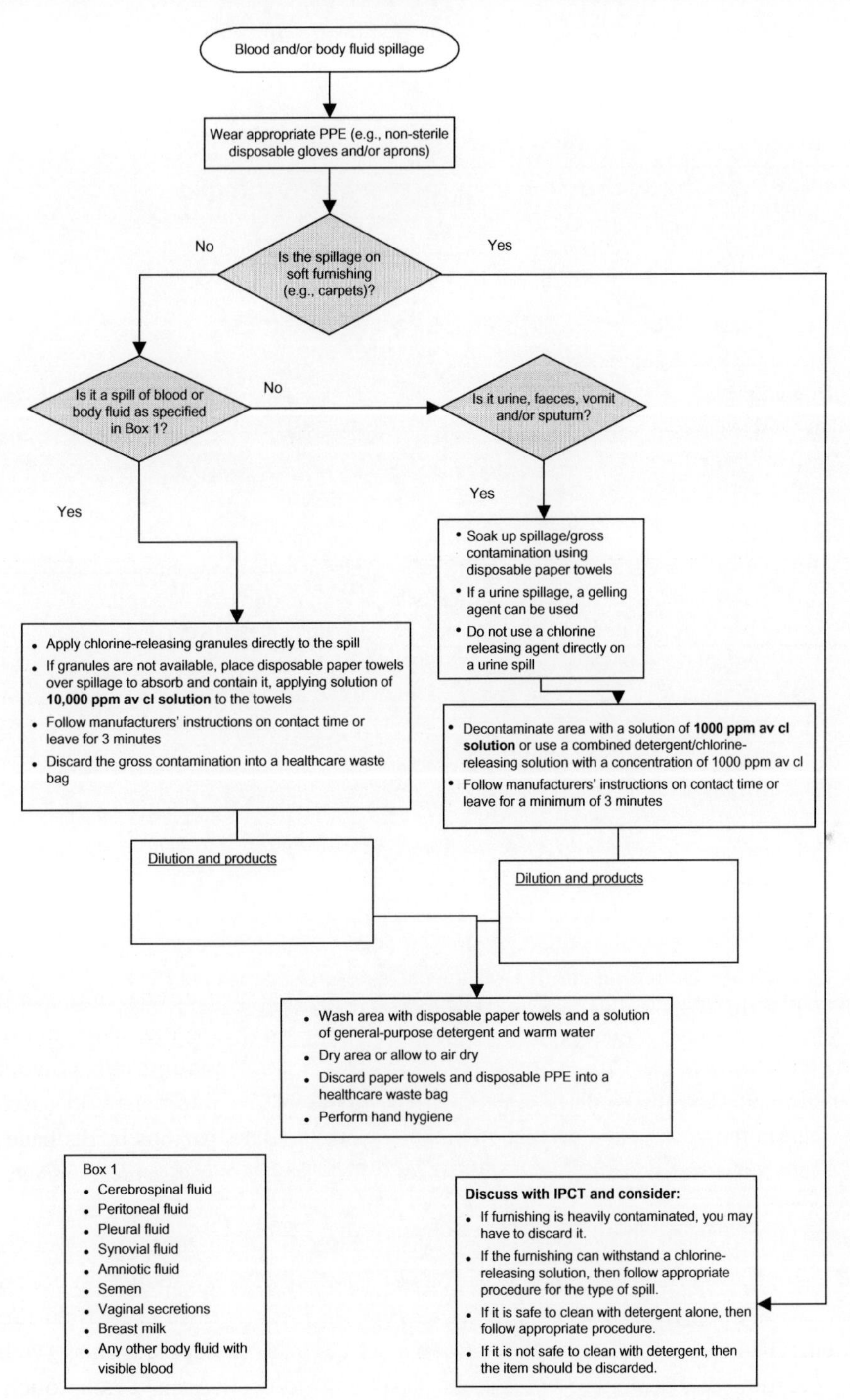

Figure 2.3 Management of blood and other body fluid spillages. *Source: Health Protection Scotland.*[9]

Exhibit 2.3 details the procedural steps for an aseptic technique as applied to the first re-dressing of a new surgical wound.

EXHIBIT 2.3 Wound Dressing Using an Aseptic Technique

Equipment

- Dressing trolley
- Alcohol-based handrub
- Sterile dressing pack
- Sachets of 0.9% saline solution
- Dressing
- Sterile gloves
- Disposable plastic apron
- Additional equipment as required

Procedure

- Explain the procedure to the patient and, where possible, gain consent.
- Offer analgesia if needed and negotiate a suitable time to undertake the procedure.
- Ensure that the dressing trolley is clean; if it is visibly dirty, clean using detergent and water or a detergent wipe.
- Gather all equipment, checking expiry dates and integrity of packaging, and place it on the bottom shelf of the trolley.
- Put on a clean, disposable apron.
- Take the trolley to the bedside, close the curtains and ensure that the patient is comfortable and positioned appropriately.
- Decontaminate hands (Hand Hygiene Moment 1).
- Loosen the dressing.
- Decontaminate hands using alcohol-based handrub (Hand Hygiene Moment 3).
- Open the outer dressing pack and drop it onto the trolley top.
- Open the sterile dressing pack, touching only the corners of the paper.
- Pour wound cleansing solution into the galley pot.
- Open any additional dressings or equipment required and drop them onto the sterile field.
- Lift the waste bag included in the sterile pack from the trolley and place one hand inside.
- Re-arrange the contents of the sterile field if necessary.
- Remove the used wound dressing from the wound site using the hand inside the waste bag. Invert the waste bag to contain the used dressing and attach it to the side of the trolley.
- Decontaminate hands (Hand Hygiene Moments 2 and 3).
- Put on sterile gloves.
- Assess the wound and use sterile equipment from the sterile field to carry out the necessary procedure, ensuring that only sterile items come into contact with the wound surface. When cleaning the wound, use each swab once; then dispose of it in the waste bag.
- Cover the wound with a sterile dressing.

- Dispose of clinical waste and sharps correctly.
- Remove gloves and dispose of them.
- Decontaminate hands (Hand Hygiene Moment 3)
- Make the patient comfortable and then open the curtains.
- Clean and return the trolley to storage.
- Decontaminate hands (Hand Hygiene Moment 5).
- Document the procedure in the patient's notes.

Source: Adapted from Randle et al. (2009).[20]

SAFE COLLECTION OF SPECIMENS

The collection of specimens for microbiological investigation enables the detection of micro-organisms in support of diagnosis, treatment and adoption of appropriate precautions. A specimen may also be required for the purpose of screening or surveillance—for example, on admission, transfer or discharge. A common request for investigation is culture and sensitivity (C&S), which means that the organism is grown (cultured) and tested to determine the most effective antimicrobial therapy (sensitivity). Most micro-organisms can be cultured within 48 hours. The use of more rapid test methods, including near-patient testing, is becoming increasingly common and may be advantageous when making risk assessments about the placement of patients.

The procedure for taking samples varies depending on the sample. Exhibit 2.4 details the procedural steps common to all microbiological sampling, which are essential to ensuring accuracy and safety. Table 2.2 provides information about specific sampling sources.

EXHIBIT 2.4 Steps for the Safe Collection of Specimens

Procedures

- Explain the procedure to the patient and, where possible, gain consent.
- Gather all relevant equipment prior to commencing the procedure, including the correct container/receptacle, request form and PPE.
- Complete the microbiology request form in full to provide as much information to the laboratory as possible.
- Decontaminate hands (Hand Hygiene Moment 2).
- Don PPE if required.
- Go directly to taking the sample from the patient and ensure that an adequate amount is collected in the appropriate container(s).

- Store the sample correctly in accordance with relevant instructions from the microbiology laboratory.
- Ensure that the patient and his or her surrounding area is decontaminated if any spillage has occurred while taking the sample.
- Remove and dispose of PPE.
- Decontaminate hands (Hand Hygiene Moment 3).
- Document in the patient's notes that the specimen has been obtained and send the sample for transportation. (If there is any contamination on the outside of the sample, it should already have been cleaned; otherwise, hand decontamination should be performed again.)

Source: Adapted from Randle et al. (2009).[20]

Table 2.2 Sampling Sources

Source	Type
Wound (e.g., surgical, traumatic, pressure ulcer, leg ulcer, burn)	Swab, fluid (e.g., pus, serous fluid), tissue
Skin or mucous membrane	Skin scrapings, swabs
Ear	Swab
Nose, throat and respiratory tract	Swab, fluid (including sputum, aspirate), tissue
Eye	Swab
Gastrointestinal tract and biliary system	Faeces, fluid, tissue
Urinary tract	Midstream urine, catheter specimen of urine, tissue
Cardiovascular system	Blood, tissue
Reproductive tract	Swab, fluid, tissue
Musculoskeletal system and soft tissue	Swab, fluid, tissue, bone
Central nervous system	Swab, fluid (e.g. cerebro-spinal fluid), tissue
Lymphatic system	Swab, fluid, tissue
Endocrine system	Swab, fluid, tissue

TRANSMISSION-BASED PRECAUTIONS

In the in-patient healthcare environment in particular, there are times when measures in addition to Standard Precautions are required in order to minimise the risk of the spread of infection. This is most often the case when a patient is either suspected or known to have a specific communicable infection that could pose a risk to others. In this situation, additional measures, known as transmission-based precautions (TBPs), may be required.

Transmission-based precautions are categorised according to the route of spread of the infectious agent—namely, airborne, droplet or contact. Use of single-room isolation may be required as part of TBPs, and this is often referred to as 'source isolation' since the patient is considered to be a potential source of infection to others. Conversely, the use of single-room isolation for a patient's own protection is referred to as 'protective isolation'. This approach has been applied in a variety of situations for patients who are recognised as being immuno-compromised as a result of an underlying illness or treatment, including those undergoing solid organ transplant, those with conditions (e.g., cystic fibrosis) and those with severe burns. However, it is generally recognised that in the majority of situations protective isolation confers no additional benefit to the use of Standard Precautions.

Current HCAI concerns, including the UK incidence of meticillin-resistant *Staphylococcus aureus* (MRSA), *Clostridium difficile* and other antimicrobial-resistant micro-organisms spread primarily via contact, mean that the elements of TBPs must be considered by those in practice on a daily basis. Specifically, the decision to isolate or cohort patients with the same infection when single rooms are unavailable is dependent on an assessment of risk, and it is important to seek the advice of the infection prevention and control team when making such decisions. Likewise, it is important to note that some communicable diseases are spread by more than one route and a combination of TBPs may be necessary depending on the risk factors for dispersal (e.g., norovirus).

As already mentioned, TBPs are required in addition to Standard Precautions for communicable infections spread by all routes of transmission: airborne, droplet and contact. The key components of airborne, droplet and contact precautions are summarised in Exhibit 2.5.

EXHIBIT 2.5 Providing Care for Patients Requiring TBPs

Placement and transfers steps

- Place the patient in a single room with en-suite facilities and keep the door closed unless other issues prevent this (e.g., patient safety).
- Communicate information on the precautions being taken to the patient, visitors and staff, without breaching confidentiality.
- Avoid ward transfers unless essential for medical reasons.
- Check the ongoing need for precautions. Only discontinue on cessation of symptoms, completion of effective treatment and/or the advice of infection prevention and control specialists.

Key Points

- The use of single-room isolation is the 'gold standard' in the management of many communicable diseases in healthcare, although it is not always possible.
- For those infections spread by the airborne route in particular, specialised, monitored isolation rooms are the best way to protect others. These include rooms with negative pressure ventilation and an ante-room.
- Cohorting is generally not recommended for those with infections spread by the airborne route.

PPE and hand hygiene steps

- Ensure that supplies of PPE and hand hygiene equipment are available at single-room/cohort areas
- Ensure that PPE is put on and removed immediately before and after care activities.
- Perform hand hygiene as per the WHO's *My Five Moments for Hand Hygiene*,[10] which includes after removal of PPE.

Key Points

Droplet precautions

- A surgical mask is one of the key precautions to be considered when providing care in close contact (within three feet).
- Masks are not expected to be routinely worn—for example, when the healthcare worker has known or proven immunity or when close-contact care is not being provided. In other seasonal situations (e.g., when there are increased numbers of influenza or respiratory syncytial virus (RSV) cases), routinely wearing masks may not be realistic. Seek advice from infection control specialists.
- Proper respiratory hygiene/cough etiquette should be encouraged as per Standard Precautions.

Airborne precautions

- Particulate respirator masks—that is, FFP3 masks (not surgical masks)—are designed to prevent inhalation of infectious airborne particles and subsequent contamination of the mucous membranes of another's respiratory tract. This is one of the key precautions to be considered when delivering care to those with infections transmissible by the airborne route.
- The use of FFP3 masks applies to care of those with, for example, active respiratory *Mycobacterium tuberculosis* (TB). They are not required routinely when caring for patients with chickenpox, measles, disseminated herpes zoster (disseminated varicella zoster/shingles). This is because the principal means of protecting healthcare workers is to ensure that only those with known immunity through vaccination/past exposure to chicken pox/measles deliver care to the infected patient.
- FFP3 masks must be fit-tested and fit-checked and changed if damaged or torn.
- Patients should never be asked to wear them.
- At times during close contact or when transfer of patients is essential, patients should be asked to wear a surgical mask if their condition allows.
- Respiratory hygiene/cough etiquette should be encouraged as per Standard Precautions.

Management of care equipment and the environment steps

- Allocate equipment to individuals; avoid sharing of equipment.
- Ensure that equipment is decontaminated before and after use and on terminal cleaning.
- Use single-use/single-patient equipment and dispose of it after use as per local policy for safe disposal of waste.
- Ensure that the environment is clutter-free, intact and clean, paying particular attention to frequently touched and horizontal surfaces.
- Ensure thorough terminal cleaning following the end of precautions.

Key Points

- There is insufficient evidence to advocate routine widespread use of disinfectants for the care environment.
- Advice for specific situations/organisms should be obtained from infection prevention and control specialists.

Safe management of linen steps

- Wear PPE when handling linen.
- Dispose of contaminated linen in an alginate bag and place in a colour-coded linen bag.

Safe disposal of waste steps

- Wear PPE when handling contaminated waste.
- Dispose of waste generated as a result of care activities as per national/local policy for healthcare waste.

Key Point

- There is no evidence to support the double-bagging of waste on removal from an isolation room.

Occupational exposure management steps

- Report occupational exposure incidents immediately.
- Ensure that occupational immunisations are up to date.

Key Point

- Consideration should be given to communicable diseases where additional immunisation is offered to healthcare professionals—for example, influenza, varicella zoster virus (VZV—i.e., chickenpox).

Source: Adapted with kind permission from Health Protection Scotland.[15]

When TBPs and, in particular, single-room isolation are required, it is important to make an effort to minimise possible adverse effects on patients, including anxiety, depression, perceptions of stigma, reduced contact with healthcare workers and increased adverse events such as falls. Visiting the patient frequently, avoiding unnecessary use of PPE, positioning the call bell within easy reach and involving visitors in finding ways to relieve boredom may help to minimise adverse effects of single-room isolation; all should be part of the routine care of the isolated patient.[18]

CONCLUSION

It is important to recognise the range of measures required to protect patients, healthcare workers and visitors from infection and to make every effort, in whichever care setting, to ensure that infection prevention and control standards are achieved.

REFERENCES

1. Royal College of Nursing. *Essential practice for infection prevention and control: guidance for nursing staff.* London; 2012.
2. Wigglesworth N, Wilcox M. Prospective evaluation of hospital isolation room capacity. *J Hosp Infect* 2006;**63**:156–61.
3. Pratt R, Pellowe C, Wilson J. Epic2: National evidence-based guidelines for preventing healthcare associated infections in NHS hospitals in England. *J Hosp Infect* 2007;**65**(Suppl. 1):S1–S64.
4. Siegel JD, Rhinehart E, Jackson M, Chiarello L, The Healthcare Infection Control Practices Advisory Committee. *Guideline for isolation precautions: preventing transmission of infectious agents in healthcare settings.* <www.cdc.gov/ncidod/dhqp/pdf/isolation2007.pdf>; 2007 [accessed 08.03.13].
5. Centers for Disease Control and Prevention *Guide to infection prevention in outpatient settings: minimum expectations for safe care.* Atlanta: CDC 24/7; 2011.
6. Slyne H, Phillips C, Parkes J. Infection prevention practice: how does experience affect knowledge and application?. *J Infect Prev* 2012;**13**:92–6.
7. Royal College of Nursing. *Tools of the trade. RCN guidance for health care staff on glove use and the prevention of contact dermatitis.* London; 2012.
8. Social Care and Social Work Improvement Scotland. *Revised health guidance on disposable gloves: use and management.* Dundee; 2011.
9. Health Protection Scotland. *Infection prevention and control manual.* Glasgow; 2013.
10. World Health Organisation. *Guidelines on hand hygiene in healthcare.* Geneva; 2009.
11. Hakko E, Rasa K, Karaman I, Enunlu T, Cakmakci M. Low rate of compliance with hand hygiene before glove use. *Am J Infect Control* 2011;**39**:82–3.
12. NICE. *Infection: prevention and control of healthcare associated infections in primary and community care.* London: National Clinical Guideline Centre; 2012.
13. Blenkharn J. Standards of clinical waste management in UK hospitals. *J Hosp Infect* 2006;**62**:300–3.
14. Haradon, C. *What is a bundle?* Institute for Healthcare Improvement, 2007.
15. Health Protection Scotland. *Transmission-based precautions.* Glasgow. <www.hps.scot.nhs.uk/haiic/ic/transmissionbasedprecautions.aspx>; 2008 [accessed 08.03.13].
16. Rowley S. Aseptic non-touch technique. *Nurs Times Plus* 2001;**97**:6–8.
17. Rowley S, Clare S. Improving standards of aseptic practice through an ANTT trust-wide implementation process: a matter of prioritisation and care. *J Infect Prev* 2009;**10**(Suppl. 1):18–23.
18. Madeo M. Infection control. The psychological impact of isolation. *Nurs Times* 2003;**99**:54–5.
19. Department of Health. *Choice framework for local policy and procedures 01-04: decontamination of linen for health and social care: management and provision*, CFPP 01-04; 2012.
20. Randle J, Vaughan N, Clarke M. Infection prevention and control. In: Randle J, Coffey F, Bradbury M, editors. *Clinical skills for adult nursing.* Oxford University Press; 2009.

CHAPTER THREE

Hand Hygiene

Claire Kilpatrick, Jacqueline Randle and Jacqui Prieto

INTRODUCTION

Hand hygiene refers to the cleaning and decontamination of hands, which in healthcare is normally performed using an alcohol-based handrub or plain soap and water.[1] It is an essential clinical action which fits into everyday practice and is considered the most effective measure in preventing healthcare-associated infections (HCAI) and, indeed, in saving lives.[2] According to Gluck et al.[3] and Allegranzi et al.,[4] HCAI are rampant in that they affect millions of people each year and can result in death and morbidity. They also cost healthcare providers and individuals additional unnecessary expense. Compliance with hand hygiene recommendations is therefore a critical component in infection prevention and control and in patient safety in every setting.[1]

Since the seminal work of Pittet et al.[5] on this lifesaving action, additional studies[6–8] have confirmed that hand hygiene is the first priority for reducing the transmission of HCAI. Although there are no large randomised control trials proving that increased compliance results in a decrease in healthcare-associated infection rates, studies have shown that improved adherence has reduced meticillin-resistant *Staphylococcus aureus* (MRSA) (1.88–0.91 cases per 10,000 bed days) and *Clostridium difficile* (16.75–9.49 cases).[9] Kirkland et al.[10] found that increasing doctors' hand hygiene compliance from 41 to 87% led to a decline in overall healthcare-associated infection rates.

As cross-transmission of micro-organisms that can be pathogenic involves five sequential steps,[11] there are many opportunities for individuals to break the chain of infection, but if these opportunities are not taken, hands become the main route of microbial transmission.[1] A review of studies between 1977 and 2007 showed the average compliance rate at 38.7%, with the range being 5 to 89%.[1] More recently, a systematic review undertaken by Erasmus et al.[12] concluded that hand hygiene non-compliance in hospitals is a universal problem despite the many interventions that allegedly have been implemented to increase and sustain compliance rates.

Biofilms in Infection Prevention and Control
Doi: http://dx.doi.org/10.1016/B978-0-12-397043-5.00003-7

Because of its importance in preventing and controlling the rate of HCAI and in order to enhance understanding of the barriers to hand hygiene so that healthcare workers can continue to address them, this chapter covers

- Issues associated with healthcare workers' hand hygiene compliance.
- Factors affecting compliance.
- Recommendations for effective hand hygiene.

ISSUES ASSOCIATED WITH HEALTHCARE WORKERS' COMPLIANCE

Although Allegranzi et al.[4] described overall compliance to be 39% worldwide, some improvements have been made. Clearly more needs to be done, but healthcare workers are working hard to ensure that hand hygiene is performed at all the right times, as described by Prieto and Kilpatrick[13] and as seen in local healthcare reporting. This is discussed in more detail later. Intervention studies have also reported sustained increases in compliance,[5,14–16] and the WHO *Guidelines on Hand Hygiene in Health Care*[1] indicates that a multimodal improvement approach can ensure change.

The systematic review by Naikoba and Hayward[17] concluded that multifaceted approaches such as those used by England's 'cleanyourhands' campaign, which provides education with written information, reminders and continuous performance feedback, have been effective. However, a systematic review by Gould et al.[18] and WHO[1] concluded that there is little robust evidence to suggest that current interventions are effective in the long term; in fact, because sustainable improvement in compliance is still rare, issues with compliance remain. Many published hand hygiene studies focus solely on the measurement of compliance rates, which provides limited information and is considered an issue in itself. Studies that address sustaining behavioural change in hand hygiene action, along with reductions in infection rates, would add value to the body of knowledge and are vital now.

One specific issue that has been underresearched but is thought to be a potential determinant for compliance is self-protection. Bahal et al.,[19] in their study of glove use and hand hygiene compliance in two countries, concluded that healthcare workers' practice is motivated by self-protection rather than protection of the patient. This is important when considering strategies for improving and sustaining compliance, as up to now the focus has been on removing barriers rather than on exploring the influential determinants of compliance that address behaviour.[20–23]

Erasmus et al.[24] concluded that self-protection explains why healthcare workers' compliance is high when hands are visibly soiled (e.g., WHO Moment 3 for hand hygiene after blood or body fluid exposure risk) or sticky. It is suggested that when this happens, it evokes an emotional response of disgust and/or discomfort, so compliance improves.[23] Further studies[25,26] also concluded that the main behavioural determinant for compliance is self-protection, and this should be taken into consideration when implementing compliance strategies.

The issue of self-protection may help explain why healthcare workers' compliance is at its lowest after contact with patient surroundings.[1] When individuals do not feel disgusted (e.g., after touching things that are not visibly dirty such as those outside of patient surroundings), they are less likely to clean their hands as they do not generally feel at risk. The role of the environment in pathogenic cross-transmission has been debated, with Dancer[27,28] arguing that environmental cleaning needs to be improved generally and specifically at near touch sites.

In a 2000 study it was found that more than half of the inanimate objects in the hospital environment were not microbiologically clean[29] and contamination of *Clostridium difficile* and vancomycin-resistant enterococci (VRE) in the hospital environment was evident.[27,30] Importantly, shared equipment can act as a vehicle for cross-transmission and can cause infections especially when hands are not cleaned at the right times.[31] However, intense emphasis on and commitment to appropriate application of all WHO's *My Five Moments for Hand Hygiene*, which are described in detail later, can and should ensure prevention of cross-transmission despite a contaminated environment, while for many reasons aspects of environmental cleaning should also be addressed.

A second issue to consider, especially in the current healthcare climate, is the specific recommended hand hygiene improvement component of compliance monitoring and feedback. This is in addition to ensuring that reliable data are available so that healthcare workers will accept the need for improvement as true and act accordingly.

Repeated monitoring of a range of indicators reflecting hand hygiene resources, such as easy access to alcohol-based handrub and sinks as well as healthcare worker practices, has been undertaken for many years now, using a range of what are usually described as audit tools. This repeated monitoring, alongside an understanding of healthcare workers' and senior managers' knowledge and perception of the problem of HCAI and the importance of hand hygiene, is a vital component of any successful hand hygiene improvement strategy.[1]

Analysis of findings and subsequent presentation at a range of levels within healthcare settings, including on wards and at senior management meetings, closes the loop on the recommended component of monitoring and feedback.[1] However, it is also important that results actively inform—for example, training and education, awareness raising and ensuring that resources for hand hygiene are always available to prevent any barriers from affecting compliance. In summary, monitoring hand hygiene compliance is important in order to

- Assess and really know the commitment of all healthcare workers, focusing on those who have patient contact, to ensure that actions, or lack of them, do not lead to avoidable harm in patients who entrust their lives to our healthcare.
- Provide feedback, as described earlier, about both poor practices and improved compliance (although the optimum compliance percentage for hand hygiene has not yet been clearly defined and remains an often discussed issue).
- Evaluate the impact of improvement interventions, including when investigating outbreaks.

Unobtrusive direct observation of hand hygiene practices by a trained observer is considered the gold standard for evaluating compliance.[1] Detailed instructions on the method for observing and recording compliance with WHO's Five Moments are included in the WHO *Hand Hygiene Technical Reference Manual*.[32] Observation forms exist for both hospitals and outpatient care settings to allow for this activity, and in some settings the information required has been included in electronic tools to make it easier to provide rapid feedback—for example, web-based data collection and entry applications using touch screen and mobile devices.

Promising innovative electronic systems for the automatic monitoring of hand hygiene compliance are also now available and can enhance data collection. They allow continuous monitoring over time as well as automatic data download and analysis. Importantly, the Hawthorne effect is thought to be significantly mitigated; the required human resources in comparison to direct observation are minimal; and these systems can serve to raise awareness and to a degree change behaviour. However, according to a recent WHO systematic review of the literature (unpublished data), limited evidence is currently available to validate the use of electronic systems, and their actual cost-effectiveness remains unknown. In most cases, these systems do not monitor standard hand hygiene indicators, and manufacturers have been strongly encouraged to integrate the Five Moments concept into their design of these tools. These new technologies present numerous advantages and may become the future approach to hand hygiene compliance monitoring when available finances permit, provided that they can reflect the Five Moments.

Consumption of hand hygiene products, in particular alcohol-based handrub, is another useful indicator. Data can be calculated easily and can be correlated with infection trends over time. However, for it to be helpful in inducing healthcare worker behavioural change, WHO recommends that this approach be used in parallel with hand hygiene compliance data—for example, those collected through direct observation.[1]

In the WHO *Guide to Implementation*,[33] you can read more about monitoring and feedback in the context of the recommended WHO Multimodal Hand Hygiene Improvement Strategy. Infection control and patient safety specialists in healthcare settings can support the most suitable approach to patient and healthcare worker monitoring.

Discussion has ensued about what level of healthcare worker compliance becomes sub-optimal, which is important in how feedback is presented, no matter how collected, for motivating staff. Although established cut-off points have been suggested—for example, 100% adherence found in NHS Trusts' hand hygiene policies or a minimum of 90% for NHS Scotland—the reality is that 100% adherence might not be strictly necessary for patient safety.[2] Cole[34] agrees and states that healthcare workers are assessed against an implausible level of best practice (i.e., 100% compliance). Voss and Widmer[35] also state that 100% compliance is unrealistic and that we must 'put an end to the reflex response that healthcare workers are neglectful of hand hygiene, which, far from helping, only demoralizes them further' (p. 208).

At the workplace of one of this chapter's authors, observations of hand hygiene compliance rates are published weekly by placing a datasheet at the entrance to wards. The rates there have never been below 98%, and it is commonplace for them to be at 100%. However, when conducting research or infection control audits, some observers (e.g., researchers and members of the infection control team) have consistently reported much lower compliance levels in the same wards. We do not think this phenomenon is specific to this geographical location but rather reflects the complexities of measuring and reporting, especially when hospital boards insist on 100% compliance, thus making the validity of the mode of data collection, as well as the reliability of all of those undertaking monitoring activities, even more critical.

The World Health Organisation provides slide sets to actually support the training of observers; without training, observation by trainers does not provide consistent results completed that people can be confident in. With monitoring and feedback a core part of what drives healthcare improvement, this issue must be carefully considered in all healthcare settings, with professionals committed to using this activity to truly support safety improvement in their patient population.

Another issue, which is topical and often overlooked, is the role of the patient's hands in the transmission of HCAI,[36] as illness and disability can often lead patients to rely on healthcare workers for hand cleaning.[37] The evidence in this area is scant, but the few studies that have been conducted imply that many patients feel reluctant to ask either for facilities to wash their hands or for staff to clean them on their behalf.[38–40]

Audits conducted by Mayers and King[37] and Whiller and Cooper[41] examined the number of times patients were offered hand hygiene by healthcare workers after using the commode; 31% stated that they had never been offered handwashing facilities.[41] A further 20% also stated that an inability to reach the facilities when they were provided prevented them from cleaning their hands.[41]

Randle et al.,[42] in their 24-hour observational study of hand hygiene compliance, found that patients' and visitors' compliance was the same as healthcare workers', although Judah et al.[43] found that more than a quarter of commuters' hands sampled had faecal matter on them. However, as few studies have involved the public, it is difficult to arrive at firm conclusions. We recommend that visitors and patients be encouraged and provided with opportunities to clean their hands in order to protect themselves and others. McGuckin[44] supports this, giving practical advice in her book *The Patient Survival Guide*.

Key Point

- Patients should always be offered hand-cleaning facilities:
  - Before eating and drinking
  - After using toilet facilities

Table 3.1 Factors Affecting Compliance

Positive impact	Negative impact
Female	Male
Nurse	Physician
Automated sink	Patient <65 years
Alcohol handrub	Interruptions
Location	Lack of knowledge of hand hygiene impact
Isolation room	Gloves
Hemodialysis unit	Understaffing
ICU	Short duration of patient contact
Administrative support	
Aware of observation	
Senior role model	

Source: Reproduced with kind permission from Gluck et al.,[3] who used data from Harrington et al.[50] to compile the table.

FACTORS AFFECTING HAND HYGIENE COMPLIANCE

Numerous studies have identified factors that can affect healthcare workers' intention to comply with hand hygiene.[8,20,45–49] A number of others have examined factors which contribute to low hand hygiene compliance levels. Table 3.1 shows several factors that either positively or negatively influence compliance.

WHO[1] provides details of all the factors which impact healthcare workers' ability to incorporate compliance into their standard clinical behaviour. As there have been numerous studies examining these factors and/or focusing on interventions to overcome them, we do not continue the debate here; instead, we recommend WHO's various publications for further reading. Needless to say, we can conclude that, accounting for these factors, it remains difficult to determine why healthcare workers choose to comply with hand hygiene or not. As previously mentioned, behavioural studies continue to be of importance.

RECOMMENDATIONS FOR EFFECTIVE HAND HYGIENE

There are a number of steps involved in ensuring effective hand hygiene in routine, everyday practice. In Exhibit 3.1 a step-by-step process for how this lifesaving action should be effectively embedded is described, with reference to a hand-cleaning technique which makes sure all areas of the hands are covered.[1] Despite scanty evidence, this is a useful step in ensuring that transient micro-organisms picked up during daily activities are removed from vital 'touch' areas. More important, outlined in this section is *when* hand hygiene should be performed (Figure 3.1), especially given what

EXHIBIT 3.1 Effective Hand Hygiene Steps during Routine Contact with and between Patients

Important factors in reliable, proper hand hygiene

- Ensure that all resources are available (i.e., alcohol-based handrub dispensers and/or soap and paper towels) by handwashing basins that are kept free from extraneous items (e.g., medicine cups, utensils)
- Ensure that hands can be adequately cleaned (i.e., keep clothing out of the way and remove all jewelry including stoned rings and wristwatches)
- Ensure that nails are short (artificial nails must not be worn)

Routine contact with patients and the right times for hand hygiene

- A nurse walks toward a patient's bed space to answer the call bell.
- As she enters the 'patient zone', the area around the patient bed where items are temporarily but exclusively dedicated to that patient, she pumps alcohol-based handrub from the dispenser at the end of the bed, performing a six-step hand cleaning technique.
- The nurse then approaches the patient and touches her arm, asking whether she is okay (performed Moment 1 action before touching a patient).
- The patient in the next bed calls the nurse over, asking her to fix her oxygen mask.
- The nurse finishes with the first patient and walks toward the next patient, pumping the alcohol-based handrub dispenser at the end of the bed as she passes it.
- The nurse fixes the mask and gets the patient's hair out of her face.
- The nurse then leaves the bed space and again pumps alcohol-based handrub from the dispenser, performing the six-step cleaning technique as she leaves (performed Moment 4).

we know about the vulnerability of patients being cared for, the times when they are most at risk and the fact that healthcare workers still do not perform this action at the right, recommended times.

WHO[1] introduced the evidence-based concept of 'My Five Moments for Hand Hygiene' based on work by Sax et al.[51] The Five Moments include hand hygiene before touching the patient, before performing aseptic and clean procedures, after being at risk from bodily fluid exposure, after touching a patient and after contact with patient surroundings. This approach is used in many healthcare settings worldwide. In other settings, including outpatient and long-term care facilities, there is often a need to adapt the emphasis on the most critical moments, but hand hygiene is equally important wherever care is being delivered.[52]

The critical point is to embed these Moments into routine workflow practices, not to have them perceived as 'add-ons' which make staff feel additionally pressured and can be a factor in non-compliance. Much work is still needed to address this, while recent publications by Health Protection Scotland have attempted to show where

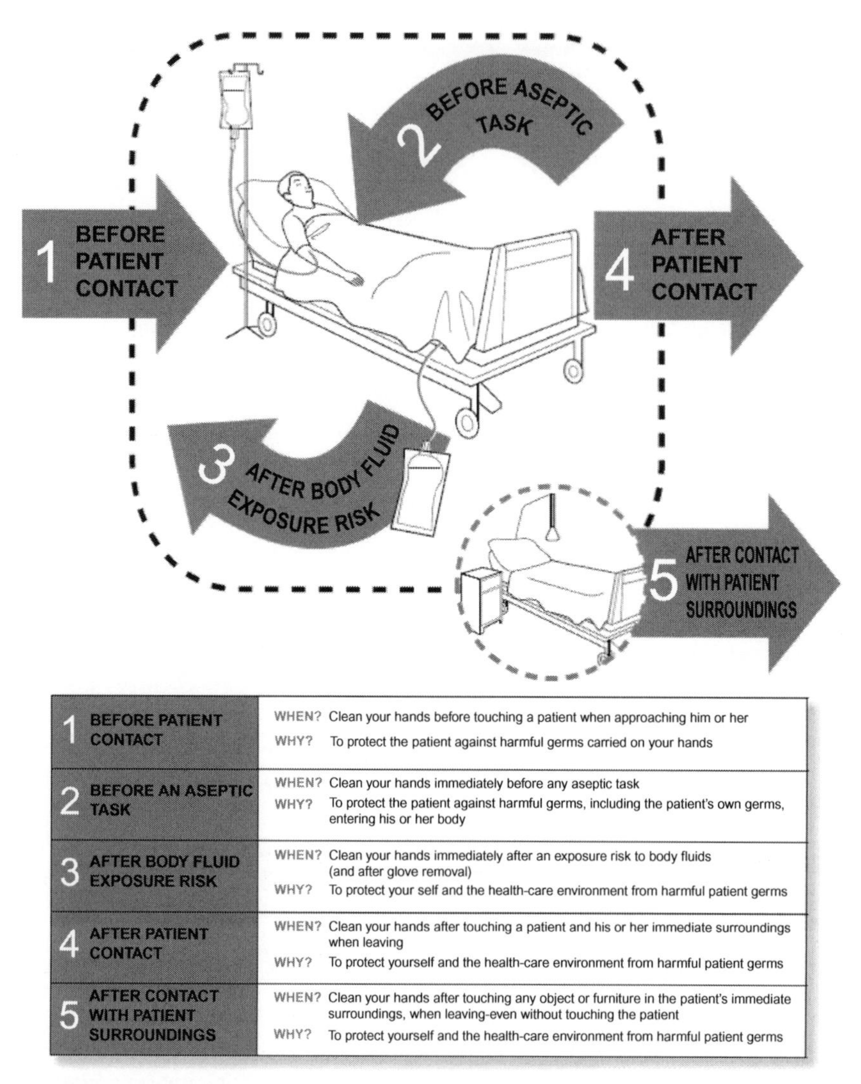

1 BEFORE PATIENT CONTACT	WHEN? Clean your hands before touching a patient when approaching him or her WHY? To protect the patient against harmful germs carried on your hands
2 BEFORE AN ASEPTIC TASK	WHEN? Clean your hands immediately before any aseptic task WHY? To protect the patient against harmful germs, including the patient's own germs, entering his or her body
3 AFTER BODY FLUID EXPOSURE RISK	WHEN? Clean your hands immediately after an exposure risk to body fluids (and after glove removal) WHY? To protect your self and the health-care environment from harmful patient germs
4 AFTER PATIENT CONTACT	WHEN? Clean your hands after touching a patient and his or her immediate surroundings when leaving WHY? To protect yourself and the health-care environment from harmful patient germs
5 AFTER CONTACT WITH PATIENT SURROUNDINGS	WHEN? Clean your hands after touching any object or furniture in the patient's immediate surroundings, when leaving-even without touching the patient WHY? To protect yourself and the health-care environment from harmful patient germs

Figure 3.1 Five Moments for Hand Hygiene. *Source: Reproduced with kind permission from the WHO Guidelines on Hand Hygiene in Health Care,*[1] *© World Health Organization.*

the Moments apply in practice when providing evidence for care bundles that aim to address common interventions that can lead to HCAI.[53]

It is critical that every healthcare worker consider the appropriate application of the Five Moments in any patient care activity, recognizing the critical times when hand hygiene should happen within a sequence of events while also ensuring that it is not undertaken unnecessarily. The steps shown in Exhibit 3.1 simply explains how hand hygiene fits into patient care activities—simple, easy and quickly undertaken but a lifesaving action that is so often neglected because the consequences of not doing it are never immediately seen.

To re-emphasise, the times when hand hygiene is required are crucially important, yet the problem of poor compliance remains. Hands should be cleaned at a range of times for general hygiene purposes. In recent years cleaning has obviously been performed at hospital entrances, and the like, as part of an attempt to role-model behaviour; however, the Five Moments highlight the most fundamental times during care delivery and daily routines that will ensure patient safety.

It should also be noted that this approach and the earlier example acknowledge that two opportunities for hand hygiene can often be met at one time. For example, when going from one patient to another, hand hygiene need not be performed after patient contact if it is being performed after the last patient contact and no other touch contact has occurred between times. Thus, this whole approach aims to make it as easy as possible for healthcare workers to comply with patient safety necessities.

The WHO *Hand Hygiene Technical Reference Manual*[32] clearly describes a number of scenarios, as do their slide sets and many other training programmes available. These resources help infection prevention professionals, trainers and all staff to properly understand implementation of the recommended approach.

Key Points

- Alcohol-based handrub is recommended as the product of choice for a number of reasons, as featured in the WHO *Guidelines*[1]
- Alcohol-based handrub is not reliable against spore-forming organisms (e.g., *Clostridium difficile*) and viral causes of gastroenteritis (e.g., norovirus—winter vomiting bug); rather, handwashing with soap and water is recommended.

Alcohol-based handrub offers a convenient alternative to plain liquid soap and water in daily practice when hands are visibly clean, but as the accompanying key points remind us, the product is not as reliable against spore-forming organisms (e.g., *Clostridium difficile*) and viral causes of gastroenteritis (e.g., norovirus). This means that for those exact situations handwashing should occur, but it does not mean that alcohol-based products have to be completely removed—a common misconception. For example, in a ward where there is a norovirus patient but not every activity involves contact with faecal matter (a Moment 3 exposure), it is often more difficult to perform hand hygiene for the other Moments.

Antiseptic soap is not required routinely for hand hygiene, but may be used as an alternative to alcohol-based handrub for procedures that require a reduction in the 'resident' flora found permanently on skin—for example, prior to a surgical or aseptic procedure. Surgical situations are specifically directed by local infection prevention and control specialists, as are outbreaks, adapting guidance appropriately to suit local needs. Every recommendation for effective hand hygiene should aim to ensure patient safety. This will remain a constant goal in healthcare for the foreseeable future.

CONCLUSION

Hand hygiene is considered the most effective way to prevent HCAI and save lives. As part of Standard Precautions it helps protect both patients and healthcare workers from exposure to infection, but it must be implemented reliably and consistently by all. Poor compliance with hand hygiene is a considerable problem in healthcare environments. Whereas measurement of compliance rates alone is ineffective in improving hand hygiene practice, there are now many resources available to support the development of a multimodal improvement strategy tailored to the needs of individual healthcare organisations.

REFERENCES

1. World Health Organisation. *WHO guidelines on hand hygiene in health care*. Geneva; 2009.
2. Allegranzi B, Pittet D. The role of hand hygiene in healthcare-associated infection prevention. *J Hosp Infect* 2009;**73**:305–15.
3. Gluck P, Nevo I, Lenchus J, Sanko J, Everett-Thomas R, Fitzpatrick M, et al. Factors impacting hand hygiene compliance among new interns: findings from a mandatory patient safety course. *J Grad Med Educ* 2010;**2**:228–31.
4. Allegranzi B, Bagheri Nejad S, Combescure C, Graafmans W, Attar H, Donaldson L, et al. Burden of endemic health-care-associated infection in developing countries: systematic review and meta-analysis. *Lancet* 2011;**377**:228–41.
5. Pittet D, Hugonnet S, Harbarth S, Mourouga P, Sauvan V, Touveneau S, et al. Effectiveness of a hospital-wide programme to improve adherence with hand hygiene. Infection Control Programme. *Lancet* 2000;**356**:1307–12.
6. Creedon S. Healthcare workers' hand decontamination practices: compliance with recommended guidelines. *J Adv Nurs* 2005;**51**:208–16.
7. Creedon S. Health care workers' hand decontamination practices. An Irish study. *Clin Nurs Res* 2006;**15**:6–26.
8. Pessoa-Silva C, Barbe-Posfay K, Pfister R, Touveneau S, Perneger T, Pittet D. Attitudes and perceptions toward hand hygiene among healthcare workers caring for critically ill neonates. *Infect Control Hosp Epidemiol* 2005;**26**:305–11.
9. Stone S, Fuller C, Savage J, Cookson B, Hayward A, Cooper B, et al. Evaluation of the national cleanyourhands campaign to reduce *Staphylococcus aureus* bacteraemia and *Clostridium difficile* infection in hospitals in England and Wales by improved hand hygiene: four year, prospective, ecological, interrupted time series study. *Br Med J* 2012 3 May.
10. Kirkland K, Homa K, Lasky R, Ptak J, Taylor E, Splaine M. Impact of a hospital-wide hand hygiene initiative on healthcare-associated infections: results of an interrupted time series. *BMJ Qual Saf* 2012;**21**:1019–26.
11. Pittet D, Allegranzi B, Sax H, Dharan S, Pessoa da Silva C, Donaldson L, et al. Evidence-based model for hand transmission during patient care and the role of improved practices. *Lancet Infect Dis* 2006;**6**:641–52.
12. Erasmus V, Daha TJ, Brug H, Richardus JH, Behrendt MD, Vos MC, et al. Systematic review of studies on compliance with hand hygiene guidelines in hospital care. *Infect Control Hosp Epidemiol* 2010;**31**:283–94.

13. Prieto J, Kilpatrick C. Infection prevention and control. In: Brooker C, Nicol M, editors. *Alexander's nursing practice*, 4th ed. Churchill Livingston/Elsevier; 2011.
14. Huggonet S, Perneger T, Pittet D. Alcohol-based handrub improves compliance with hand hygiene in intensive care units. *Arch Intern Med* 2002;**162**:1037–43.
15. Randle J, Clarke M, Storr J. Hand hygiene compliance in healthcare workers. *J Hosp Infect* 2006; **64**:205–9.
16. Huang J, Jiang D, Wang X, Liu Y, Fennie K, Burgess S, et al. Changing knowledge, behaviour and practice related to universal precautions among hospital nurses in China. *J Contin Educ Nurs* 2002;**33**:217–24.
17. Naikoba S, Hayward A. The effectiveness of interventions aimed at increasing handwashing in healthcare workers—a systematic review. *J Hosp Infect* 2001;**47**:173–80.
18. Gould D, Chudleigh JH, Moralejo D, Drey N. Interventions to improve hand hygiene adherence in patient care. *Cochrane Database Syst Rev* 2007;**2**:CD005186.
19. Bahal A, Karamchandani D, Fraise A, McLaws M. Hand hygiene compliance: universally better post-contact than pre-contact in healthcare workers in the UK and Australia. *Br J Infect Control* 2007;**8**:24–8.
20. Pittet D, Mourouga P, Perneger TV. Compliance with handwashing in a teaching hospital. Infection Control Program. *Ann Intern Med* 1999;**130**:126–30.
21. Whitby M, McLaws ML. Hand washing in healthcare workers: accessibility of sink location does not improve compliance. *J Hosp Infect* 2004;**58**:247–53.
22. Braun V, Clarke V. Using thematic analysis in psychology. *Qual Res Psychol* 2006;**3**:77–101.
23. Porzig-Drummond R, Stevenson R, Case T, Oaten M. Can the emotion of disgust be harnessed to promote hand hygiene? Experimental and field-based tests. *Soc Sci Med* 2009;**68**:1006–12.
24. Erasmus V, Brouwer W, van Beeck EF, Oenema A, Daha TJ, Richardus JH, et al. A qualitative exploration of reasons for poor hand hygiene among hospital workers: lack of positive role models and of convincing evidence that hand hygiene prevents cross-infection. *Infect Control Hosp Epidemiol* 2009; **30**:415–9.
25. Borg M, Benbachir M, Cookson B, Redjeb S, Elnasser Z, Rasslan O, et al. Self-protection as a driver for hand hygiene among healthcare workers. *Infect Control Hosp Epidemiol* 2009;**30**:578–80.
26. Jang JH, Wu S, Kirzner D, Moore C, Youssef G, Tong A, et al. Focus group study of hand hygiene practice among healthcare workers in a teaching hospital in Toronto, Canada. *Infect Control Hosp Epidemiol* 2010;**31**:144–50.
27. Dancer S. The role of environmental cleaning in the control of hospital acquired infection. *J Hosp Infect* 2009;**73**:378–85.
28. Dancer S. Hospital cleaning in the 21st century. *Eur J Clin Microbiol Infect Dis* 2011, 17 April.
29. Griffith CJ, Cooper RA, Gilmore J, Davies C, Lewis M. An evaluation of hospital cleaning regimes and standards. *J Hosp Infect* 2000;**45**:19–28.
30. Lemmen SW, Häfner H, Zolldann D, Stanzel S, Lütticken R. Distribution of multi-resistant Gram-negative versus Gram-positive bacteria in the hospital inanimate environment. *J Hosp Infect* 2004;**56**:191–7.
31. Arias K. Contamination and cross contamination on hospital surfaces and medical equipment. Initiatives in Safe Patient Care. Saxe Healthcare Communications; 2010. <https://docs.google.com/viewer?a=v&q=cache:Nn_goQBXhkYJ:www.initiatives-patientsafety.org/Initiatives4.pdf+c-an+shared+equipment+cause+cross+transmission&hl=en&gl=uk&pid=bl&srcid=ADGEESgMUBOewmowGI2lnXEDOrMvzd47kbgZGKzm7qLvJpAwnxtXFCRv6XnMzVorUjbZwvVvoffrB0fmHFhoBe1EW4aVuKgbMHGWpSGh7TOstcjSWl1Gu0Zsy9LF4gcuusvDx1hYJ8e-&sig=AHIEtbTJ755YUxzJ_eo0bF9DzHQ_QiU2Zw> [accessed 14.02.13].
32. World Health Organisation. *WHO hand hygiene technical reference manual*. Geneva; 2009a.
33. World Health Organisation. *WHO guide to implementation*. Geneva; 2009b.
34. Cole M. Adherence and infection control guidelines: a complex phenomenon. *Br J Nurs* 2008; **17**:700–4.
35. Voss A, Widmer AF. No time for hand washing? Hand washing versus alcoholic rub: can we afford 100% compliance? *Infect Control Hosp Epidemiol* 1997;**183**:205–8.

36. Ward D. Improving patient hand hygiene. *Nurs Stand* 2003;**17**:39–42.
37. Mayers D, King S. Hand hygiene for patients with rheumatic disease. *Nurs Times* 2000;**96**:47–8.
38. Lawrence M. Patient hand hygiene: a clinical inquiry. *Nurs Times* 1983;**79**:24–5.
39. Ouschan R, Sweeney JC, Johnson LW. Dimensions of patient empowerment. *Health Market Q* 2000;**18**:99–114.
40. Duncanson V, Pearson LS. A study of the factors affecting the likelihood of patients participating in a campaign to improve staff hand hygiene compliance. *Br J Infect Control* 2005;**6**:26–30.
41. Whiller J, Cooper T. Clean hands: how to encourage good hygiene by patients. *Nurs Times* 2000; **96**:37–8.
42. Randle J, Arthur A, Vaughan N. A 24-hour observational study of hospital hand hygiene adherence. *J Hosp Infect* 2010;**76**:252–5.
43. Judah G, Donachie P, Cobb E, Schmidt W, Holland M, Curtis V. Dirty hands: bacteria of faecal origin on commuters' hands. *Epidemiol Infect* 2010;**38**:409–14.
44. McGuckin M. *The patient survival guide 8: simple solutions to prevent hospital or healthcare-associated infections*. Demos Medical Publishing; 2012.
45. Harris AD, Samore MH, Nafziger R, DiRosario K, Roghmann MC, Carmel Y. A survey on hand washing practices and opinions of healthcare workers. *J Hosp Infect* 2000;**45**:318–21.
46. van de Mortel T, Bourke R, McLoughlin J, Nonu M, Reis M. Gender influences hand washing rates in a critical care unit. *Am J Infect Control* 2001;**29**:395–9.
47. Roberto M, Mearns K, Silva S. Social and moral norm differences among Portuguese 1st and 6th year medical students towards their intention to comply with hand hygiene. *Psychol Health Med* 2012;**17**:408–16.
48. Girou E, Chai SHT, Oppein F, Legrand P, Ducellier D, Cizeau F, et al. Misuse of gloves: the foundation for poor compliance with hand hygiene and the potential for microbial transmission? *J Hosp Infect* 2004;**57**:162–9.
49. Rosenthal V, Guzman S, Safdar N. Reduction in nosocomial infection with improved hand hygiene in intensive care units of a tertiary care hospital in Argentina. *Am J Infect Control* 2005;**33**:392–7.
50. Harrington L, Lesh K, Doell L, Ward SK. Reliability and validity of hand hygiene measures. *J Healthc Qual* 2007;**29**:20–9.
51. Sax H, Allegranzi B, Uckay I, Larson E, Joyce J, Pittet D. My five moments for hand hygiene: a user-centred design approach to understand, train, monitor and report hand hygiene. *J Hosp Infect* 2007;**67**:9–21.
52. World Health Organisation. *Hand hygiene in outpatient and home-based care and long-term care facilities*. Geneva; 2012.
53. Health Protection Scotland. *Evidence for care bundles and other quality improvement tools*; 2012. <www.hps.scot.nhs.uk/haiic/ic/evidenceforcarebundles.aspx?subjectid> [accessed 03-13].

CHAPTER FOUR

Decontamination

Tracey Cooper, Geoff Sjogren, Jacqueline Randle and Steven L. Percival

INTRODUCTION

Decontamination is an essential part of any infection prevention and control programme and has been documented to reduce the risk of healthcare-associated infection (HCAI).[1] Although it remains controversial as to the extent that the environment contributes to HCAI,[2,3] it is evident that every patient has the right to clean, safe care in an environment where clinicians deal daily with micro-organisms, blood, body fluids, dust and other 'soil'.[4–6]

Because both opportunistic and strict pathogenic micro-organisms can contaminate, persist and be disseminated in a viable state for weeks and months in a healthcare environment,[7–9] it is vital that equipment and the environment in which it is used are clean and fit for purpose, thus reducing patients' risk of infection. Decontamination in a hospital environment is particularly important to prevent the contamination and spread of high-risk problematic bacteria, including as examples *Acientobacter baumannii*, meticillin-resistant *Staphylococcus aureus* (MRSA) and *Clostridium difficile*.[10–12]

Whilst the procedural details of many associated decontamination practices, such as hand hygiene, clearing of spills and environmental cleanliness, are detailed in other chapters, the general principles of decontamination are examined here to highlight the aforementioned procedures and explicate the principles underpinning them.

It is very important that individuals undertaking decontamination be appropriately trained. The Department of Health has published and made available guidance documents to aid in this training. These include, among others, *Choice Framework for Local Policy and Procedures 01-01: Management and Decontamination of Surgical Instruments (Medical Devices) Used in Acute Care* (CFPP 01-01)[13] and *Choice Framework for Local Policy and Procedures 01-06: Decontamination of Flexible Endoscopes—Operational Management* (CFPP 01-06).[14]

This chapter is concerned with the fundamental principles of decontamination and focusses mainly on the decontamination of medical devices; however, it makes some reference to environmental cleaning and decontamination. The range of practices which fall under the general term of 'decontamination', and areas and tasks where decontamination may take place, are described. Overall, the chapter covers (1) the

Biofilms in Infection Prevention and Control
Doi: http://dx.doi.org/10.1016/B978-0-12-397043-5.00004-9

terms associated with decontamination, (2) a risk matrix to assist in choosing appropriate processes, and (3) principles underpinning decontamination of reusable medical devices, skin and the environment.

USEFUL DEFINITIONS

For the purpose of this chapter and the book as a whole, *decontamination* can be used generically for the process, or series of processes, which renders an object free of contaminants and safe for use.[15] The time spent on cleaning a hospital, however, does not necessarily correlate to the level of cleanliness.[16] The level of decontamination required depends on the item and the risk posed by its use. Any item which is decontaminated undergoes a combination of cleaning, disinfection and/or sterilisation according to the level of risk identified and regularly evaluated.[17]

Cleaning is defined as the physical removal of organic matter, which may include dirt, dust or body fluids. In the removal of this organic matter a large number of micro-organisms are also removed, but cleaning should be considered more than routine.[18] It is a pre-requisite to any further processing and can be used in isolation for items which are considered to be low risk (Table 4.1). Cleaning can be manual or automated, using reusable equipment or single-use wipes depending on local policy.

Disinfection is the next level of decontamination and kills micro-organisms through a number of either heat or chemical processes. However, it does not necessarily kill bacterial spores.[19,20] Items which are suitable for disinfection are those considered to be of medium risk (see Table 4.1) or those which may be low risk but contaminated with an organism of concern (e.g., a blood spill on a vinyl floor and/or on hospital textiles).[21] Heat or chemicals can achieve disinfection, but heat is generally more reliable and therefore preferable.[22]

Table 4.1 Spaulding Classification of Risk and Method of Decontamination

Risk	Definition	Process of choice
Low	Items which come into contact with intact skin	*Cleaning* Disinfection may be indicated if there is concern regarding particular pathogens.
Medium	Items which come into contact with broken skin or intact mucous membranes	*Disinfection* Sterilisation may be considered where there is concern regarding certain pathogens; the item may not need to be sterile at point of use.
High	Items which come into contact with broken mucous membranes or enter sterile body cavities	*Sterilisation* Some items which are not heat-tolerant may undergo high-level chemical disinfection for extended periods of time according to the assessment of risk.

Sterilisation is the highest level of decontamination, rendering an item free from all microbial contamination including spores. It is used for high-risk items (see Table 4.1), which in the main are reusable invasive medical devices such as surgical instruments. However, it is very important to report that sterilisation does not inactivate prion proteins, the causative agents of transmissible spongiform encephalopathies such as CJD.[23]

CHOOSING THE APPROPRIATE PROCESS

The factors which influence the method of decontamination include the risk of infection from the contaminant, the risk from the use of the item and whether the item needs to be sterile or only clean at the point of use. A risk matrix commonly used to identify the level of risk posed and choose the appropriate decontamination method is the adapted Spaulding Classification of Risk.[24] However, in some circumstances it has been suggested that for the cleaning of theatre equipment the Spaulding classification should be extended.[25]

Key Point

- Additional factors which need to be considered in the choice of a decontamination method include tolerance to heat/chemicals/moisture, availability of processing methods and risks to staff handling the items or the chemical agents. For example, some medical equipment (e.g., fibre-optic endoscopes) are not tolerant of heat so cannot withstand sterilisation or autoclaving, while certain items are less compatible with some chemicals, some plastics and alcohol. The manufacturer's decontamination instructions must be followed at all times.

Both fresh water and seawater have the capacity to remove chemical agents not only through mechanical force but also via slow hydrolysis. The effect of water and water/soap solutions is the physical removal or dilution of agents. A very good chemical decontamination reaction is oxidative chlorination, which includes agents such as hypochlorite.

Generally cleaning is carried out using detergent and water. Detergents break down grease molecules, allowing debris to be removed from surfaces. Cleaning generally should be carried out using water at no more than 30°C to prevent proteins from adhering and fixing in place, and items should be rinsed to remove traces of detergent. Any item which is cleaned should also be dried to reduce the risk of recontamination from various pathogens that flourish in moist environments. In many areas of healthcare, cleaning is a manual activity, although there are more and more automated cleaning systems available, in particular floor cleaners and automated endoscope washer-disinfectors (EWD) for endoscopes and control of biofilm.[26]

Flexible endoscopes with channels must be cleaned and disinfected in an EWD following manual pre-cleaning, as highlighted in the recent Department of Health (England) publication, *Choice Framework for Local Policy and Procedures 01-06: Decontamination of Flexible Endoscopes—Operational Management* (CFPP 01-61, p. 4). Automated washer-disinfectors must also be used to clean reusable invasive medical devices such as surgical instruments.[13]

Key Points

- It is essential that any equipment used for cleaning be maintained in a clean and dry condition between uses, and that single-use items, such as cloths, be used when appropriate.
- Single-use detergent wipes provide a quick and easy solution and are increasingly available, though care must be taken to ensure compatibility with the item to be cleaned.[27]
- Some items (e.g., endoscopes and surgical instruments) require automated cleaning in addition to any required manual cleaning to ensure that as much physical soiling as possible is removed.

Principles Underpinning Manual Cleaning

When manual cleaning is undertaken, it is important to follow the manufacturer's decontamination instructions. The following example procedure highlights a standard protocol that is often employed for manual cleaning to help the healthcare professional.

1. Wash hands.
2. Prepare necessary cleaning equipment or use single-use detergent wipes if appropriate.
3. Put on gloves and apron.
4. Move from areas of least contamination to areas of greater contamination.
5. If using reusable cloths, rinse cloth regularly during use—single-use wipes should be changed as necessary.
6. Cleaning water should be changed if it becomes visibly dirty, and thrown away as soon as cleaning is completed; depending on the item(s) being cleaned, it may be necessary to change the water after each item.
7. Reusable equipment must be cleaned and dried or allowed to dry and stored dry.
8. Disposable cloths must be thrown away after use.
9. Hand hygiene must always be performed after gloves and apron are removed post-cleaning.

Key Point

- There are a number of proprietary disposable detergent cloths which can be used for cleaning. Reference should be made to the manufacturer's decontamination instructions and local policies regarding their use, or discussions should be held with the infection prevention and control team.

Principles Underpinning Automated Cleaning

In general all automated cleaning systems should have specialist instructions for use. In addition, reference should be made to local policy and procedure and instruction given by an individual competent and trained in the use of automated cleaning systems.

In the use of automated cleaning systems, the general principles—including the use of personal protective equipment (PPE) and appropriate decontamination of cleaning equipment—should always apply. Additionally all equipment for automated cleaning must be used in accordance with the manufacturer's instructions. This includes the choice of detergent, rinse aids and, where necessary, water softeners. The equipment must be maintained as part of a planned preventative maintenance programme to ensure that it is functioning properly. This includes periodic testing using proprietary soil kits.

A number of national guidance documents exist on the use of automated systems for the decontamination of medical devices (e.g., *Choice Framework for Local Policy and Procedures 01-01: Management and Decontamination of Surgical Instruments (Medical Devices) Used in Acute Care* (CFPP 01-01)[13] and *Choice Framework for Local Policy and Procedures 01-06: Decontamination of Flexible Endoscopes—Operational Management* (CFPP 01-06).[14] The reader may find these useful.

Principles Underpinning Disinfection

There are a number of different types of disinfectants, either using heat or chemicals; the choice depends on the item to be disinfected and the nature of the disinfecting agent.[1,28] Some specialist equipment requires processing with certain chemical disinfectants, particularly endoscopic equipment and other medical devices.[29] In such circumstances it is imperative that the user refer to local policy and procedure.

It is important that when using chemical disinfectants a control of substances hazardous to health (COSHH) assessment be carried out and guidance made available on its use. For chemical spillages, appropriate kits are required in the event of an accidental spill. Risk analysis and checks should always be made and the licence should be checked to ensure that any product is suitable for its intended use.

All items being disinfected must be pre-cleaned to remove soiling. Where heat disinfection is used, such as a bedpan washer[30] or automated endoscope washer-disinfector, the disinfection unit must be maintained according to the manufacturer's instructions. Periodic tests should be conducted to ensure that the temperatures are within the required limits for thermal disinfection (e.g., 70–95°C, according to CFPP 01-01: Part D, p. 2). All equipment should be maintained following guidelines set out in relevant national guidance documents—for instance, CFPP 01-01 and CFPP 01-06.

When using chemical disinfectants, it is always important for the user to wear PPE; in certain circumstances facial protection should be employed. In particular, some chemical disinfectants require special controls and should be used in a well-ventilated area. In addition all items' surfaces must be in contact with the chemical disinfectant for the time recommended by the manufacturer to ensure that the contact time is sufficient to kill any micro-organisms present.

Furthermore it is important to ensure that where chemicals are diluted this is done to the correct strength recommended by the manufacturer and using the appropriate diluents. Some manufacturers produce test strips to confirm the correct dilution. Therefore it is important that the user ensure that solutions are changed in accordance with the manufacturer's instructions.

Following the period of disinfection, it is important to rinse the item well to remove all chemical residues and then dry it. Some items may require rinsing with sterile water, and manufacturer's instructions or local policy should be followed. In general, items that are required to be sterile at the point of use should be rinsed in sterile water. Equipment used for disinfecting should be cleaned, rinsed and stored dry. Chemical disinfectants should be used only if recommended by the manufacturer and in conjunction with advice from the infection prevention and control team.

Major advances are being made in hospital cleaning and disinfection as reported by Carling and Huang.[28]

Key Point

- There are a number of proprietary disposable disinfectant wipes which can be used for both environmental and device disinfection for low-risk items. Their use should be discussed with the infection prevention and control team.

Principles Underpinning Sterilisation

Sterilisation is usually a process reserved for reusable invasive, high-risk medical devices, including surgical instruments.[31,1] It is best achieved using steam under pressure in an autoclave—in the United Kingdom typically generating temperatures of 134 to 137°C for at least 3 minutes at a pressure of 2.2bar. In some countries a temperature of 121°C is used with a longer hold time at that temperature. Other methods of sterilisation are available for equipment which requires sterilisation but is heat labile, such as ethylene oxide and gas plasma sterilisation. However, these processes require very specialised equipment, are usually expensive and are only carried out following the advice of a specialist.

Sterilisation requires careful monitoring to ensure that the correct parameters have been achieved to destroy all micro-organisms. These are set out in national guidance (e.g., CFPP 01-01) documents. However, it is important that only staff specifically trained in sterilisation carry out this function. Wherever possible, a designated sterile supplies department (SSD), which is in compliance with the European Union Medical Devices Directive 93/42/EEC (updated as Directive 2007/47/EC), should be used.

For reusable invasive medical devices it is essential that there be an audit trail which connects instrument trays to individual patients and to the effective decontamination processes. Instruments that are considered high risk in terms of variant Creutzfeldt-Jakob disease (vCJD) have to be kept separate from other procedural trays.[32,33] This is a specialist area, so additional advice is often sought and followed. Guidance documents are provided by The Advisory Committee on Dangerous Pathogens, Transmissible Spongiform Encephalopathy Risk Management Subgroup (ACDP TSE), which the reader may find very useful.

Instruments for sterilisation must be pre-cleaned, usually in an automated washer-disinfector, before sterilisation. This is because automated washer-disinfectors achieve high temperatures as part of their processes and therefore achieve both cleaning and disinfection.

At all times, it is important for the healthcare professional to follow local instructions and standard operating procedures. Also, handwashing is very important in the prevention of intravenous bacterial injection from healthcare providers' hands.[34]

DECONTAMINATION OF MEDICAL DEVICES

For the decontamination of a medical device, it is essential that the entire life cycle of the device be considered, taking into account procedures from its acquisition through use to disposal at the end of its useful life.[35] This life-cycle procedure is particularly important when considering decontamination.*

An assessment should be performed which includes processes for cleaning and disinfection and, where appropriate, sterilisation. An example of a pre-purchase questionnaire can be found on the Institute of Decontamination Sciences' website at *www.idsc-uk.co.uk/news.php* (MDA/2013/019).[36]

Many items require servicing or repair on a regular basis. It is essential that prior to devices' being sent for examination, service or repair, a suitable process for decontamination be carried out to make an item safe for others to handle. For many devices this may be as simple as cleaning the outside. A certificate of decontamination should accompany the device to indicate the process that it has been subjected to and to provide assurance for maintenance/repair staff working on it.

Key Points

- It is essential that prior to purchasing any medical device an evaluation be undertaken to ensure that the process for decontamination recommended by the manufacturer can be carried out.
- Many organisations utilise a pre-purchase questionnaire which has to be completed and approved by the infection prevention and control team and the local decontamination advisor to ensure this has been done correctly.

*See Decontamination: Health Technical Memorandum—01-01, Decontamination of reusable medical devices, Figure 2, p. 8 (2007).

Single-Use and Reusable Equipment

In recent years there has been a shift toward increased use of single-use disposable equipment: both medical devices and instruments as well as cloths and wipes for decontamination processes. Paper maceratable bedpans and urinals have become popular in healthcare facilities in recent years, yet recently there has been a move back to washer-disinfectors and reusable bedpans, as some feel this is more economically beneficial. However, in considering which type is appropriate a full risk and economic assessment should be carried out, including reprocessing costs versus disposal costs in conjunction with a specialist's infection prevention advice.

Also, all instruments and devices must be confirmed to be of good clinical quality, as examples of poor-quality single-use instruments have been identified which may impact the success of a procedure—for example, forceps that do not fit together well.

If single-use products are adopted, they must be used once only and then disposed of. If a single-use device is reprocessed, then full liability for the functioning of that instrument and the associated infection risks are taken on by the user. Devices which are classed as single use are marked with the following symbol (meaning 'Do not use twice').

Key Point

- Some devices, including nebulisers, are considered to be for single-patient use. This means that they can be used more than once on the same patient but not reused for different individuals. In these circumstances, the device must be decontaminated and stored between uses, in line with the manufacturer's recommendations.

DECONTAMINATION OF THE ENVIRONMENT

A clean environment generally promotes confidence in the care that is delivered. Patient surveys in England have repeatedly shown that the provision of 'clean, safe care' is a priority for patients.[37,38]

While designated, trained staff usually undertake many cleaning tasks, there are frequently items in the environment which healthcare workers are responsible for cleaning.[39] It is essential that all staff are clear on who is responsible for cleaning which items and that no items or areas of the environment are overlooked so that no one is responsible for cleaning them.[40,41]

Healthcare workers in England must ensure that they are happy with the standard of cleaning in their area, and know how to report concerns if not. In hospitals, nurses in general and ward sisters in particular (and matrons where this role exists) are seen as the guardians of cleanliness standards in wards or departments. They must routinely check that environmental cleaning standards meet the needs and expectations of patients and service users.

Cleaning standards are set out in national guidelines, which often include cleaning methods, frequencies and a recommendation for colour coding to reduce the risks of cross-infection in ward areas (e.g., the National Patient Safety Agency guidance documents[40,42,43]).

Innovative technologies have been implemented for environmental cleaning in recent years, including micro-fibre cleaning products, steam cleaning, chemical agents (e.g., detergent-chlorine) and products which fog the area using hydrogen peroxide vapour to achieve proper disinfection. Evidence of the importance of the environment as a source of patient infection is re-emerging, and it is predicted that technological innovation will continue in response. Local policy should be followed, and new technology should be introduced in a planned manner to ensure that all relevant staff are trained in new processes and techniques.

One area of concern is water quality and risk of cross-contamination.[44] In addition, as part of any decomtamnation procedure, it is very important for health implications that control procedures be effective against spores[45] and biofilms.[46]

Key Point

- It is essential to ensure that everyone is aware of their cleaning responsibilities and that there are no items which no one is responsible for cleaning. For instance, it is generally accepted, in relation to bed cleaning, that nurses clean mattresses and the upper bedframe while domestics clean the lower bed parts. However, if workers do not discuss this together as a team, parts may be missed and equipment will be dirty.

DECONTAMINATION OF BLOOD SPILLAGE

In reference to the decontamination of blood spillage, it is important that local policies be followed. There are a number of key principles involved, including the following:

- Standard Precautions must be adopted at all times—personal protective equipment must be worn and hands must be decontaminated after the procedure.
- Designated colour-coded equipment must be used.
- Warning signs should be used to make the area safe.

- For blood:
  - Decontaminate the area using 10,000 ppm chlorine solution and rinse and dry
    OR
  - Use a powdered chlorine product to soak up the spill and mop the area using detergent and water and then dry.
    OR
  - Place paper towels over the spill and flood the area with a 10,000 ppm chlorine solution, mop up the towels and put them in clinical waste disposal, mop the area with detergent and water and then dry.

Key Points

- Chlorine disinfectants should **never** be used on urine spills; the principle of cleaning first with detergent and water and then disinfecting with a 10,000 ppm chlorine solution should be followed.
- Local policy for the method of choice should be consulted.
- All cleaning equipment which is reusable must be cleaned and rinsed and stored in a manner that allows it to dry between uses—this includes inverting mops and buckets.

CONCLUSION

Decontamination is an essential part of any infection prevention and control programme.[15] As discussed throughout this chapter, it is very important that appropriate protocols be employed for the decontamination of areas at high risk of microbial contamination in the hospital environment. In particular, in reducing the incidence and prevalence of HCAI the correct decontamination of medical devices is fundamental. Of growing relevance to hospital cleanliness and effectiveness are the evidence of biofilms. Biofilms should always be considered significant as they have a role to play in influencing the effectiveness of decontamination procedures in hospitals.[46] The reasons for this are discussed in further detail in Part 2 of this book.

REFERENCES

1. Rutala WA, Weber DJ. Disinfection and sterilization: an overview. *Am J Infect Control* 2013;**41**(Suppl. 5):S2–S5.
2. Hota B. Contamination, disinfection, and cross-colonization: are hospital surfaces reservoirs for nosocomial infection? *Clin Infect Dis* 2004;**39**:1182–9.
3. Otter JA, Yezli S, Perl TM, Barbut F, French GL. A request for an alliance in the battle for clean and safe hospital surfaces. *J Hosp Infect* 2013:24.
4. Kehoe B. Cleaning evaluation pays. New rewards coming for reducing HAIs. *Health Facil Manage* 2010; **23**(12):30–3.
5. Otter JA, Yezli S, Salkeld JA, French GL. Evidence that contaminated surfaces contribute to the transmission of hospital pathogens and an overview of strategies to address contaminated surfaces in hospital settings. *Am J Infect Control* 2013;**41**(Suppl. 5):S6–11.

6. Hughes JA. Infection control practices: they're not just suggestions. *Dent Assist* 2012;**81**(6):40–64.
7. Kramer A, Schwebbe I, Kampf G. How long do nosocomial pathogens persist on inanimate surfaces? A systematic review. *BMC Infect Dis* 2006;**6**:130.
8. Sexton T, Clarke P, O'Neill E, Dillane T, Humphreys H. Environmental reservoirs of *Staphylococcus aureus* in isolation rooms: correlation with patient isolates and implications for hospital hygiene. *J Hosp Infect* 2006;**62**:187–94.
9. Bergen LK, Meyer M, Høg M, Rubenhagen B, Andersen LP. Spread of bacteria on surfaces when cleaning with microfibre cloths. *J Hosp Infect* 2009;**71**(2):132–7.
10. Passaretti CL, Otter JA, Reich NG, Myers J, Shepard J, Ross T, et al. An evaluation of environmental decontamination with hydrogen peroxide vapor for reducing the risk of patient acquisition of multidrug-resistant organisms. *Clin Infect Dis* 2013;**56**(1):27–35.
11. Barbut F, Yezli S, Mimoun M, Pham J, Chaouat M, Otter JA. Reducing the spread of *Acinetobacter baumannii* and methicillin-resistant *Staphylococcus aureus* on a burns unit through the intervention of an infection control bundle. *Burns* 2013;**39**(3):395–403.
12. Hughes GJ, Nickerson E, Enoch DA, Ahluwalia J, Wilkinson C, Ayers R, et al. Impact of cleaning and other interventions on the reduction of hospital-acquired *Clostridium difficile* infections in two hospitals in England assessed using a breakpoint model. *J Hosp Infect* 2013;**84**(3):227–34.
13. Department of Health. *Choice Framework for Local Policy and Procedures 01-01: Management of Surgical Instruments (Medical Devices) Used in Acute Care* (CFPP 01-01)—*Part A: The Formulation of Local Policy and Choices; Part B: Common Elements*; *Part C: Steam Sterilization*; *Part D: Washer Disinfectors*; *Part E: Alternatives to Steam for the Sterilization of Reusable Medical Devices*; 2013a.
14. Department of Health. *Choice Framework for Local Policy and Procedures 01-06: Decontamination of Flexible Endoscopes* (CFPP 01-06)—*Operational Management*; *Choice Validation and Verification*; *Testing Methods*; 2013b.
15. Meredith SJ, Sjorgen G. Decontamination: back to basics. *J Perioper Pract* 2008;**18**(7):285–8.
16. Rupp ME, Adler A, Schellen M, Cassling K, Fitzgerald T, Sholtz L, et al. The time spent cleaning a hospital room does not correlate with the thoroughness of cleaning. *Infect Control Hosp Epidemiol* 2013;**34**(1):100–2.
17. Miller KL. Techniques for cleaning outside the box. *Health Facil Manage* 2012;**25**(10):46.
18. Griffith CJ, Obee P, Cooper RA, Burton NF, Lewis M. Evaluating the thoroughness of environmental cleaning in hospitals. *J Hosp Infect* 2007;**67**(4):390.
19. Dubberke ER, Reske KA, Noble-Wang J, Thompson A, Killgore G, Mayfield J, et al. Prevalence of *Clostridium difficile* environmental contamination and strain variability in multiple health care facilities. *Am J Infect Control* 2007;**35**(5):315–8.
20. Weber DJ, Rutala WA. Assessing the risk of disease transmission to patients when there is a failure to follow recommended disinfection and sterilization guidelines. *Am J Infect Control* 2013;**41**(Suppl. 5):S67–71.
21. Fijan S, Turk SŠ. Hospital textiles, are they a possible vehicle for healthcare-associated infections? *Int J Environ Res Public Health* 2012;**9**(9):3330–43.
22. Wilson J. *Infection control in clinical practice*, 3rd ed. Bailliere Tindall; 2006.
23. Fernie K, Hamilton S, Somerville RA. Limited efficacy of steam sterilization to inactivate vCJD infectivity. *J Hosp Infect* 2012;**80**(1):46–51.
24. Spaulding EH. Chemical disinfection of medical and surgical materials. In: Block SS, editor. *Disinfection, sterilization and preservation*, 4th ed. Lea & Febiger; 1968.
25. Lewis T, Patel V, Ismail A, Fraise A. Sterilisation, disinfection and cleaning of theatre equipment: do we need to extend the Spaulding classification? *J Hosp Infect* 2009;**72**(4):361–3.
26. Balsamo AC, Graziano KU, Schneider RP, Antunes Jr M, Lacerda RA. Removing biofilm from a endoscopic: evaluation of disinfection methods currently used. *Rev Esc Enferm USP*; 2012: 46, Spec No. 91-8.
27. Medicines and Healthcare Products Regulatory Agency. *Medical Device Alert*—MDA/2013/019: Detergent and Disinfectant Wipes Used on Reusable Medical Devices with Plastic Surfaces; 2013.
28. Carling PC, Huang SS. Improving healthcare environmental cleaning and disinfection: current and evolving issues. *Infect Control Hosp Epidemiol* 2013;**34**(5):507–13.

29. Kehoe B. Surface safety. Part 4. Best practices in surface and medical device disinfection. *Health Facil Manage* 2013;**26**(1):2p.
30. Bryce E, Lamsdale A, Forrester L, Dempster L, Scharf S, McAuley M, et al. Bedpan washer disinfectors: an in-use evaluation of cleaning and disinfection. *Am J Infect Control* 2011;**39**(7):566–70.
31. Rutala WA, Weber DJ. Sterilization, high-level disinfection, and environmental cleaning. *Infect Dis Clin North Am* 2011;**25**(1):45–76.
32. NICE, 2008. www.nice.org.uk/nicemedia/pdf/ip/IPG196guidance.pdf
33. National Institute for Health and Care Excellence. Patient safety and reduction of risk of transmission of Creutzfeldt–Jakob disease (CJD) via interventional procedures; 2006, November. Intervention Procedures IPG196, London.
34. Loftus RW, Patel HM, Huysman BC, Kispert DP, Koff MD, Gallagher JD, et al. Prevention of intravenous bacterial injection from health care provider hands: the importance of catheter design and handling. *Anesth Analg* 2012;**115**(5):1109–19.
35. Alfa MJ. Monitoring and improving the effectiveness of cleaning medical and surgical devices. *Am J Infect Control* 2013;**41**(Suppl. 5):S56–9.
36. Institute of Decontamination Sciences. MDA/2013/019 and MDA/2013/019; 2003. www.idsc-uk.co.uk/news.php.
37. Mitchell BG, Dancer SJ, Shaban RZ, Graves N. Moving forward with hospital cleaning. *Am J Infect Control* 2013:28 June.
38. Weber DJ, Anderson D, Rutala WA. The role of the surface environment in healthcare-associated infections. *Curr Opin Infect Dis* 2013;**26**(4):338–44.
39. Dancer SJ. Hospital cleaning in the 21st century. *Eur J Clin Microbiol Infect Dis* 2011;**30**(12):1473–81.
40. Jansen I, Murphy J. Environmental cleaning and healthcare-associated infections. *Healthc Pap* 2009;**9**(3):38–43.
41. Kleypas Y, McCubbin D, Curnow ES. The role of environmental cleaning in health care-associated infections. *Crit Care Nurs Q* 2011;**34**(1):11–17.
42. National Patient Safety Agency. *Notice 15: Colour Coding Hospital Cleaning Materials and Equipment*; 2007, 10 January.
43. Thrall TH. Complete cleaning: improved cleaners, disinfectants, monitoring systems and training help close the loop on infection prevention. *Health Facil Manage* 2013;**26**(4):43–6.
44. Bosley M. Water quality—how to ensure a pure supply. *Health Estate* 2009;**63**(7):53–5.
45. Otter JA, French GL. Survival of nosocomial bacteria and spores on surfaces and inactivation by hydrogen peroxide vapor. *J Clin Microbiol* 2009;**47**(1):205–7.
46. Yezli S, Otter JA. Does the discovery of biofilms on dry hospital environmental surfaces change the way we think about hospital disinfection? *J Hosp Infect* 2012;**81**(4):293–4.

CHAPTER FIVE

Challenges to Healthcare Providers

Natalie Vaughan and Jacqueline Randle

INTRODUCTION

The *Code of Practice,*[1] which forms part of the Health and Social Care Act of 2008,[2] is unequivocal in its message that good infection prevention and control are essential in order for healthcare and social care service users to receive safe and effective treatment. Thus, all healthcare workers have a responsibility to become actively involved in preventing and reducing healthcare-associated infections (HCAI). Numerous guidelines, procedures and policy documents have been issued in an attempt to promote safe and effective infection prevention and control practice, and patients and their families rely on healthcare workers to consistently implement best practice in order to provide a safe environment for them.

This chapter provides information concerning three micro-organisms that are of ongoing concern to healthcare providers. By concentrating on just three, we do not wish to negate the sometimes devastating effect of other micro-organisms nor do we wish to trivialise other less-known (in the United Kingdom) ones such as Carbapenemase-resistant *Enterobacteriaceae*. Instead, the intention is to provide up-to-date information that will benefit both clinical practice and patient care.

CLOSTRIDIUM DIFFICILE

The *Operating Framework* for the National Health Service (NHS) in England 2012-2013[3] identifies national priorities for delivering improved quality and cost-efficient care, with one priority being the reduction of *Clostridium difficile* (*C. difficile*) cases. Significant activity has resulted in progress whereby the number of cases has been reduced, but new ambitions have been set. The NHS is being asked to continue year-on-year reductions in *C. difficile* infections, and if the target of a 17% reduction is achieved by the end of 2013, it will reduce these infections from 19,754 to 16,100.[4]

Clostridium difficile is a Gram-positive anaerobic bacterial infection which produces toxins A and B.[5] Toxin A is a cytotoxin and a potent enterotoxin which causes

Biofilms in Infection Prevention and Control
Doi: http://dx.doi.org/10.1016/B978-0-12-397043-5.00005-0

fluid secretion, mucosal damage and intestinal inflammation, while toxin B is a potent cytotoxin which causes its characteristic explosive diarrhoea by stimulating muscle contractions.[6]

C. difficile is the most common cause of hospital-acquired infectious diarrhoea.[7] Patients excrete large numbers of organisms which rapidly form spores to survive. These spores, which are extremely resilient, contaminate the patient's environment and can persist for many months, thus providing a source of infection for future patients. Up to 25% of patients suffer a relapse of diarrhoea following successful treatment.

Rates of colonisation increase significantly at the age of 65 and older,[6] but it is possible for younger patients to suffer with *C. difficile* infection.[8] The most important pre-disposing factor for susceptibility to *C. difficile* infection, however, is the use of a broad-spectrum antibiotic.[9] The gastrointestinal flora, which inhibits growth of the organism, is altered by antibiotic use, enabling *C. difficile* to grow and multiply, producing excess toxins.[10] It is this alteration that causes harm, although the specific mechanism is still not fully understood.[6] The result is inflammation, fluid and mucous secretion and damage, which leads to diarrhoea, colitis, pseudomembranous colitis and occasionally death.[11] These symptoms are known as *C. difficile*-associated disease (CDAD) or *C. difficile* infection (CDI). This should be considered as a diagnosis in its own right.

The Health Protection Agency[12] has identified the following additional factors predisposing patients to *C. difficile*:

- Proton pump inhibitors
- Severe underlying disease
- Length of stay in hospital/nursing home
- Nasogastric intubation
- Older patients
- Non-surgical gastrointestinal procedures
- Long antibiotic course
- Administration of multiple antibiotics or multiple courses

A report compiled by the Healthcare Commission[13] investigating outbreaks of *C. difficile* at Maidstone and Tunbridge Wells NHS Trust, UK, found evidence of avoidable deaths. The report concluded the following:

- Inadequate care was received by 80% of patients.
- Medical care was poor in a third of the cases.
- There was a lack of specialist care.
- There was a delay in stopping antibiotics and/or treatment with inappropriate antibiotics.

The following list contains more details about the preceding from the Healthcare Commission report:

Inadequate care: There was a complete lack of nutritional assessment, use of fluid and stool charts, and inaccurate assessments using the Bristol Stool Chart.

Medical care: Medics did not regularly review the patient's *C. difficile* status, and at times there was no mention in the case notes that patients had this diagnosis.

Specialist care: In a fifth of cases specialists were not involved and so, although microbiologists were involved in a third of the cases, infection prevention and control nurses were only involved in 8% of cases; the intensive care unit outreach team was involved in only 12% of cases.

Antibiotic use: There was a delay in antibiotic treatment of up to two weeks, mainly due to lack of communication about positive test results or failure to repeat negative test results even in the presence of diarrhoea. Additionally, patients continued to receive broad-spectrum antibiotics when simple antibiotics less prone to causing *C. difficile* would have sufficed. Excessive antibiotics were used for simple infections and, in some patients, even in the absence of significant infections. They were often used for excessive periods of time, and broad-spectrum antibiotics were sometimes not discontinued even when *C. difficile* had been diagnosed.

The report's findings,[13] along with those of reports from other hospitals with *C. difficile* outbreaks, formed the basis for current recommendations for care of patients suspected of CDI.

Diagnosis and Treatment

To help promote early identification, the Department of Health and the Health Protection Agency[8] recommend the use of the acronym SIGHT, as follows:

- **S**uspect that a patient may be infective where there is no clear alternative for the cause of diarrhoea.
- **I**solate the patient and consult with the infection prevention and control team while determining the cause of the diarrhoea.
- **G**loves and aprons must be used for all contact with patients and their environment.
- **H**andwashing with soap and water should be carried out before and after each patient and environment contact.
- **T**est the stool for toxin by immediately sending a specimen.

In 2009, concerns were raised regarding the accuracy and effectiveness of testing kits used for diagnosing *C. difficile*.[4] Consequently, new guidance[4] aims to promote more effective and consistent diagnosis, testing and treatment of CDI. Diagnosis is by a stool specimen from one episode of diarrhoea, which ranges from mild, self-limiting to severe and frequent episodes of watery, green, foul-smelling stools every day.[14] Once diarrhoea is suspected, the Bristol Stool Chart is used as part of the diagnostic process. Heaton[15] developed the chart, which is depicted in Figure 5.1, to visually assess the likelihood of infection; the figure shows the seven types of stools.

Specimens that are not types 5 through 7 should not be processed, and the stool has to be liquid enough to take the shape of the container. It is also important to note that many patients may be embarrassed by diarrhoea and may try to hide this from clinical

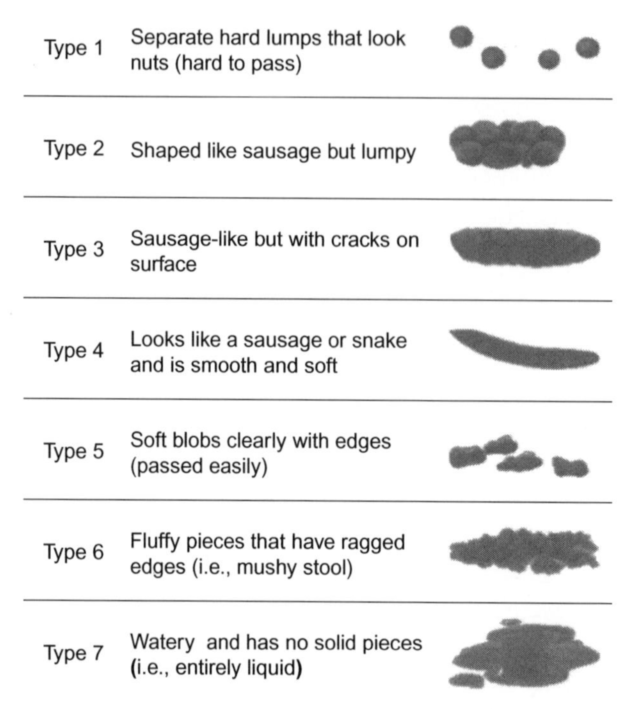

Figure 5.1 ***Bristol Stool Chart***. *Source: Adapted from image in Heaton.*[15]

workers. It is vital that the staff clearly explain the importance of patients reporting episodes of diarrhoea and the potential adverse consequences of not doing so.

The stool specimen should be sent promptly to the microbiology laboratory; it should be placed in a refrigerator if there is to be a delay, but must be received by the laboratory within 24 hours. If the result is a positive toxin, the diarrhoea may be due to CDI and the patient is then a cross-infection risk. Enteric precautions (described in Chapter 2) must be implemented. If the result is a positive polymerase chain reaction (PCR), the diarrhoea may not be due to CDI, but the patient is carrying *C. difficile* and may make the toxin. Therefore the patient is a cross-infection risk and enteric precautions must be implemented. If the first stool sample is negative and the patient still has diarrhoea, another sample should be sent for examination before the patient is classified as CDI-negative.

Interventions

All *C. difficile* interventions should follow local policies and be patient-focussed in order to provide safe patient care. The following interventions are required:[6,16]

- Strict adherence to local antibiotic policy, with all antibiotics that are not required stopped. Broad-spectrum antibiotics, as well as any other medication that might cause diarrhoea, are to be avoided.

- Consultation with a medical microbiologist and a *C. difficile*/diarrhoea Gastroenterology Nurse Specialist.
- Implementation of enteric precautions. Isolation, unless patient safety is compromised, or cohorting. The infection prevention and control team should be consulted.
- Staff adherence to local hand hygiene policy.
- Implementation of a bare-below-elbow policy and handwashing with soap and water. Alcohol-based handrub alone is not effective against *C. difficile*.
- Effective communication with patient, relatives and visitors.
- Effective communication with other healthcare workers.
- Management of other symptoms such as malnutrition.
- Maintenance of hydration. Some organisations prescribe oral rehydration therapies routinely for patients with *C. difficile* to prevent dehydration.
- Assessment of the patient's skin condition and appropriate intervention to prevent skin damage.
- Assessment of the patient's psychological status and appropriate intervention, if needed.
- Accurate documentation such as the Bristol Stool Chart.
- Enteric precautions to be discontinued when the patient is diarrhoea-free for 72 hours. However, the patient should open his or her bowels with a formed stool and/or exhibit back-to-normal bowel habit before enteric precautions are discontinued.
- Thorough decontamination of the room or bed space daily, and on discharge/discontinuation of enteric precautions, with a chlorine-containing cleaning agent (at least 1000 ppm available chlorine). Use of hydrogen peroxide decontamination should also be considered.
- Decontamination of patient equipment as in the previous bullet point.
- Clean bed linen and night wear or clothing if the patient is being moved to a new room or bed space.

Assessment of Nutrition and Malnutrition

As the preceding indicates, all patients should have a nutritional screening and assessment to identify malnutrition,[17] which is very common in hospitalised patients. Better nutrition can play a critical role in reducing HCAIs.[18] Numerous screening tools have been developed, with the most commonly used being the *Malnutrition Universal Screening Tool* (MUST), a national directive in England for adults only.[19] However, any nutritional screening and assessment tool should include the following:[20]

- *Dietary history*—frequency of eating and habits, preferences, meal pattern and portion sizes
- *Living environment*—facilities for storing and preparing food
- *Socioeconomic status*—resources for purchasing food and reliance on other people for mealtime support
- *Psychosocial factors*—depression, anxiety, bereavement and dietary knowledge

- *Disability*—ability to purchase, prepare and eat food
- *Disease*—acute or chronic disease which might influence diet, appetite, energy expenditure or swallowing
- *Gastrointestinal symptoms*—anorexia, nausea, vomiting, diarrhoea and constipation
- *Significant weight change*—comparison of current weight with usual weight (rapid weight loss can indicate fluid loss and dehydration as a result of diarrhoea)
- *Drug therapy*—many drugs can induce gastrointestinal symptoms
- *Dental health*—poor dentition, denture suitability, mouth ulceration or gum disease
- *Mobility and activity level*—comparison of energy expenditure with energy intake
- *Physical appearance*—emaciation, cachexia (physical wasting with loss of weight and muscle wastage), obesity or loose clothing
- *Blood biochemistry*—plasma proteins, haemoglobin, serum vitamin and mineral levels and immunological competence
- *Physical assessment of nutritional status*—patient's height and weight

As part of the assessment process, the patient's weight and height should be established; do the following to establish the patient's height:[20]

- Explain the procedure and gain consent.
- Decontaminate your hands.
- Ask the patient to remove his or her shoes.
- Ask the patient to stand against a wall and measure height from the floor to the top of the crown of the head.
- If the patient cannot stand, lay him or her flat and measure height from the base of the foot to the crown of the head.
- Decontaminate your hands.
- Document the patient's height in his or her record.

To establish the patient's weight[20] do the following:

- Explain the procedure and gain consent.
- Decontaminate your hands.
- Decontaminate the weighing scales.
- Ask the patient to remove his or her shoes and any heavy articles of clothing.
- Ensure that the scales are calibrated according to the manufacturer's instructions.
- Ask the patient to get on the weighing scales.
- Ask the patient to keep as still as possible.
- Take the weight measurement.
- Ask the patient to get off the scales.
- Decontaminate the weighing scales.
- Decontaminate your hands.
- Document the patient's weight in his or her record.

Once the patient's weight and height measurements have been taken, calculate the patient's body-mass index (BMI) using the following formula[20] as an indicator of nutritional status in patients 18 years of age and older.

$$BMI = weight(kg)/(height[m])^2$$

BMI < 20: Underweight
BMI 20–24.9: Desirable range
BMI 25–29.9: Grade I obesity (overweight)
BMI > 30: Grade II obesity (obese)
BMI > 40: Grade III obesity (morbidly obese)

After the patient's usual weight has been ascertained, the following formula can be used to determine whether she or he is malnourished:

$$\%\,\text{weight loss} = \text{usual weight} - \text{actual weight} \times 100 / \text{usual weight}$$

Unintentional weight loss of 10% over a period of 6 months represents malnutrition
Unintentional weight loss of 20% over a period of 6 months represents severe malnutrition

METICILLIN-RESISTANT *STAPHYLOCOCCUS AUREUS*

Staphylococcus aureus is a Gram-positive bacterium that colonises the nose, skin and perineum. About 90% of such infections are *meticillin-resistant Staphylococcus aureus* (MRSA), which is the term used for those that have acquired resistance to meticillin.[21] The micro-organism does not usually cause harm in a healthy person, and most people with MRSA are colonised without suffering harm.[22] If it causes infection, it can result in boils, pimples, skin sepsis, scalded skin syndrome, wound infection and pneumonia. If it enters the bloodstream (bacteraemia), it can result in osteomyelitis, liver abscesses, kidney and liver endocarditis, toxic shock syndrome and death.[23,24]

The presence of the bacteria becomes more problematic if the patient develops an ulcer or wound or is being treated with an invasive procedure/device. Infection occurs when bacteria enter the soft tissues, multiply and cause tissue damage, and/or when bacteria enter the bloodstream. Infections due to MRSA are associated with a greater risk of treatment failure, increased patient mortality and higher medical costs.[25] The emergence of MRSA can be blamed on the overprescription of antibiotics.[26]

Additionally, poor compliance with hand hygiene is a major cause of the spread of MRSA, and although environmental hygiene is important in helping reduce HCAIs, its importance in relation to MRSA is often overemphasised.[27] However, prevention and control relies on a high standard of infection prevention and control precautions at all times (including single-room isolation or cohorting if there are several cases), appropriate screening policies and good compliance with topical therapies aimed at reducing or removing carriage of the organism.[28]

Key Point

- MRSA is an important pathogen because, although it is no more infectious than some other micro-organisms, it is resistant to some antimicrobial medications.

The Current Trend

The incidence of MRSA continues to rise globally and so presents an ongoing challenge to healthcare workers.[29] The NHS is aiming to reduce the number of MRSA infections in 2012–2013 by a further 29%, and if this target is achieved, annual numbers of MRSA bloodstream infections will decrease.[30] The following are the vehicles for transmission:

- A colonised patient
- An infected patient
- A colonised healthcare worker
- An infected healthcare worker
- Contaminated equipment and/or environment

Attempts to control the spread of MRSA have focussed on these three areas:[25]

1. Hand hygiene
2. Restriction of antibiotics
3. Detection and isolation of infected or colonised patients

Key Points

- It is rare for healthcare workers who have no underlying disease to become infected with MRSA.
- Patients are at greater risk due to underlying illness, clinical procedures, invasive devices *in situ*, antibiotic therapy and being cared for in environments where MRSA exists.

In a MRSA infection, the micro-organisms cause clinical signs of infection.[21] The signs and symptoms are not specific to MRSA and are the same as the general signs and symptoms expected for that type of infection.

Patients can be classified as those at high, medium or low risk of contracting MRSA or those for whom the result of infection may have severe consequences,[25] as shown in Table 5.1. Additionally, patients at high risk of MRSA[25] are those who are

- Known to have been infected or colonised in the past.
- Frequently admitted to healthcare facilities.
- Transferred between units in hospital or between hospitals.
- Inpatients at hospitals that have or are likely to have a high MRSA prevalence rate.
- Residents of nursing homes where there is a known or likely high prevalence of MRSA carriage.

Table 5.1 Patient Risk of Contracting MRSA

High risk—any patient in	Medium risk—any patient in	Low risk—any patient in
Intensive care	Admissions ward	Psychiatric
Neonatal intensive care	General surgical	Long-term facilities
Burns units	Urological	
Transplantation	Paediatric	
Cardiothoracic	General medical	
Orthopaedic	Elderly medicine	
Trauma		
Vascular surgery		
Renal		
National or international referral centres		
Specialist areas (as determined by the infection prevention and control team)		

- Injecting drug users (especially PVL-producing strains).
- HIV patients.

Some types of *Staphylococcus aureus* produce a toxin called Panton-Valentine leukocidin (PVL-SA), which destroys white blood cells and is associated with virulent and transmissible strains of *S. aureus*, including CA-MRSA.[31] PVL-SA is easily spread by close contact and through sharing of personal items, which explains why it has become a major public health problem in the United States and is rapidly spreading across Europe.

Unlike those infected with MRSA, healthy people are equally at risk of acquiring a PVL-SA infection. The Health Protection Agency[31] reports that rates remain 'modest' in the United Kingdom and Europe, but warns that this toxin is highly transmissible with a high attack rate and if untreated, can lead to illnesses such as severe bone and joint infections, endocarditis and necrotising pneumonia. It is recommended that community and practice nurses pay close attention to patients with recurring skin infections, particularly in high-risk areas (e.g., care homes), where the infection can easily spread.[31]

The PVL-SA micro-organism has no defined set of symptoms; instead, symptoms are general and often ambiguous and are those that are common to infections caused by other bacteria. The micro-organism can cause a wide range of infections and is highly adaptable. Superficial and minor skin lesions are the primary infections caused by *Staphylococcus aureus*, and although acquiring a serious *Staphylococcus aureus* infection is difficult, it can be devastating for the patient and his or her caregivers. The impact of every infection should not be underestimated.

If MRSA (or any *Staphylococcus aureus*) spreads from a local site into the bloodstream, it can lodge at various sites in the body (e.g., lungs, kidneys, bones, liver, spleen) and cause one or more deep abscesses distant from the original site. These can be painful and can result in the following:

- High fever
- High white blood cell count
- Signs of inflammation near the infection
- General feeling of being very unwell
- Disturbance of blood clotting with a tendency to bleed
- Rigors (shivers)
- Low blood pressure (shock)

Over a period of time, the body enters a catabolic state with breakdown of tissue, loss of weight and failure of essential organs. This is usually linked with an associated septicaemia.

Screening and Treatment

MRSA and PVL-SA can be detected by routine screening swabs (nose and perineum), but specialist reference laboratory testing is required to confirm the presence of the latter. However, a new diagnosis of MRSA and PVL-SA is often determined from a clinical specimen. MRSA screening is routine for emergency, elective and day case patients.

MRSA and PVL-SA can be difficult to eradicate because they persistently colonise; however, topical treatment can be used to eradicate skin and nasopharyngeal colonisation, depending on the individual patient. It should be remembered that some patients (e.g., older patients with friable skin or those with a pre-existing skin condition) are unlikely to be able to tolerate antiseptic skin washes.[23] However, products are available that are less likely to cause damage to fragile skin; it is recommended that these be used as it is important for the patient to still receive decolonisation therapy.

The following are general guidelines for eradication:

- Bathe or shower for five days with an antiseptic body wash that contains chlorhexidine octenidine.
- Wash hair with antiseptic body wash on days 1 and 4 of treatment.
- Apply mupirocin 2% nasal ointment (Bactobran) three times a day for five days; however, some strains are resistant to this drug and so an alternative (e.g., naseptin) may be prescribed.
- Apply ointment to the inside of the nostril in a 'match head' or 'pea-sized' amount.
- Administer antimicrobial therapy if the patient becomes systemically unwell, and seek the advice of a medical microbiologist.

Intervention

All interventions should follow local policies and be patient-focussed in order to provide safe patient care. This list contains interventions that should be implemented in the context of local policies:[32,33]

- Wound and skin precautions should be implemented.
- Isolation is recommended, unless patient safety is compromised, or cohorting (seek advice from the infection prevention and control team).
- A patient with a highly resistant strain of MRSA (e.g., Mupirocin resistance) should be accorded the highest priority for single-room accommodation and should be managed with the strictest wound and skin precautions.
- Healthcare workers must strictly adhere to the local hand hygiene policy.
- Communicating effectively with patients, visitors and relatives is essential.
- Communicating effectively with other healthcare workers is essential.
- The patient's psychological status and appropriate intervention must be assessed.
- Documentation must be accurate.
- Screening and decolonisation must be carried out per local policy.
- Wound and skin precautions should be discontinued only on the advice of the infection prevention and control team and only after three screens' results are negative. Screening must be done at least 48 hours after completion of treatment and at weekly intervals.
- The room or bed space should be thoroughly decontaminated daily, and on discharge/discontinuation of wound and skin precautions, with chlorine-containing cleaning agents (1000 ppm available chlorine).

Screening of Healthcare Workers

In some countries, such as Denmark and Sweden, healthcare workers are routinely screened for MRSA carriage. In the 1990s many trusts in the United Kingdom included routine screening, but review of the evidence on its effectiveness demonstrated that it was not necessary as part of basic control measures. This remains controversial, and in the United Kingdom although screening of some healthcare workers does occur, it is not routine. Instead, Coia et al.[34] recommend a risk-based approach to individual cases, rather than mass healthcare worker screening; thus, perform a screen only when basic control measures have failed or when there is evidence of a cluster of cases that could be related to an individual healthcare worker.

Treatment in otherwise healthy individuals is usually very effective, and healthcare workers who are found to be MRSA-positive may not be excluded from work while receiving treatment. Those who have to be away from work (e.g., they may have infected/colonised hand lesions) or who work in a high-risk area (e.g., a neonatal unit) may have their work modified during treatment. They should be assessed by their occupational health department in consultation with the infection prevention and control team.

A healthcare worker with a proven PVL-SA infection should not work in clinical areas until the acute infection has resolved (e.g., affected skin is intact) and 48 hours of a five-day decolonisation regimen has been completed.[33] Healthcare workers should not be discriminated against in terms of their current employment or future employment prospects by a diagnosis of MRSA.[35]

NOROVIRUS

The burden of norovirus on patients and healthcare providers is increasing,[36] and outbreaks of this infection can cause considerable disruption to service delivery.[37] Acute onset of vomiting and/or diarrhoea is the prominent symptom, and others include nausea, abdominal cramps, headache, myalgia, chills and fever.[38] However, it is estimated that 30% of infections may be asymptomatic.[38] Noroviruses are non-enveloped viruses belonging to the *Caliciviridae* family and used to be commonly known as the winter vomiting bug. Table 5.2 provides clinical information about the virus.

Because norovirus is highly infectious, immediate action is required once a patient or patients become ill with clinical signs of the aforementioned symptoms. Isolation/cohorting of affected patients and the instigation of enteric precautions should occur as soon as symptoms start. A confirmed laboratory diagnosis is not a pre-requisite; however, it is required for identification and surveillance. Ideally patients are accommodated in single rooms or can be cohorted in bays.

Declaration of an outbreak occurs when there is a trigger point for the activation of an organisational response. This may not require a rigid definition and can be tailored to suit the organisation and resource limitations. Isolation of a patient is not dependent on the declaration of an outbreak.[37] Health Protection Scotland[38] provides a very useful decision tree to help clinicians decide whether an outbreak is occurring and, if so, provide algorithms for helping clinicians to decide when to close the ward or bay and when to reopen it (see Exhibit 5.1).

Table 5.2 Clinical Information about Norovirus

Incubation period	**Usually 24–48 hours[39]**
Common clinical features	Symptoms of norovirus most commonly begin around 12–48 hours after becoming infected. The illness is self-limiting and symptoms last for 24–60 hours.
Reservoir	Human gastrointestinal tract
Transmission	Principal route of transmission of enteric pathogens is via the faecal–oral spread[40] and by vomiting probably caused by widespread aerosol dissemination of virus particles, environmental contamination and subsequent indirect person-to-person spread.[41] The virus also can be transmitted via contaminated water and food.

EXHIBIT 5.1 Outbreak Control Measures

Ward

- Close affected bay(s) to admissions and transfers.
- Keep doors to single-occupancy room(s) and bay(s) closed.
- Place signage on door(s) informing all visitors of the closed status and restricting visits to essential staff and essential social visitors.
- Place patients in the ward for the optimal safety of all patients.
- Prepare for re-opening by planning the earliest date for a terminal clean.

Healthcare Workers

- Ensure that all staff are aware of the norovirus situation and how norovirus is transmitted.
- Ensure that all staff are aware of the work exclusion policy and the need to go off duty at first symptoms.
- Allocate staff to duties in either affected or non-affected areas of the ward but not both unless unavoidable (e.g., therapists).

Patients and Relatives

- Provide all affected patients and visitors with information on the outbreak and the control measures they should adopt.
- Advise visitors of the personal risk and how they might reduce it.

Continuous Monitoring and Communications

- Maintain an up-to-date record of all patients and staff with symptoms.
- Monitor all affected patients for signs of dehydration and correct as necessary.
- Regularly brief organisational management, public health organisations and the media office.

Personal Protective Equipment

- Use gloves and aprons to prevent personal contamination with faeces or vomitus.
- Consider face protection with a mask only if there is a risk of droplets or aerosols.

Hand Hygiene

- Use liquid soap and warm water as per WHO's *My Five Moments for Hand Hygiene*.
- Encourage patients to perform hand hygiene and assist them with it.

Environment

- Remove exposed foods (e.g., fruit bowls) and prohibit eating and drinking by staff in clinical areas.
- Intensify cleaning, ensuring that affected areas are cleaned and disinfected; toilets used by affected patients must be included.
- Decontaminate frequently touched surfaces with detergent and disinfectant containing 1000 ppm available chlorine (or its equivalent).
- Discard all waste as healthcare waste.
- Remove PPE and wash hands with liquid soap and warm water.

Equipment

- Use single-patient-use equipment wherever possible.
- Decontaminate all other equipment immediately after use.

Linen

- While clinical area is closed, discard linen from the closed area in a water-soluble (alginate) bag and then a secondary bag.

Spillages

- Wearing PPE, decontaminate all faecal and vomit spillages.
- Remove spillages with paper towels and then implement outbreak control measures.

The definition of the end of an outbreak is when terminal cleaning has been successfully undertaken. Patients with continuing symptoms should be moved to side rooms if doing so expedites terminal cleaning (if their condition allows).

Clinical Care

In addition to the preceding, rehydration and avoidance of dehydration are the main priorities of care. These are to be undertaken in conjunction with treatment for the patient's underlying condition. Anti-emetics and anti-diarrhoeal agents are discouraged because of their effect on the gut, which may predispose the patient to *C. difficile* infection.

Discharge and Transfer of Patients

Patients can be discharged to their own home when their condition allows. For transfer to continuing care homes and community-based institutions, they should be symptom-free of any norovirus for at least 48 hours. Patients should not be transferred between affected areas in hospitals.

CONCLUSION

This chapter covers three micro-organisms as separate topics and provides a comprehensive approach to each. As already stated, other micro-organisms, which can have devastating effects, are also common in healthcare environments. Specialist advice should be sought from the infection prevention and control team.

REFERENCES

1. Department of Health. *The Health and Social Care Act 2008: Code of Practice on the prevention and control of infections and related guidance*. London; 2010.
2. Health and Social Care Act. legislation.gov.uk; 2008.
3. Department of Health. *NHS Operating Framework for 2012/2013 (England)*. London; 2011.
4. Department of Health. *Updated guidance on the diagnosis and reporting of Clostridium difficile*. London; 2012.
5. National *Clostridium difficile* Standards Group. Report to the Department of Health. *CDR Weekly* 2003;13:40.
6. Hall J, Horsley M. Diagnosis and management of patients with *Clostridium difficile*-associated diarrhoea. *Nurs Stand* 2007;**21**:49–56.

7. Simor A, Bradley S, Strausbaugh L, Crossley K, Nicolle L. *Clostridium difficile* in long-term facilities for the elderly. *Inft Cont Hosp Epidemiol* 2002;**23**:696–702.
8. Department of Health and the Health Protection Agency. *Clostridium difficile infection: How to deal with the problem. Advice on the most effective methods of prevention and control of this infection and the management of outbreaks*. London; 2009.
9. Wiegand PN, Nathwani D, Wilcox MH, Stephens J, Shelbaya A, Haider S. Clinical and economic burden of *Clostridium difficile* infection in Europe: a systematic review of healthcare-facility acquired infection. *J Hosp Infect* 2012;**81**:1–14.
10. Jenkins L. The prevention of *Clostridium difficile* associated diarrhoea in hospital. *Nurs Times* 2004; **100**:56–9.
11. Sunenshine R, Clifford L. *Clostridium difficile*-associated disease: new challenges from an established pathogen. *J Med* 2006;**73**:187–97.
12. Health Protection Agency Regional Microbiology Network. *A Good Practice Guide to Control of Clostridium difficile*. London; 2007.
13. Healthcare Commission. *Investigation into Outbreaks of Clostridium difficile at Maidstone and Tunbridge Wells NHS Trust*. London; 2007.
14. Brazier J. The diagnosis of *Clostridium difficile*-associated disease. *Antimicrobial Chemother* 1998: **41**(suppl):67–69.
15. Heaton KW. The Bristol Stool Form Scale. Practical procedures for nurses. *Nurs Times* 1999;**95**:25.
16. Nottingham University Hospitals NHS Trust. *Clostridium difficile* Diarrhoea Policy. Nottingham; 2012.
17. British Association for Parenteral and Enteral Nutrition. *A Toolkit for Commissioners and Providers in England: Malnutrition Matters—Meeting Quality Standards in Nutritional Care*. Redditch, Worcester; 2010.
18. Curtis L. Prevention of hospital-acquired infections: a review of non-pharmacological interventions. *J Hosp Infect* 2008;**69**:204–19.
19. British Association for Parenteral and Enteral Nutrition. Malnutrition Universal Screening Tool (MUST); 2012. www.bapen.org.uk/screening-for-malnutrition/must/musttoolkit/the-must-itself [accessed 15.03. 2013]
20. Stayt L, Randle J. Gastrointestinal system. In: Randle J, Coffey F, Bradbury M, editors. *Clinical skills in nursing*. Oxford University Press; 2009.
21. Winter G. A bug's life. *Nurs Stand* 2005;**19**:16–18.
22. Gill J, Kumar R, Todd J, Wiskin C. Methicillin-resistant *Staphylococcus aureus*: awareness and perceptions. *J Hosp Infect* 2006;**62**:333–7.
23. Rayner D. MRSA: An infection control overview. *Nurs Stand* 2003;**17**:47–53.
24. Mims C, Dockrell H, Goering R, Roitt I, Wakelin D, Zuckerman M. *Medical microbiology*, 3rd Ed. Elsevier Mosby; 2004.
25. Xue Y. *MRSA: Screening*. Joanna Briggs Institute-COnNECT; 2008.
26. Washer P, Joffe H, Solberg C. Audience readings of media messages about MRSA. *J Hosp Infect* 2008;**70**:42–7.
27. MRSA Working Group. *Separating Fact from Fiction: The Role of Treatment within MRSA Management*. London; 2008.
28. Perry C. *Infection prevention and control*. Blackwell Publishing; 2007.
29. Department of Health. *Getting Ahead of the Curve: A Strategy for Combating Infectious Diseases (including other aspects of health protection*; 2002. www.doh.gov.uk/cmo/idstrategy [accessed 19.03.2013].
30. Department of Health. New objectives set to reduce MRSA and *C difficile*; 2012. www.dh.gov.uk/health/2012/01/mrsa-cdifficile-objectives/ [accessed 19.03.2013].
31. Health Protection Agency. *Guidance on the Diagnosis and Management of PVL-associated Staphylococcus aureus Infections (PVL-SA in England). Report by the PVL of the Steering Group on Healthcare Associated Infection Subgroup*. London; 2008.
32. Nottingham University Hospitals NHS Trust. Methicillin-resistant *Staphylococcus aureus* (MRSA) policy. Nottingham; 2010.
33. Nottingham University Hospitals NHS Trust. Panton-Valentine Leukocidin Associated *Staphylococcus aureus* (PVL-SA) policy. Nottingham; 2013.
34. Coia J, Duckworth G, Edwards D, Farrington M, Fry C, Humphreys H, et al. Guidelines for the control and prevention of methicillin-resistant *Staphylococcus aureus* (MRSA) in healthcare facilities. *J Hosp Infect* 2006;**63**(Suppl 1):S1–44.

35. RCN. *Methicillin-resistant Staphylococcus aureus (MRSA): Guidance for Nursing Staff.* London: Royal College of Nursing; 2005.
36. Tam C, Rodrigues L, Vivani L, Dodds J, Evans M, Hunter P, et al. Longitudinal study of infectious intestinal disease in the UK (11D2 study): incidence in the community and presenting to general practice. *Gut* 2012;**61**:69–77.
37. Norovirus Working Party. *Guidelines for the management of norovirus outbreaks in acute health and social community settings.* Department of Health, London; 2012.
38. Health Protection Scotland. *Norovirus Guidance: Preparedness, Control Measures and Practical Considerations for Optimal Patient Safety and Service Continuation in Hospitals.* Glasgow; 2012.
39. Licorish J, Mohamed H, Taylor B, Chipwete J, Robinson R, Docherty A, et al. *Norovirus Prevention and Control. A toolkit for planning and providing high quality services in the West Midlands.* Health Protection Agency; 2008. www.hpa.org.uk [accessed 19.03.2013].
40. Irving W, Boswell T, Ala'Aldeen D. F8: Gastroenteritis and food poisoning in BIOS instant notes. In: *Medical microbiology.* Taylor and Francis; 2005.
41. Chadwick PR, Beards G, Brown D, Caul EO, Cheesborough J, Clarke I, et al. Management of hospital outbreaks of gastro-enteritis due to small round structured viruses. *J Hosp Infect* 2002;**45**:1–10.

CHAPTER SIX

Changing Practice

Jacqueline Randle and Mitch Clarke

INTRODUCTION

Infection prevention and control are high on the healthcare agenda. In response to the escalating costs identified in previous chapters and because of the harm healthcare-associated infections (HCAI) can cause to patients and their families,[1] the Department of Health in England has produced several guidance documents—*Getting Ahead of the Curve: A Strategy for Combating Infectious Diseases (Including Other Aspects of Health Protection)*,[2] *Winning Ways: Working Together to Reduce Health Care Associated Infection in England—Report from the Chief Medical Officer*,[3] *A Matron's Charter: An Action Plan for Cleaner Hospitals*[4] and *Clean Safe Care: Reducing Infections and Saving Lives*[5]—in attempts to reduce HCAI. However, the implementation of this guidance has been described as inconsistent and patchy.[6,7]

In an attempt to remedy this inconsistency, the *Health and Social Care Act 2008: Code of Practice on the Prevention and Control of Infections and Related Guidance* for registered providers in England was introduced.[8,9] Its ultimate aim is to improve and to embed high standards of infection prevention and control practice and thereby reduce the burden of HCAI. Importantly, in contrast to previous guidance, the *Code of Practice* provides punitive actions, including improvement notices, fines, 'special measures' for registered providers found not to be compliant, and in the most serious cases criminal prosecution. Since its publication in 2008 the code's scope has widened to include all registered providers of healthcare and adult social care, primary dental care and medical care, as well as independent-sector ambulance providers.

Along with the *Code of Practice*, the *Report of the Mid-Staffordshire NHS Foundation Trust Public Inquiry*—known as the *Francis Report*[10]—has detailed its recommendations for healthcare practice, following the discovery of numerous failings which resulted in avoidable patient deaths and unacceptable levels of care. As infection prevention and control practice does not occur in a vacuum, it is reasonable to suggest that a wider perspective which encompasses healthcare contexts is required because poor practice occurs in poor environments.[11] Consequently, this chapter looks at the following:

- The *Code of Practice* for the prevention of healthcare-associated infection and the Care Quality Commission

Biofilms in Infection Prevention and Control
Doi: http://dx.doi.org/10.1016/B978-0-12-397043-5.00006-2

- Commissioning of infection prevention and control services
- The *Francis Report*
- Empowerment and power
- Whistleblowing

THE *CODE OF PRACTICE*

The *Code of Practice*[9] is divided into the following sections:

- The first is aimed at registered providers, organisations and their environment, describing the structures and processes providers should follow for compliance.
- The second details the criteria against which registered providers are judged on their compliance with the requirements of cleanliness and infection control for registration.
- The third and fourth sections relate to guidance compliance and represent the basic steps required to ensure that the code's criteria are being met. Information aims to help registered providers interpret the criteria and develop their own risk assessments.

The ten duties that registered providers are expected to comply with are as follows:

1. Establish systems to manage and monitor the prevention and control of infection. These systems use risk assessments and consider how susceptible service users are and any risks that their environment and other users may pose to them.
2. Provide and maintain a clean and appropriate environment in managed premises that facilitates the prevention and control of infection.
3. Provide suitable accurate information on infections to service users and their visitors.
4. Provide suitable accurate information on infections to any person concerned with providing further support or nursing/medical care in a timely fashion.
5. Ensure that people who have or who develop an infection are identified promptly and receive the appropriate treatment and care to reduce the risk of passing on the infection to other people.
6. Ensure that all staff and those employed to provide care in all settings are fully involved in preventing and controlling infection.
7. Provide or secure adequate isolation facilities.
8. Secure adequate access to laboratory support as appropriate.
9. Have and adhere to policies designed for the individual's care and provider organisations that help to prevent and control infections.
10. Ensure, so far as is reasonabe, that healthcare workers are free of and protected from exposure to infections that can be caught at work, and that all staff are suitably educated in the prevention and control of infection associated with the provision of health and social care.

The ten criteria are judged by the Care Quality Commission (CQC) and taken into account when the commission decides about providers' registration. Not all criteria apply to every provider; however, providers cannot cherry-pick the criteria but rather have to consider them as a whole. Health commissioning groups have the responsibility to commission services that are meeting infection prevention and control standards and so are likely to work closely with the CQC but not replace or duplicate it.[9]

The Department of Health[9] states that the main purpose of the *Code of Practice* is to make requirements for registration clear so that they can be met in terms of cleanliness and infection control; provide guidance for CQC staff to make judgements about compliance; and provide information on who uses the services of the registered provider.

Registered providers do not have to comply with the code by law, but to become and stay registered they have to meet the full range of requirements. However, by law the CQC must account for the code when it makes decisions about cleanliness and infection control.[9] The aim of the CQCs' regulatory model[12] is to ensure that providers and managers who carry on regulated activities are registered, and that the care people receive from these providers and managers meets the essential standards of quality and safety in terms of cleanliness and infection control. The aim of the CQCs' regulatory model[12] is to ensure that providers and managers who carry on regulated activities are registered, and that the care people receive from them meets the essential standards of quality and safety in terms of cleanliness and infection control.

Before the *Code of Practice* and the CQC's revised regulatory model existed, many government documents, such as *Saving Lives: A Delivery Programme to Reduce Healthcare-Associated Infection Including MRSA*,[13] identified the processes and measures that should be in place. Although these documents had a profound impact[7] and raised the profile of infection control in the National Health Service (NHS),[14] the guidance they offered was poorly implemented.[6] At this point trusts' performance was monitored via the self-reporting 'star rating' system that included generic performance indicators for infection control, primarily around the management and provision of infection prevention and control teams.[15,7]

The National Audit Office (NAO)[6] reviewed the self-reporting systems, concluding that they were inadequate, and therefore recommended that the Healthcare Commission develop suitable performance indicators to measure outcomes rather than simply systems and processes. Consequently, the CQC published *Judgement Framework: How We Judge Providers' Compliance with the Section 20 Regulations of the Health and Social Care Act 2008*, which helps CQC staff to reach decisions about whether a provider or a manager is meeting the essential standards and to decide their regulatory response if not. The framework is split into four stages, as shown in Table 6.1.

By consulting the *Judgement Framework*, it is envisaged that staff can judge whether providers or managers are either compliant or non-compliant with one or more of the

Table 6.1 Stages of Action from the *Judgement Framework*

Stage	Action
1	Determining whether there is enough evidence to make a judgement.
2	Determining whether the evidence demonstrates non-compliance with one or more of the regulations.
3	Determining the level of impact of non-compliance on people.
4	Determining the regulatory response (which includes referring to the Enforcement Policy).

regulations. When they are judged to be non-compliant, the level of non-compliance is assessed as one of the following:

Minor: People who use the service experience poor care that has an impact on their health, safety or welfare, or there is a risk of this happening. The impact is not significant and the matter can be managed or resolved quickly.

Moderate: People who use the service experience poor care that has a significant impact on their health, safety or welfare, or there is a risk of this happening. The matter may need to be resolved quickly.

Major: People who use the service experience poor care that has a serious current or long-term impact on their health, safety and welfare, or there is a risk of this happening. The matter needs to be resolved quickly.

Once it has been determined that providers or managers are non-compliant, the Regulatory Response Escalator is used to help staff select a proportionate response. The Enforcement Policy sets out the range of regulatory responses that can be taken to promote change (Table 6.2).

Despite its aims, the CQC has already been criticised for not following its own recommendations by intervening on a failing hospital (i.e., University Hospitals of Morecombe Bay Foundation Trust). It had the highest mortality rate and neonatal deaths, some of which were seen to be avoidable, of any trust in England.

Similarly, there has been discussion of the nature, implementation and evidence base for the *Code of Practice* by professionals, and it has been suggested that the code will have little effect on actual infection rates if other factors are not taken into account, including

- Bed occupancy
- Patient turnover
- Staffing levels and workload

All of these have been identified as pivotal to improving compliance and reducing HCAI.[6,16]

Dancer and Simmons[17] state that if it is not resourced, no amount of improvement notices or sanctions can improve HCAI rates. It could be that while the *Code of Practice* puts pressure on registered providers, infection prevention and control activities still

Table 6.2 Enforcement Policy Indicating Regulatory Response and Outcome

Response	Outcome
Formal regulatory action	
Compliance action	Provider and/or manager submits a robust report detailing the actions they intend to take to become compliant within an agreed timeframe. • Update required from the provider and/or manager on progress and upon completion of actions to become compliant. • Follow-up is either through a desktop review or an inspection.
Enforcement action	
Warning notice	Provider and/or manager implements the necessary changes to become compliant within the timeframe imposed. • Follow-up visit to check compliance.
Criminal law	
Fixed penalty notice	Registered person pays a fine and makes the necessary changes.
Simple caution	Registered person admits offence and a caution is accepted.
Prosecution	Registered person is prosecuted.
Civil enforcement	
Conditions/urgent conditions	To restrict activity to ensure the health, welfare and safety of people. • Follow-up visit to check compliance.
Suspension of registration/ urgent suspension of registration	Provider and/or manager implements the necessary changes to become compliant so temporary restrictions can be lifted. • Follow-up visit to check compliance.
Cancellation/urgent cancellation of registration	Provider and/or manager is no longer registered to carry on regulated activities.

have to compete with other hospital requirements such as waiting lists and diminishing resources.[17,18] However, increased penalties and sanctions on registered providers may have a real impact on prioritisation and result in tangible improvement. Storr[19] predicts that with a competent workforce, the right regulation and regulators who understand infection control, users can receive safe care.

COMMISSIONING INFECTION PREVENTION AND CONTROL SERVICES

The Conservative government introduced the concept of commissioning in order to improve healthcare provisioning, and health services have recently undergone another major reorganisation. The commissioning of health services is a cycle of activities intended to secure the best possible health and well-being outcomes at the best value for local citizens and taxpayers. Starting in 2013 every primary care trust (PCT) is responsible for buying services for its population, and PCT members work with practice-based commissioners (PCBs) to achieve this. The latter consists of front-line clinicians and GP practices that hold and manage a delegated budget from which PCB members commission prescribing, acute, community and emergency care services that meet the health needs of their local population.

The Infection Prevention Society and the Royal College of Nursing have published a toolkit which incorporates guidance for nursing and commissioning staff[20] in relation to infection prevention and control services. The toolkit is explicit that commissioners should aspire to achieve zero-tolerance of HCAI and should not accept that healthcare-associated infections are inevitable or an acceptable risk that is an unavoidable part of health and social care.

Provider organisations must

- Be registered with the CQC to provide care that meets the requirements of the *Code of Practice*.[9]
- Have their own local infection prevention and control strategy and assurance framework that reflects their local commissioning of a HCAI reduction plan and contractual requirements, and that provides evidence of their compliance with the code.[9]
- Undertake assessments of their compliance, at intervals agreed to with the commissioning organisation, with the code.[9]
- Actively engage the processes for HCAI prevention and control performance and quality monitoring and be active members of any relevant cluster health economy infection prevention group (or other forum as appropriate).

The toolkit consists of a basket of indicators that can be considered during the commissioning process.

The achievement of indicators helps commissioners buy services that comply with the *Code of Practice*.[9] The indicators have been developed to help commissioning organisations understand, compare and predict outcomes and improve care. They should not be seen in isolation but should comply with the Department of Health's *Operating Framework for the NHS in England 2012-13*[21] and *Healthy Lives, Healthy People: Improving Outcomes and Supporting Transparency*.[22]

Presently, the framework has only two mandatory key performance indicators: reduce *meticillin-resistant Staphylococcus aureus* (MRSA) bacteraemias and *Clostridium difficile* cases. The toolkit's basket of indicators can be used by local commissioning groups depending on their local compliance reports, surveillance data and intelligence.[20] Examples of suggested indicators are provided in Table 6.3.

The following are the four main requirements for effectively commissioning infection prevention and control:

1. Development and leadership of the healthcare and social care economy
2. Contracting to include the setting of clear expectations of achievement
3. Performance monitoring against the contract (gaining assurance)
4. Organisational accountability

Table 6.3 Suggested Indicators for Commissioners

Quality requirement	Threshold/ expectation	Measurement method	Breach
Patients isolated as per agreed local policy/advice from IPCT	100% compliance to agreed local policy	Confirmation of percentage of compliance (including exceptions of variation to policy) Quarterly audit—terms of audit to be determined locally	Escalation through appropriate clause of contract
IPC training programme adhered to as per locally agreed plan for each staff group	100% compliance to agreed local plan	Quarterly confirmation of percentage of compliance	Escalation through appropriate clause of contract
100% compliance with national cleaning standards for areas of very high risk, high risk, significant risk, and low risk	100% achievement of national standards for cleaning	Monthly confirmation of percentage of achievement of standard in areas of very high risk, high risk, and significant risk	Escalation through appropriate clause of contract

It is envisaged that commissioning services will result in a culture of continuous improvement which uses a whole-system approach.[23]

THE *FRANCIS REPORT*

In February 2013 the *Report of the Mid-Staffordshire NHS Foundation Trust Public Inquiry*, known as the *Francis Report*,[10] was published in response to documented failings in care at the Mid-Staffordshire Trust. Although the report focusses on this specific trust, it is widely believed Mid-Staffordshire was not unique in its failings. The latest claims arising in March 2013 about the Morecombe Bay Foundation Trust support this belief. Some highlighted issues from the inquiry are likely to affect the majority of healthcare providers; for instance, it was found that a chronic shortage of staff, particularly nursing staff, was largely responsible for the reported substandard care. However, the RCN reported an inconsistent level of staffing levels which makes a negative difference to patient outcomes.[24] Therefore the inquiry has lessons for all healthcare providers and not just for one trust.

The inquiry chairman, Robert Francis, QC, concluded that in the Mid-Staffordshire case, patients were routinely neglected by a trust that was preoccupied with cost-cutting, targets and processes and lost sight of its fundamental responsibility to provide safe care. Targets for the NHS were first introduced by the Conservative government during the 1990s, but they are probably most associated with the Labour Government which from 1997 on vigorously introduced a volume of centrally managed targets.[25] Safe care in terms of cleanliness and infection control were not upheld; instead, it is reported that the standards of hygiene were at times awful, with families forced to remove used bandages and dressings from public areas and clean toilets themselves for fear of becoming infected.

The *Francis Report*[10] is vast and has identified many issues such as standard training and education for nurses and the regulation of healthcare assistants. It also noted that morale at the trust was low, while many staff did their best in difficult circumstances, others showed a disturbing lack of compassion towards patients. Staff who did speak out felt ignored, bullied and fearful if they continued to highlight inadequacies. Patient groups and visitors were also routinely ignored and complaints were mismanaged.

Staff, patients and visitors were routinely ignored as the trust's board was found to be disconnected from what was actually happening in the hospital and chose instead to rely on apparently favourable performance reports by outside bodies such as the Healthcare Commission, rather than on effective internal assessment and feedback from staff and patients. Such deficiencies were considered to be systemic and deep-rooted.

The report[10] made 290 recommendations but concluded that

> *People must always come before numbers. Individual patients and their treatment are what really matters. Statistics, benchmarks and action plans are tools not ends in themselves. They should not come before patients and their experiences. This is what must be remembered by all those who design and implement policy for the NHS.*

Francis is now considered to be at the cornerstone of root-and-branch reform for healthcare providers; however, Wanless[26] had already stated that the government's focus on narrow targets was like a blunt instrument and that a step change would be required to move to secure good health for all of the population. This would require strong leadership and sound organisation.

Darzi[27] had also called for reforms where the patient, rather than the achievement of centralised targets, would be at the heart of service provision. This meant that the focus would not be on the quantity of NHS services but rather on their quality. It was Darzi's report that explicitly stated that high-quality care should be as safe and effective as possible, with patients treated with compassion, dignity and respect. Darzi made clear that as well as clinical quality and safety, quality means care that is personal to each individual. It should be remembered that calls, like those of Wanless and Darzi for high-quality patient care delivered with dignity and compassion, were being made at the same time as the events at Mid-Staffordshire were happening.

Since its inception, the NHS has been viewed as more centralised than any other European healthcare system, with change the result of a complex and interrelated array of political, social and economic factors.[28] Politicisation and intensive media interest have influenced leadership, causing turbulence and organisational defensiveness.[29,30] Historically, the NHS and the government have had a top-down hierarchical management approach that has been slow to change, with critical events rather than management initiatives forcing change.[31,32] Unfortunately, this seems to have happened in Mid-Staffordshire, with the *Francis Report*[10] stating that the problem is healthcare providers that often provide care characterised by negative aspects of culture, including a lack of criticism, a lack of consideration for patients and defensiveness.

The characteristics reported by the *Francis Report*[10] may have been worsened by the implementation of 'managerialism', where the NHS adopted models based on the private sector and more recently 'lean' models such as those developed by commercial organisations.[28] Also, it focussed on the achievement of targets such as the four-hour waiting time for patients being treated by Accident and Emergency departments. For some, the changes have been viewed as imposing a stricter hierarchical and bureaucratic organisational culture.[33] Pressure from central government to deliver key targets rapidly increases the likelihood that this model of managerialism will prevail.[34] The change in leadership from professionals to managers in the NHS has had a major impact on NHS healthcare workers, and they have had to adapt.

Research by the Department of Health shows that managers and management came in for significant criticism from staff working within the NHS because of a lack of clear leadership.[35] Managers often default to a dictatorial style that fails to consult and engage.[35] This can result in staff not being empowered, which can lead to them having a negative outlook and feeling disenfranchised.

For a variety of reasons, leadership, as opposed to management, is consistently discussed in recent Department of Health literature and reports—for example, Darzi[27]—and the *Francis Report*,[10] with continued emphasis on its importance. The Department of Health[36,37] considers engagement with healthcare staff to be essential in shaping the workforce and developing and delivering high-quality care. Leadership is again identified as a core skill that should be integrated into student programmes and beyond, to equip healthcare staff with the clinical and managerial skills they need to improve care and lead change.[36,37]

The Department of Health developed the Leadership Qualities Framework (LQF), which identifies behaviours that illustrate the core leadership skills required by staff at all levels. Fifteen qualities for leadership are identified, as shown in Table 6.4.

EMPOWERMENT AND POWER

Despite the necessity for empowerment of healthcare workers, it is often difficult to define empowerment and be certain whether it is taking or has taken place. There

Table 6.4 Leadership Qualities Based on the NHS *Judgement Framework*

Personal qualities These are at the core of the Framework. The complexity and scale of the change in agenda and the level of accountability means that NHS leaders need to draw deeply on their personal qualities to see them through the demands of the job.	Self-belief and -awareness Self-management Drive for improvement Personal integrity
Setting direction The outstanding leader sets a vision for the future, drawing on his or her political awareness of the social and healthcare context. This political awareness is underpinned by intellectual flexibility and this, with the drive for results and seizing the future, is key to inspiring and motivating others to work with him or her.	Seizing the future Intellectual flexibility Broad scanning Political astuteness Drive for results
Delivering the service High-performing leaders provide leadership across the organisation and the social and healthcare context in order to deliver results. They use a range of styles which challenge traditional organisational boundaries and ways of working, and emphasise integration and partnership.	Leading change through people Holding to account Empowering others Effective and strategic influencing Collaborative working

Source: Adapted from the NHS Leadership Qualities Framework.[12]

are several views on what 'empowerment' is, appearing to mean different things to different people.[31] Murphy[33] defines empowerment as

- Encompassing power sharing
- Delegating authority
- Returning control and power from bureaucrats to nurses (and other healthcare professionals)

The Department of Health has repeatedly emphasised the importance of 'empowering staff', and that healthcare workers must be positioned for success in leadership and management.[36] In recent years, empowerment of NHS Trust employees has been given substantial political support, with government documents clearly extolling its virtues.[36,37]

Corbally et al.[38] found managerial style to be important in relation to views of empowerment, with supportive managers contributing to empowerment and authoritarian managers being counterproductive because, having previously had a greater sense of control, they cannot experience real 'empowerment'.

Empowerment requires there to be well-developed leadership skills in order to facilitate change. The *NHS Plan: A Plan for Investment, a Plan for Reform*[39] promotes the importance of leadership, asking nurses 'to take the lead in the way local services are organised and in the way that they are run'. One small-scale study relating to infection prevention and control was conducted by Koteyko and Nerlich,[40] who assessed the impact of empowerment and transformational leadership on 'modern matrons', following the publication of *A Matron's Charter*,[4] and its impact on infection prevention and control.

The study describes cases of difficulty in fulfilling leadership requirements because of organisational barriers to empowerment despite arguments to the contrary. Unless a significant budgetary responsibility is made part of the modern matron's role, personal skills (communication, problem solving) alone may not be sufficient to sustain it and may not lead to achieving control over infection, which was the initial trigger for instituting this role.

Koteyko and Nerlich examined matrons' accounts to assess how far the rhetoric of empowerment affected the realities of day-to-day management and infection prevention and control practice.[40] Their findings suggest the following:

- Empowerment is more symbolic than real.
- Matrons have little or no control over medical staff, describing them as 'challenging'. While the term *empowerment* is used, it is more aspirational compared to the reality, and this can result in compliance with infection prevention and control practice not being embedded.
- There is difficulty in fulfilling clinical leadership because of the organisational barriers to empowerment.

This study relating to infection prevention and control identifies the challenges that are faced when attempting to deliver high-quality care in a complex organisation such as the NHS. Unless real changes are made to assert the patient as central to the provision of healthcare, it is likely that failings will continue.

WHISTLEBLOWING

Findings from the *Francis Report*[10] show the catastrophic effects when concerns are not listened to or acted upon. The Public Interest Disclosure Act of 1998[41] is a channel for healthcare workers and others to raise concern in the public interest while being protected by law.[9] The question of whether it is fit for the purpose has now been raised.

In relation to whistleblowing about infection prevention and control issues, both of the Healthcare Commission's reports[42,43] on failings at Stoke Mandeville[42] and later at Maidstone and Tunbridge Wells[43] found that clinical staff repeatedly expressed their concerns about moving patients to different wards because of the likely spread of infection, but they were not listened to. The reasons for this were seen to be a result of the trust's decision to prioritise so that the achievement of the government's target for a maximum waiting time in Accident and Emergency of four hours took precedence over safe care.[42]

Consequently, patients with diarrhoea were placed in open wards rather than in isolation facilities and the shortage of nurses meant that they were too rushed to

Table 6.5 Policy for Raising Concerns in a NHS Organisation

Step 1	Raise the issue with your line manager either verbally or in writing.
Step 2	If you feel unable to raise the issue with your line manager, contact the organisation's designated officer.
Step 3	If these channels have been followed or you feel unable to raise the issue with those in Steps 1 and 2, contact the medical director, director of nursing or the chief executive or the person who is responsible for the organisation's whistleblowing policy.

Source: Social Partnership Forum (2010).[44]

adhere to basic Standard Precautions such as hand hygiene. Similarly the impact of financial pressures[43] took priority over any concerns that were raised. Earlier in 2010, the publication of *Speaking up for a Healthy NHS*[44] provided guidance for NHS organisations in building trust and confidence so that tragedies like the Bristol Royal Infirmary and the Harold Shipman case could be avoided. For individuals who have concerns about risk, malpractice or wrongdoing in the workplace, this document recommends they follow three steps, as shown in Table 6.5.

Despite the guidance embedded in each trust's policy, there seems to be a hesitance in healthcare workers to blow the whistle. A RCN survey in 2011 reported that 84% of nurses would be reluctant to do so. In light of this and in the post–*Francis Report* era, when whistleblowers have been allegedly gagged by their previous employers, not only may guidance have to be rewritten but a change in cultural norms should also occur to signify a shift so that the patient is truly at the heart of healthcare delivery.

CONCLUSION

For more than a decade we have witnessed healthcare failings and scandals, with reports calling for high-quality, safe patient care. Those who work directly with patients are at the sharp end of healthcare delivery and are very much influenced by the wider social, cultural and political healthcare context. The information detailed in this chapter is just a small sample of the range of centralised guidance deemed necessary to shape our healthcare practice and deliver safe, individualised care to patients.

REFERENCES

1. Department of Health. *The third prevalence survey of health care associated infections in acute hospitals.* London; 2007.
2. Department of Health. *Getting ahead of the curve: a strategy for combating infectious diseases (including other aspects of health protection).* London; 2002.
3. Department of Health. *Winning Ways: working together to reduce health care associated infection in England—Report from the Chief Medical Officer.* London; 2003.
4. Department of Health. *A Matron's Charter: an action plan for cleaner hospitals.* London; 2003.
5. Department of Health. *Clean, safe care: reducing infections and saving lives.* London; 2006.

6. National Audit Office. *Improving patient care by reducing the risk of hospital acquired infection: a progress report—report by the Comptroller and Auditor General*. London: The Stationery Office; 2004.
7. Weaving P, Cooper T. Infection control is everyone's business. *Nurs Manage* 2006;**12**:18–22.
8. Department of Health. *The Health and Social Care Act 2008*. London; 2008.
9. Department of Health. *The Health and Social Care Act 2008: Code of Practice on the prevention and control of infections and related guidance*. London; 2010.
10. Francis R. *Report of the Mid-Staffordshire NHS Foundation Trust Public Inquiry*. London: The Stationery Office; 2013.
11. Rafferty AM. Opinion. *Nurs Times* 2013;**109**(6):7.
12. Care Quality Commission. *Judgement Framework: how we judge providers' compliance with the section 20 regulations of the Health and Social Care Act 2008*. London; 2012.
13. Department of Health. *Saving Lives: a delivery programme to reduce health care associated infection including MRSA*. London; 2005.
14. Gould DJ, Hewitt-Taylor J, Drey NS, Gammon J, Chudleigh J, Weinberg JR. The cleanyourhands campaign: critiquing policy and evidence base. *J Hosp Infect* 2007;**65**:95–101.
15. Bradshaw PL, Bradshaw G. *Health policy for health care professionals*. SAGE Publications; 2004.
16. Halwani M, Solaymani-Dodaran M, Grundmann H, Coupland C, Slack R. Cross-transmission of nosocomial pathogens in an adult intensive care unit: incidence and risk factors. *J Hosp Infect* 2006;**63**:39–46.
17. Dancer SJ, Simmons NA. MRSA behind bars (editorial). *J Hosp Infect* 2006;**62**:261–3.
18. Dancer SJ, Simmons NA. MRSA behind bars (authors' reply). *J Hosp Infect* 2006;**63**:356–7.
19. Storr J. How should *Francis* change nursing practice? *Nurs Times* 2013;**109**:12.
20. Infection Prevention Society and Royal College of Nursing. *Infection prevention and control commissioning toolkit. Guidance and information for nursing and commissioning staff in England*. London; 2012.
21. Department of Health. *The operating framework for the NHS in England 2012-13*. London; 2011.
22. Department of Health. *Healthy lives, healthy people: improving outcomes and supporting transparency*. London; 2012.
23. NICE. Prevention and control of healthcare-associated infections. Quality Improvement Guide. *NICE Public Health Guidance* 36; 2011.
24. Royal College of Nursing. *Mandatory nurse staffing levels*. London; 2012.
25. Kings Fund. *Have targets improved NHS performance?* <www.kingsfund.org.uk/projects/general>; 2010.
26. Wanless D. *Securing good health for the whole population*. London: The Stationery Office; 2004.
27. Darzi A. *High quality care for all. NHS next stage review final report*. London; 2008.
28. Brooks I, Brown RB. The role of the ritualistic ceremonial in removing barriers between subcultures in the National Health Service. *J Adv Nurs* 2002;**38**:341–52.
29. Goodwin N. Leadership in the UK health service. *Health Policy* 2000;**51**:49–60.
30. Allen P. New localism in the English National Health Service: what is it for? *Health Policy* 2006; **79**:244–52.
31. Hewison A, Stanton A. From conflict to collaboration? Contrasts and convergence in the development of nursing and management theory. *J Nurs Manage* 2003;**11**:15–24.
32. Truman P. A question of style. *Nurs Manage* 2000;**7**:10–12.
33. Murphy L. Transformational leadership: a cascading chain reaction. *J Nurs Manage* 2005;**13**:128–36.
34. Davies P, Gubb J. *Putting the patient last. How the NHS keeps the ten commandments of business failure*. CIVITAS, Institute for the Study of Civil Society; 2009.
35. Department of Health. *What matters to staff in the NHS: research study conducted for the Department of Health*. London; 2008.
36. Department of Health. *A high quality workforce: NHS next stage review*. London; 2008.
37. Department of Health. *A consultation on the NHS constitution*. London; 2008.
38. Corbally M, Scott P, Mathews A, Gabhann L, Murphy C. Irish nurses' and midwives' understanding and experiences of empowerment. *J Nurs Manage* 2007;**15**:169–79.
39. Department of Health *The NHS plan: a plan for investment, a plan for reform*. Crown Publishing; 2000.
40. Koteyko N, Nerlich B. Modern matrons and infection control practices: aspirations and realities. *Br J Infect Control* 2008;**9**:18–22.

41. Legislation Government. *The Public Interest Disclosure Act*. <www.Legislation.Gov.uk/>; 1998 [accessed 20.03.13].
42. Healthcare Commission. *Investigation into outbreaks of Clostridium difficile at Stoke Mandeville Hospital, Buckinghamshire Hospitals NHS Trust*; 2006.
43. Healthcare Commission. *Investigation into outbreaks of Clostridium difficile at Maidstone and Tunbridge Wells NHS Trust*; 2006.
44. Social Partnership Forum. *Speaking up for a healthy NHS*. <www.socialpartnershipforum.org/News/Pages/SpeakingupforahealthyNHS.aspx>; 2010 [accessed 20.03.13].

CHAPTER SEVEN

Invasive Devices

Tracey Cooper, Lauren Tew, Jacqueline Randle and Steven L. Percival

INTRODUCTION

There are a number of well-defined principles that underpin the use of invasive devices. The selection of a medical device is and should be determined on the basis of treatment effectiveness and cost. Risk in relation to infection is a significant concern; however, the best way to reduce the risk of infection from any invasive device is by not inserting it at all, although in most cases this is not possible. Consequently, there should always be a clear indication for insertion of any invasive device.

Except in an emergency, or when the patient is not competent to consent, prior to insertion of any invasive device consent should be sought. This is usually a verbal consent, following discussion between the patient and the healthcare worker of the need for the device, the relative risks associated with its use or non-use, and the likely duration of device use. Insertion must always be in a manner that does not introduce micro-organisms into the body either aseptically or through an aseptic non-touch technique.[3] The choice of technique employed depends on the invasiveness of the device, the site of insertion and the patient's vulnerability.

To ensure compliance, documentation of the insertion and removal of an invasive device is essential. This should include the person who inserted it; the date, time and site of insertion; and level of asepsis used for insertion or removal. In addition, some devices considered high risk should be regularly monitored for signs of infection.[3] All of these observations and interventions should be clearly documented.

In addition, every device must be cared for in a way that minimises infection; the best way to do this is to implement practices that are in accordance with the international evidence base. Whenever possible, patients, their family and caregivers should be given appropriate information about the care of an invasive device, about what they can do to reduce the risk of infection, and about signs of infection that they should report if they develop. The need for an invasive device must be regularly reviewed and the device removed as soon as it is no longer required. Some devices require removal

Biofilms in Infection Prevention and Control
Doi: http://dx.doi.org/10.1016/B978-0-12-397043-5.00007-4

after a pre-determined period of time in order to prevent a significantly increased risk of infection. All reviews and the removal of devices must be documented.

SHORT- AND LONG-TERM URINARY CATHETERS

It is estimated that 25% of people admitted to hospital will have a urinary catheter inserted as part of their treatment.[4,5] These are hollow tubes inserted into the bladder via the urethra.[5] Their purpose is usually to drain urine from the bladder, but they can also be used to insert medication and to facilitate recovery from urethral surgery. They are increasingly inserted through the abdominal wall, above the pubic bone (supra-pubic catheterisation). Foley catheters have a balloon which is inflated with sterile water to retain them in the bladder. Otherwise, a simple Nelaton catheter can be used for intermittent urine drainage.

Infection Risks

Invasive devices have been found to increase the patient's risk of hospital-acquired infection by seven times.[6] Pratt and colleagues[7] described the risk of urinary tract infection (UTI) from the insertion and use of urinary catheters as 'significant', and Bond and Harris[8] stated that urinary tract infections are the most common form of healthcare-associated infections (HCAI). There are increasing numbers of reports highlighting methodologies that can be employed in the prevention of UTIs.[9,10]

In hospitals, 80% of urinary tract infections are related to the use of urinary catheters.[11] In the community, there are fewer studies that have attempted to establish the rate of infection as a consequence of urinary catheterisation, and Getcliffe and Newton[12] concluded that there is a lack of data on the prevalence and economic impact of catheterisation in the community. However, Getcliffe[13] had earlier estimated that patients who have long-term urinary catheters *in situ* occupy 4% of the community nurse's workload in the United Kingdom alone. Sorbye and colleagues' study[14] of elderly community patients in 11 European countries found that the risk of acquiring a urinary tract infection is nearly seven times greater for those patients who are catheterised. Disturbingly, Landi et al.[15] found that elderly and frail female patients who are catheterised are more likely to die within a year when compared to their non-catheterised counterparts.

It is therefore essential that these devices be used only if absolutely necessary and once alternatives have been considered and determined to be inappropriate. Dingwall and McLafferty[16] found that many nurses still consider catheterisation for the elderly patient as 'a habit'. They reported that many patients are catheterised before it is established whether they can regain continence by other methods or strategies. More recently, Hart[17] reported that as many as 50% of patients are catheterised when they should not be.

Generally most urinary tract infections are caused by bacteria, although *Candida* has been documented to cause them as well.[18] The most common bacteria reported as causing a catheter-related infection include *Escherichia coli*, *Klebsiella*, *Proteus mirabilis*[19] and *Enterococci*. Of particular concern is when these bacteria form biofilms in the urinary catheters which significantly increase the risk of infection.[20] This is discussed in further detail in Part 2 of this book.

To reduce infection risks the urinary catheter should be removed from the patient as soon as is feasibly possible.[21] This rationale is significant as evidence has shown that leaving a catheter in place for more than six days constitutes the greatest infection risk factor to the patient.[22] Stamm[18] reported that all patients with permanent in-dwelling catheters will eventually become infected. There are a number of opportunities available to help in these situations. For example, the use of supra-pubic catheters is recommended by NICE[23] to reduce the risk of urinary tract infections as these remain *in situ* for much shorter periods of time.

Opportunistic pathogens enter the catheterised bladder in a number of ways, but most specifically as a result of poor technique during insertion so that they enter on the external surface of the catheter and cause interruption to the closed drainage system by travelling upwards inside the catheter lumen. It is imperative that indwelling urinary catheters be used when necessary.[24,25]

General Principles

There are a number of general principles and guidelines to help significantly reduce the risk of infection,[3,26–28] including the following:

- A protocol should be designed to reduce risk of infection.[29]
- Urinary catheterisation should be undertaken only by healthcare workers who have been appropriately trained and are competent to perform it. Extra training is necessary for changing or inserting supra-pubic catheters.
- Medical consent should be required for the first catheterisation procedure; subsequent changes of a catheter are not generally necessary.
- Patients must give their informed consent for this procedure.[3]

In addition, appropriate records must be collated and written in the patient's care record and should include, as an example, justification for the catheterisation; a review date for its removal/replacement; the patient's informed consent; the date and time of catheterisation; the size, material and length of the catheter; the batch number and expiry date; the route of catheterisation; and the name and signature of the accountable professional, patient's experience and findings. Volume and description of urine drained, difficulties during procedure and breach of asepsis should also be documented.

Choice of Urinary Catheter for Adults

A comprehensive assessment of the patient should always be undertaken to select the appropriate catheter and drainage system to maximise the patient's comfort and safety while the catheter is in place. Numerous aspects should always be considered, including the following:

Diameter of catheter: The smallest-diameter catheter should always be used to allow drainage of urine[31]; sizes 12 and 14 are usually large enough for routine drainage of clear urine; urology patients may require special catheters with larger dimensions.

Balloon size: A 10 ml balloon is used for routine drainage; 30 ml balloons are used for post-operative urology cases.

Length of catheter: A female-length catheter (20–26 cm) may be suitable for an ambulant female. However, if the woman is likely to be immobile or is obese, a standard-length catheter (40–45 cm) may be more appropriate. Never use a female-length catheter on a male as the self-retaining balloon may inflate in the urethra and cause severe bleeding.

Drainage system: Closed urinary drainage systems are the most common in the healthcare environment. These systems must be kept closed to reduce the risk of micro-organisms entering them and causing infection. Link systems can be used overnight when the patient uses a leg bag during the day; some patients prefer to use a valve to open their catheter and allow drainage intermittently rather than be attached to a drainage bag.

Catheter material: Latex-free systems are available for patients with latex allergy; no particular catheter material (e.g., silicone or latex) is associated with fewer UTIs; however, silver-alloy hydrogel-coated latex catheters have been shown to reduce infection rates.[32]

Preference and experience: The patient's own preference and healthcare workers' clinical experience influence the choice of catheter.[31]

Key Points

- The larger the catheter lumen, the more likely it is to increase bladder spasm and bypassing of urine.[8]
- Too large a lumen may cause pain.[33]

Insertion Procedure

The equipment that is necessary for the insertion of an indwelling catheter includes an alcohol-based handrub, a disposable plastic apron, sterile disposable gloves, non-sterile disposable gloves, patient hygiene equipment, a catheter, sterile 0.9% sodium chloride, a dressing pack (according to local policy), sterile water and syringe (unless a catheter with a pre-filled balloon is used), anaesthetic gel/lubricant supplied in a single-use sterile container, a drainage system, and the patient's healthcare record. Preparation prior to insertion is very important, as detailed in the following:

1. Explain the procedure to the patient and obtain informed consent.

2. Ensure privacy and warmth.
3. If the patient has not done so already, assist him or her in washing the genital area with soap and water and drying it thoroughly, while wearing a disposable apron and non-sterile gloves. The hygiene equipment should then be removed.
4. Place a pad under the patient to absorb any accidental spillage during the catheterisation.
5. Remove gloves and perform hand hygiene.
6. Assemble the equipment on a clean surface; open the dressing pack to create a sterile field; open the catheter and anaesthetic gel/lubricant packaging and empty it onto the sterile field. Pour the sterile saline into a gallipot. Open the sterile catheter drainage bag and place it on the sterile field.
7. If the catheter does not have a pre-filled balloon, fill the syringe with the appropriate volume of sterile water and place it to one side in a clean receiver.

Key Points

- Healthcare organisations are required to have policies to reduce the risks to their patients from invasive devices.[34] Local policies must be followed during this procedure.
- Catheterisation is an aseptic technique.[31]
- Additional training and assessment are required before performing this procedure.

For female catheterisation, the following is the recommended procedure:

1. Clean your hands; put on an apron and sterile gloves.
2. Expose the tip of the sterile catheter and place it on the sterile field.
3. Assemble the anaesthetic gel/lubricant.
4. Separate the labia minora with the non-dominant hand using low-linting swabs and identify the urethral meatus.
5. Maintain labial separation with one hand and clean the meatus with 0.9% sodium chloride, using single downward strokes.
6. Instil the anaesthetic gel/lubricant and allow its effect to develop (5 minutes).
7. Using the dominant hand, gently insert the tip of the catheter into the urethral meatus; use no force. If resistance is felt, stop and seek assistance.
8. Inflate the balloon once urine is draining freely and attach the balloon to the drainage system.
9. Secure the catheter to prevent pulling on the trigone or irritation of the urethral meatus.
10. Ensure the patient's comfort.
11. Dispose of all equipment according to policy.
12. Clean your hands.
13. Record the procedure fully (as described previously).

Key Points

- The drainage bag should be kept below the level of the bladder to prevent backflow of urine up the tubing. If this occurs, urine from the bag, along with colonising micro-organisms, will be transferred up the tubing, increasing the risk of infection.
- All manipulations and disconnections of the catheter and its drainage system provide opportunities for contamination of the system with micro-organisms that could cause infection (e.g., *E.coli*, the commonest cause of UTI), so must be avoided. Wear an apron and non-sterile disposable gloves for such procedures, which must be undertaken aseptically. Hands must be decontaminated before and after use of gloves.
- Remove the catheter at the earliest opportunity.
- Daily personal hygiene with soap and water is sufficient cleaning. Antiseptic cleaning is not necessary.
- Wherever possible, encourage patients with long-term catheters to empty their catheter hygienically themselves.
- Provide the patient with appropriate information.

Maintenance

The equipment required for emptying a urinary catheter bag includes an alcohol-based handrub, a disposable plastic apron, non-sterile disposable gloves, a suitable container for collecting drained urine (e.g., a disposable or reusable jug or urinal) and two 70% isopropyl alcohol wipes. The recommended procedure for emptying a catheter bag follows:

1. Collect the equipment.
2. Perform hand hygiene.
3. Put on an apron and gloves.
4. Place the container under the drainage bag outlet.
5. Clean the drainage outlet with a 70% isopropyl alcohol swab; allow 30 seconds for drying.
6. Drain the urine into the receptacle.
7. Close the drainage outlet and clean it with the other 70% isopropyl alcohol swab.
8. Dispose of used wipes and wrappings in a foot-operated clinical waste bin.
9. Measure the volume of urine drained if the patient is on a fluid balance chart.
10. If the container is disposable, place it in a macerator.
11. If the container is reusable, place it in a bedpan washer.
12. Remove gloves and apron and perform hand hygiene.
13. Record the volume of urine on the fluid balance chart.

Key Points

- Urinary catheter drainage bags form part of the closed system of the urinary catheter. Each time the catheter drainage bag is opened, it breeches the closed system and potentially allows micro-organisms present in the external part of the drainage port to migrate upwards into the bag. Once present in the bag, these organisms can migrate upwards and eventually enter the bladder, leading to potential infection.
- The bag should not be opened more frequently than necessary to maintain urine flow and to prevent reflux of urine up the tubing; routine emptying at pre-set times must be avoided.
- A drainage bag must be chosen that allows sufficiently close monitoring of urine output yet minimises the need to open the drainage port.

Key Points

- If the patient is to be discharged from hospital with a catheter ***in situ***, this must be communicated to appropriate community healthcare workers. A thorough summary of the information documented about the patient and the catheter must be shared with them to ensure seamless care in the community setting.
- The patient should be given details about whom to contact should she or he have any concerns and provided written advice about living with a catheter.
- Catheter manufacturers often supply booklets to record information for patients.

FAECAL MANAGEMENT SYSTEMS

The equipment required for a faecal management system includes an alcohol-based handrub; a disposable plastic apron; non-sterile disposable gloves; a 50 ml syringe; water for inflating the retaining balloon (volume and type per the manufacturer's instructions); a lubricant; a clinical waste bag; an incontinence pad; patient hygiene equipment; a faecal management system, including a drainage bag; and the patient's healthcare record, including stool chart. The recommended procedure for faecal management is shown in Exhibit 7.1.

EXHIBIT 7.1 Faecal Management

Procedure

1. When faecal management systems are available, consider their use when a patient develops diarrhoea that is frequent or uncontrollable.[35]
2. Determine the type of diarrhoea and document it using a system such as the Bristol Stool Chart.[36]

3. If an infectious cause of diarrhoea cannot be excluded, take a faecal specimen and send it to the laboratory for investigation.
4. Before use of the system, perform an individual patient assessment to ensure that it is suitable and that the patient has no contra-indications for use.
5. Perform hand hygiene and assist the patient into position, lying on his or her left side with knees bent upwards as far as comfortable. Ensure privacy using bedclothes.
6. Position an incontinence pad under the patient's buttocks in case of leakage during insertion.
7. Perform hand hygiene and open the items of equipment onto a clean surface.
8. Put on gloves and the apron.
9. Draw up the correct amount of water into the syringe and place it to one side in a clean receiver.
10. Apply lubricant to the tip of the faecal management system.
11. Insert the system per the manufacturer's instructions.
12. Monitor the nutrition and hydration status of the patient to ensure that he or she does not suffer dehydration or malnutrition due to diarrhoea because this can be life-threatening. A fluid balance chart should be used, and careful notes should be made of food intake. Daily weighing of the patient helps to identify weight loss indicative of nutritional or hydration problems.
13. Where there may be difficulty maintaining hydration, consider the use of oral rehydration solutions when the patient is able to take oral fluids. Rehydration using this approach avoids the use of indwelling intravascular devices and therefore avoids the infection risks associated with them. Use of subcutaneous fluids or IV fluids may be required, however, if the patient cannot take oral therapy. These should all be prescribed.
14. Consider the use of sip-feeds and other nutritional supplements when there is concern about a patient's nutritional status. Have the patient reviewed by a dietician immediately, especially when he or she is already frail or malnourished prior to the development of diarrhoea, to ensure that nutritional intake can be optimised. Malnutrition increases vulnerability to infection, and all possible measures must be taken to prevent it.
15. Remove the faecal management system when diarrhoea reduces or stops, or per the manufacturer's instructions.

Key Points

- Faecal management systems are a relatively recent innovation and have been introduced as a way to manage liquid or semiliquid faecal incontinence. They can be ideal for managing persistent diarrhoea.
- Such systems retain diarrhoea and thus prevent the environmental contamination from faeces and faecal organisms that often results. This benefit may be particularly helpful in preventing environmental contamination with *Clostridium difficile* spores and other bowel organisms.
- By mid-2008, two products had been evaluated by the Rapid Review Panel of the Health Protection Agency and a recommendation for category 1 was given. This means that the products had been evaluated by a national expert panel and demonstrated benefits that the panel recommended should be available to the National Health Service (NHS) to include in its infection control protocols.

- Faecal management systems also reduce the pain and skin excoriation caused by frequent diarrhoea and the washing and drying of the skin required for cleaning, therefore reducing distress to the patient.[37]
- There are some contra-indications for the use of faecal management systems, and each patient must be assessed individually to ensure that these systems are suitable.

INTRAVENOUS ACCESS DEVICES

Intravascular (IV) access devices have been used since 1830.[38] There are a range of such devices, which are inserted through the skin into vessels of the circulatory system. Depending on their design and placement, they can deliver medication, fluids or nutritional products such as parenteral nutrition; monitor pressures within the vascular system such as blood pressure or pulmonary artery pressures; and allow access for withdrawal of blood samples for analysis (e.g., arterial blood gas monitoring).

Local infection can occur at the insertion site of an IV device, or it can track up the vein and surrounding tissue to cause phlebitis. Once micro-organisms are introduced, the bloodstream can act as a 'sterile motorway', carrying them throughout the body and potentially leading to systemic sepsis. The presence of any medical device material increases the risk of infection considerably, even with micro-organisms that are considered to have low virulence.[39]

Some IV devices placed for long-term use are inserted through a 'tunnel' made in the skin before entering a central vein. This allows the skin to heal around the device, reducing the risk of infection by preventing the migration of micro-organisms along the outside of the device. Additional training and assessment are required for many of the activities relating to IV devices.

Key Points

- The skin is naturally colonised with a range of micro-organisms,[40] many of which can cause infection if introduced into the bloodstream. Micro-organisms from the skin, from the environment and from the hands of healthcare workers inserting the device can enter the bloodstream at the following times:[41]
  - During insertion as a result of poor hand hygiene, poor skin preparation or poor technique, causing accidental contamination of the device.
  - Through the lumen of the device during use; via contaminated infusion fluid and bolus injections; or via extension sets, hubs and bungs.
  - Along the outside of the device while it is *in situ*; particularly if dressings are loose or the site becomes contaminated.

Use of Intravascular Devices

It is always important to consider whether an IV device is really needed, and the question should always be asked as to whether there is an alternative way to achieve the therapeutic outcomes required. In particular, an IV device should not be inserted unless it is needed.

Verbal consent must be sought from the patient, and it must be ensured that any risks of insertion have been fully explained. This is particularly important for IV devices that are inserted into the central vascular system, as the risks associated with the procedure, depending on the route and site of insertion, may include pneumothorax, haemothorax, malpositioning and infection.[42]

It is very important that insertion always be performed aseptically. For central vascular devices this must include sterile drapes, sterile gown and gloves with a full aseptic technique. In an emergency—for instance, during a cardiac arrest—in order to save a life, a peripheral IV device may be inserted without appropriate asepsis. In such circumstances it is important that the device be removed or replaced as soon as possible and always within 48 hours.[43] This significantly reduces the risk of biofilm formation and infections.[44]

The insertion of any IV device must be clearly documented, including type and size of device, insertion site, date and time of insertion, aseptic technique used and name of the person who inserted the device. The device must be cared for in accordance with the international evidence base. This should include choice of dressing, solution used for site care during dressing changes, and length of time the device is left *in situ*. Three-way taps should not be used, as they act as a reservoir for the multiplication of micro-organisms which can then be injected into the bloodstream. There are a large number of procedures available for guidance to help reduce infections.[45]

Biofilm formation is a significant problem with IV catheters, increasing the risk of bloodstream infections.[46–48] Wound dressings that are employed must remain intact and protect the insertion site due to detachment of bacteria from the biofilm. If they become loose or soiled, or if there is pooling of sweat or blood underneath, they must be changed using an aseptic technique. Any IV device insertion site must be observed regularly for signs of infection or phlebitis, and its condition must be documented. The site must always be observed prior to use of the device. Several systems exist for scoring the condition of the insertion site. These provide an objective scale that can be used consistently by different healthcare workers, and they help ensure consistency of care.

Needle-less connectors can be used to reduce the risk of sharps injury and lessen the need to completely break the closed system. They must be used in accordance with the manufacturer's instructions and changed when recommended. Devices must be removed when they are no longer needed. Peripheral IV cannulae must be removed within 72 hours of insertion, as the risk of infection increases significantly beyond

72 to 96 hours. Removal must be documented and the site observed daily until healed in order to detect any signs of infection post-removal.

Peripheral IV Cannulae

The Department of Health report *Winning Ways; Working Together to Reduce Healthcare Infection in England*[49] details specific guidance on the use of peripheral intravenous cannulae. The Infection Control Nurses Association audit tool[50] has produced similar guidance in an attempt to reduce meticillin-resistant *Staphylococcus aureus* (MRSA) and other HCAI.

Insertion

Equipment that is suggested for insertion of a peripheral IV cannula includes an alcohol-based handrub, a disposable plastic apron, sterile disposable gloves, non-sterile disposable gloves, peripheral IV cannulae (two or three in case of failed attempts), a dressing pack/cannula pack (according to local policy), a syringe and needle, sterile 0.9% sodium chloride, an applicator or solution of 2% chlorhexidine gluconate in 70% isopropyl alcohol, a sharps bin, a transparent polyurethane dressing with a high moisture–vapour transfer rate, a needle-less or other connector if required, a clinical waste bag and the patient's healthcare record. Exhibit 7.2 details the procedure recommended for the insertion of a peripheral IV cannula.

EXHIBIT 7.2 Insertion of a Peripheral IV Cannula

Procedure

1. Ensure that all items required for insertion are collected, including sharps bin, dressing pack or IV pack if available, cannula, skin-cleaning solution, gloves, apron, handrub and trolley.
2. Administer skin-cleaning solution either through an applicator or as a solution using sterile gauze. The most effective solution is 2% chlorhexidine gluconate in 70% isopropyl alcohol.
3. Whenever possible, use safety cannulae to protect healthcare workers from accidental sharps injury. Devices that activate the protective mechanism automatically are the safest.
4. Perform hand hygiene before contact with the patient.
5. Obtain verbal consent from the patient whenever possible; assist the patient into a comfortable position.
6. If the patient is uncooperative or aggressive and the IV device is required for urgent treatment, obtain sufficient help so that the risk of sharps injury to healthcare workers, and harm to the patient, is minimised.
7. Don an apron, clean the hands with handrub and open the cannula pack or sterile field using a non-touch technique; all equipment should be opened on the sterile field in a manner that does not lead to accidental contamination.

8. Position the sharps container so that the used sharp can be placed directly into it during the procedure.
9. If needle-less connectors are to be used, prepare them correctly, flushing with 0.9% sodium chloride to prevent any risk of air embolism.
10. Once the site for insertion is selected following review and palpation of available veins, apply a tourniquet to the arm, perform hand hygiene using alcohol-based handrub and prepare the site of insertion using the skin preparation solution.
11. When 2% chlorhexidine gluconate in 70% isopropyl alcohol is used in a prepared applicator, sterile gloves are not necessary. If another solution is used from a gallipot or the skin is to be re-palpated after the skin preparation solution is applied, sterile gloves need to be put on first.
12. Once the skin is prepared, allow the solution to dry over the insertion site, as it is this process of drying that leads to bacterial death and a reduced infection risk.
13. If hands are ungloved, clean them again with alcohol-based handrub and put on non-sterile gloves.
14. Pick up the cannula, taking care to touch only the parts that will remain outside the patient after insertion; the skin should not be re-palpated, unless sterile gloves are being worn; insert the cannula through the skin.
15. Withdraw the cannula if not placed correctly. A second attempt must not be made using the same cannula. A new sterile cannula should be opened for a second attempt.
16. Once correct placement has been confirmed, release the tourniquet and remove the introducer, dropping it into the sharps container without contaminating the hands.
17. Immediately connect the primed connector or bung to the open end of the cannula and flush it through to ensure that there is no blood pooled in the cannula.
18. Secure the cannula using a semi-permeable transparent polyurethane dressing, ensuring that the actual insertion site is securely covered and sealed. The dressing selected should help prevent build-up of sweat underneath it.
19. Place all used items in the clinical waste bag, remove gloves into the bag, remove the apron into the bag and perform hand hygiene using alcohol-based handrub.
20. Document the insertion, including date, time and name of the person inserting the device. If a label is included on the dressing, the date and time of insertion should also be recorded on it.
21. Advise the patient not to try to lift the dressing and to report any soreness, redness or discharge at the insertion site.

Key Point

- Insertion of a peripheral IV cannula is an extended role which requires additional training and assessment.

Management of Peripheral Intravenous Cannulae

The insertion of a peripheral intravenous cannula is one of the most common invasive procedures performed on hospitalised patients[51]; therefore healthcare workers are likely to come across a patient with a peripheral intravenous cannula *in situ*. In intensive care settings the need for intravenous therapy is 100%.[39] Peripheral intravenous catheters are used therapeutically for a variety of reasons, including administration of fluid, nutrients, medications and blood products and monitoring of the patient's haemodynamic status.[52] The longer the patient has a peripheral intravenous cannula *in situ*, the greater the opportunity for micro-organisms to multiply.[53] This is because medical devices are an easy way for bacteria in particular to spread into a patient's tissues[54] as they are foreign objects breaking the skin.

The incidence of bacteraemia from peripheral intravenous cannulae is low.[55,59] However, if it does occur, it may be life-threatening, especially in immuno-compromised patients, and the consequences of intravenous device infection can be multiple.[39] Most cases of *Staphylococcus epidermidis* bacteraemia are associated with infected intravascular devices.[57,58] For a minority of patients, peripheral intravenous cannulae have the potential to cause harm because of the extracellular slime that *Staphylococcus epidermidis* produces when it grows as a biofilm that adheres to the devices and then multiplies and penetrates tissue or invades cells.[59] The growth of biofilms on intravenous cannulae is a major problem and needs to be considered when anything enters the human body.[60]

A common catheter-related infection is phlebitis, which is largely a physicochemical or mechanical phenomenon rather than an infectious one.[53,61] Wilson[62] defines phlebitis as the inflammation that occurs in the vein where the catheter is positioned, and Horton and Parker[63] describe how micro-organisms colonising the skin can be introduced into the break of the skin caused by the cannula tip. When phlebitis occurs, there may be an increased risk of a local catheter-related infection. The overall incidence for phlebitis is 2.3 to 43%.[55] Symptoms are tenderness, redness, pyrexia with unknown cause and exudates.[64]

The following are factors which influence the incidence of phlebitis[55,65]:

- Age (affects the older person)
- Gender (affects men more than women)
- Patient diagnosis (e.g., diabetes)
- Infection at another body site

Additional process factors include[66]:

- Site preparation
- Frequency of tubing changes
- Catheter material
- Osmolarity and dose of the drug/diluent

- Use of filters
- Hand hygiene
- Aseptic or venipuncture technique
- Number of entries into the system
- Type and size of the catheter

Another very significant factor affecting the likelihood of phlebitis is the experience of staff in inserting and maintaining the catheter.[56] The presence of an IV team has been shown to decrease rates of phlebitis and other complications.[67,68]

Key Points

- Any patient with a peripheral IV device *in situ* is at an increased risk of infection, and healthcare workers must therefore be alert to any signs of infection.
- Bandages should not be used to aid the security of a peripheral IV device, as they obscure the insertion site and can pull tight, obstructing the venous return and sometimes even pulling on the IV device and dislodging it. If additional security is required, an appropriately sized tubigrip can be applied. This has the advantage of being easy to lift for observation of the site and applying consistent pressure over the entire area.
- The insertion site should be observed prior to each use and, if not in regular use, a minimum of twice in every 24-hour period.[69]
- The site should be observed for heat, redness, pain, discharge, cellulitis or hard, palpable venous chord. Several observation rating scales have been developed for determining signs of infection, including the Visual Infusion Phlebitis (VIP) score system.[70]
- When the site is observed, the condition of the dressing should also be inspected. If the dressing is not intact or is soiled, or there is pooling of sweat or blood underneath the dressing, it should be replaced, as detailed in Exhibit 7.3. This must be done as a aseptic technique using a dressing pack, and the site should be cleaned using 2% chlorhexidine gluconate in 70% isopropyl alcohol before applying a new dressing.
- When an IV device is no longer required, it should be removed as soon as possible.
- The risk of infection due to a peripheral IV device increases significantly after it has been *in situ* for 72 to 96 hours. Therefore all peripheral IV devices should be removed after they have been *in situ* for 72 hours.[71]
- It is helpful to document, at the time of insertion, the date and time the device should be removed and also to inform the patient of this.

EXHIBIT 7.3 Re-dressing a Peripheral IV Device

Equipment

- Alcohol-based handrub
- Disposable plastic apron
- Sterile disposable gloves or non-sterile disposable gloves—depending on choice of skin-cleaning product and palpation technique used

- Applicator or solution of 2% chlorhexidine gluconate in 70% isopropyl alcohol
- Dressing pack
- Patient's healthcare record
- Sharps bin
- Transparent semi-permeable polyurethane dressing
- Clinical waste bag

Procedure

1. Collect all equipment required: sterile gloves, apron, handrub, dressing pack, cleaning solution and sterile gauze or cotton wool; alternatively, cleaning solution in an applicator, clinical waste bag, replacement sterile dressing and trolley.
2. Perform hand hygiene before contact with the patient, using the alcohol-based handrub or soap and water.
3. Inform the patient that the device is to be re-dressed, and assist her or him into a comfortable position.
4. Perform hand hygiene.
5. Open the dressing pack and sterile items onto the sterile field using a non-touch technique. The glove pack should be opened last and placed on the sterile field nearest the nurse, as this is required first.
6. Don the apron and carefully loosen the dressing from the skin; remove it without directly touching the insertion site. Dispose of it into a clinical waste bin.
7. Perform hand hygiene using the handrub. Sterile gloves should be put on if a cleaning solution is to be used with gauze or if touching the site is unavoidable. If an applicator cleaning solution is used and the site does not need to be touched, non-sterile gloves can be worn instead.
8. Carefully clean the insertion site and allow it to dry.
9. Once the site is dry, apply the replacement dressing to it; take care to fully cover it and do not trap the gloves underneath the dressing.
10. Once the site is secure, dispose of the waste in the waste bag and then the gloves and apron.
11. Perform hand hygiene and document the dressing change, along with the condition of the insertion site.

Key Points

- The use of the procedure in Exhibit 7.3 requires additional training. If an IV dressing is not intact or is soiled, or if there is pooling of sweat or blood underneath it, it should be replaced. This is because any of the situations described can result in increased proliferation of micro-organisms under the dressing and lead to an increased risk of infection.
- The dressing change must be performed with an aseptic non-touch technique using a dressing pack, and the site should be cleaned using 2% chlorhexidine gluconate in 70% isopropyl alcohol and allowed to dry before applying a new dressing.

Removal of Devices

It has been established that peripheral intravenous cannulae should be removed every 72 to 96 hours (depending on the type of therapy) or sooner if complications are suspected.[56,72–74] From a microbiological perspective, we know that devices (e.g., plastics) allow the adhesion of bacteria such as *Staphylococcus epidermis*, which are very avid biofilm formers. Adherence of *Staphylococcus epidermidis* to the surface of the device is not a one-time phenomenon but rather an evolving process. Initially, there is a rapid attachment that is mediated by non-specific factors.[61] This indicates the importance of either removing or replacing peripheral intravenous cannulae within a specific time range for the majority of patients.

However, the question of routine replacement of peripheral intravenous cannulae which have been well cared for and are not showing signs of phlebitis, should be reviewed.[53,56] Curran et al.[53] suggest that the benefits of removing peripheral intravenous cannulae must outweigh the risks and discomfort of re-siting. Reasons for extended dwell time include poor venous access, expected site continuation within 24 hours and use of a saline lock. Current UK policy is to remove the device after 72 to 96 hours; however, we are bringing it to your attention that evidence suggests a risk assessment can be appropriate.

Equipment required for the removal of peripheral IV devices include an alcohol-based handrub, a disposable plastic apron, non-sterile disposable gloves, the patient's healthcare record, a clinical waste bag, sterile cotton wool balls or gauze and sterile dressing or plaster (check with the patient for allergy to sticking plaster before selecting the product to be used).The procedure is detailed in Exhibit 7.4.

EXHIBIT 7.4 Removal of Peripheral IV Devices

Procedure

1. Collect all equipment required: non-sterile gloves, apron, handrub, sterile gauze or cotton wool, yellow clinical waste bag and sterile dressing or sticking plaster. If signs of infection are present, a swab for laboratory testing should also be taken at the bedside.
2. Perform hand hygiene before contact with the patient, using handrub or soap and water.
3. Inform the patient that the device is to be removed and assist him or her into a comfortable position.
4. Place the waste bag close by so that items can be disposed of while the healthcare worker is near the patient.
5. Perform hand hygiene using alcohol-based handrub or soap and water.
6. Don gloves and aprons and loosen the dressing carefully from the skin without directly touching the insertion site.

7. Observe the insertion site for signs of infection; if discharge or pus is present, swab the site and send the swab for culture and sensitivity.
8. Using sterile gauze or cotton wool, place light pressure over the insertion site, at the same time sliding the plastic cannula backwards and removing it.
9. Maintain pressure on the insertion site in order to stop any bleeding, and drop the used cannula and dressing into the receiver or waste bag. As there is no sharp left in the patient following insertion, a sharps bin is not needed for disposal of the removed cannula.
10. Maintain pressure until any bleeding or oozing has stopped; then drop the sterile gauze or cotton wool into the receiver or waste bag, along with the gloves.
11. Perform hand hygiene using the alcohol-based handrub.
12. Open the sterile plaster using a non-touch technique, and protect the insertion site using the sterile plaster.
13. After disposing of the waste products, document the cannula removal and record the condition of the insertion site at removal. If a swab is sent, this must also be recorded.
14. Check the condition of the insertion site daily after removal, until healed, in order to detect any signs of infection.

Managing IV Infusions

The procedure for the management of IV infusions is as follows:

1. Perform hand hygiene using the alcohol-based handrub.
2. If the bag of fluid is to replace an existing bag, check the IV administration set to see whether it also requires replacement.
3. If blood or blood products are being administered, change the set when the infusion is complete, with a maximum hanging time of 12 hours.
4. If a crystalloid solution is being administered, leave the set for up to 72 hours, but replace it if the IV device is replaced during that time period.
5. When parenteral nutrition solutions containing lipid emulsions are infused, change the bag and administration set every 24 hours.
6. Always check containers of IV fluid to ensure that the packaging has not been damaged, as any damage to the packaging will breach the sterility of the product. If there is any doubt about the condition of the packaging, reject the product.
7. Open the outer packaging and leave the inner administration port cover in place until the spike of the administration set is ready to be connected.
8. Connect the bag to the sterile spike of an IV administration set without touching either the spike or the port on the IV bag. Push the spike firmly into the port, ensuring that it pierces the internal membrane.
9. If the administration set is new, prime and label it with the date and time of opening.

10. Connect the set to the patient connector using an aseptic non-touch technique.
11. Clean the membrane of a needle-less connection system if used or the injection port of the IV device before accessing it using a wipe containing 2% chlorhexidine gluconate and 70% isopropyl alcohol. Allow it to dry before accessing it.

Key Points

- The procedures for removal of peripheral IV devices (see Exhibit 7.4) and the management of IV infusion are extended functions which require additional training and assessment.
- Contaminated IV infusions can rapidly lead to systemic infection, though this risk has largely been eliminated by the central production of IV infusions by commercial manufacturers and local pharmacy aseptic services. It is still possible, however, to introduce micro-organisms into these products as a result of damage to the packaging or poor handling that allows contamination of the drug additive port with skin or environmental organisms.
- Once introduced, micro-organisms can multiply and be infused into the patient. This can lead to rapid onset of rigors, pyrexia, sepsis and collapse.

Key Points

- The use of the procedure in Exhibit 7.5 is an extended function which requires additional training and assessment.
- Intravenous infusion fluids are produced with a range of electrolytes added, most notably potassium chloride. However, the need to add more potassium or medication to bags of infusion fluids is not uncommon.
- Care must be taken when preparing and adding to IV fluid bags or preparing large syringes of drugs for infusion, as accidental introduction of microbial contamination can result in patient infection.
- Because these solutions usually have a relatively long hanging time at room temperature, it is possible for bacterial multiplication and growth to occur over a number of hours, increasing the risk further. For this reason, additions to IV infusion fluids should be performed in a clean area (e.g., in a clean treatment room).
- Although rare, the possibility of infusion-related infection should be considered for any patient who rapidly develops pyrexia, rigors and other signs of infection shortly after a new IV infusion is commenced.

EXHIBIT 7.5 Preparing Drugs or Electrolytes for IV Infusions

Equipment

- Alcohol-based handrub
- Syringes and needles
- Sterile 0.9% sodium chloride or sterile water as a diluent

- Drugs and additives as required
- Sharps bin
- Wipe containing 70% alcohol
- Injection tray
- Waste bag

Procedure

1. Collect the required equipment: drug, appropriate diluents if required, syringe, needle, injection tray, alcohol handrub, wipe impregnated with 70% isopropyl alcohol and sharps bin.
2. Perform hand hygiene.
3. Open the syringe and needle and connect them together without allowing the open ends to come into contact with anything else; then place the devices on the injection tray.
4. Check the drug to be drawn up; if it comes from a multi-use vial (e.g., insulin), check that it is still in date and within the permitted re-use time period, and clean the membrane thoroughly using the alcohol-impregnated wipe. Allow it to dry.
5. Open the vial of diluent if required and draw it up, taking care not to contaminate the needle by accidental contact with the outside of the vial. The cap can be carefully replaced on the needle to allow expulsion of air and to ensure that the correct amount of fluid has been drawn up.
6. Draw up the drug, removing the cap on single-use drug vials first. Care must be taken not to touch the membrane on the vial or to contaminate the needle by accidental contact with the outside of the vial.
7. Once the drug is mixed, draw it up. The needle must then be carefully removed from the vial and the needle cover carefully replaced to allow expulsion of air. The drug can then be placed back on the injection tray.
8. The previous step does not constitute re-sheathing, as the needle is a clean one. Any accidental injury to a staff member does not put him or her at risk of blood-borne viruses and does not need follow-up as a contaminated sharps injury; however, the drug, the tray and all equipment must to be discarded to prevent any risk to the patient.
9. Remove the outer packaging of the IV fluid bag, first taking care to check that the packaging is intact.
10. The injection membrane on the IV fluid bag is sterile; place the bag on the work surface so that the membrane does not come into contact with any surface.
11. Pick up the syringe and inject the drug through the injection membrane, holding the outer surface of the bag.
12. On removal of the syringe and needle, drop them into the sharps bin, taking care not to touch the bin with the hands.
13. Vigorously agitate the IV bag, with the added drug, to ensure thorough mixing. A label should be attached to the bag, stating the date and time of addition as well as details of the drug added.
14. Return any multi-use vial to the drug fridge for correct storage. All rubbish should then be disposed of in the sharps bin and domestic waste bag.
15. Perform hand hygiene again, and connect the bag to an administration set and to the patient.

Key Points

- The procedure described in Exhibit 7.6 is an extended function which requires additional training and assessment.
- Microbial contamination can be introduced into a patient via contaminated bolus drugs. It is therefore important to ensure that drugs are prepared using a non-touch technique in order to maintain drug sterility.

EXHIBIT 7.6 Preparation of IV Drugs: Bolus

Equipment

- Alcohol-based handrub
- Syringes and needles
- Sterile 0.9% sodium chloride or sterile water as a diluent
- Drugs and additives as required
- Sharps bin
- Wipe containing 70% isopropyl alcohol
- Wipe containing 2% chlorhexidine gluconate and 70% isopropyl alcohol
- Injection tray
- Waste bag

Procedure

1. Collect the required drug, appropriate diluents if required, and other equipment.
2. Perform hand hygiene.
3. Open the syringe and needle and connect them together without allowing the open ends to come into contact with anything else; then place the devices on the injection tray.
4. Check the drug to be drawn up; if it comes from a multi-use vial (e.g., insulin), check that it is still in date and within the permitted re-use period, and clean the membrane thoroughly using the wipe impregnated with 70% isopropyl alcohol. Allow it to dry.
5. Open the vial of diluent if required and draw it up, taking care not to contaminate the needle by accidental contact with the outside of the vial. The cap can be carefully replaced on the needle to allow expulsion of air and to ensure that the correct amount of fluid has been drawn up.
6. Draw up the drug, removing the cap on single-use drug vials first. Take care not to touch the membrane on the vial or contaminate the needle by accidental contact with the outside of the vial.
7. Draw up the drug once mixed. The needle must be carefully removed and the needle cover carefully replaced to allow expulsion of air. The drug can then be placed back on the injection tray.
8. Return any multi-use vial to the drug fridge for correct storage. All rubbish should be disposed of in the sharps bin and domestic waste bag.
9. Take the prepared drugs and flushes, plus the wipe impregnated with 2% chlorhexidine gluconate and 70% isopropyl alcohol, sharps bin and handrub, to the patient.

10. Perform hand hygiene at the point of care.
11. Check the IV insertion site for signs of infection; if signs are present, the cannula should be removed and re-sited.
12. If a needle-less connector is used, clean the membrane using the wipe impregnated with 2% chlorhexidine gluconate and 70% isopropyl alcohol and allow it to dry; then remove the needle from the syringe and connect the syringe carefully to the connector. The drug can then be injected slowly.
13. If an extension set is used, clean the membrane using the wipe impregnated with 2% chlorhexidine gluconate and 70% isopropyl alcohol and allow it to dry before removing the needle from the syringe and replacing it with a new sterile needle. The needle can then be pushed carefully into the sterile lumen of the connector and the drug injected slowly.
14. If the drug is to be injected directly into the cannula, open the cap on the injection port and clean the port using the the wipe impregnated with 2% chlorhexidine gluconate and 70% isopropyl alcohol; allow it to dry. The needle can then be removed from the syringe, the syringe connected carefully to the port and the drug injected slowly.
15. Once the drug(s) have been administered, dispose of the sharps at the bedside using the sharps bin; also dipose of rubbish and then perform hand hygiene.
16. Record the administration of the drugs, along with the condition of the insertion site, preferably using a scoring systems such as VIPS.

Central Vascular Catheter Insertion

The central vascular catheter (CVC) is the medical device most likely to cause sepsis because it is long in length and remains *in situ* for extended periods of time.[75,76] Most pathogens associated with CVCs are coagulase-negative *Staphylococci*, *Enterococci*, *Enterobacter* species, *Pseudomonas aeruginosa* and *Candida* species.[39,77] It is vital that effective handwashing be adhered to as well as meticulous aseptic technique throughout the procedure for CVC insertion, as detailed in Exhibit 7.7. It has been shown that hospitals with healthcare workers who are well-trained and experienced have the lowest rates of CVC-related sepsis.[39,78–80]

EXHIBIT 7.7 CVC Insertion Procedure

Equipment

- Alcohol-based handrub
- Disposable plastic apron
- Non-sterile disposable gloves
- Sterile disposable gloves

- Clinical waste bag
- Sterile gown
- Cap and mask if required by local policy
- Large sterile drapes
- CVC insertion pack or large dressing pack
- CVC device
- Needle-less or other connectors, sufficient for the number of lumens
- Syringes and needles
- Sterile 0.9% sodium chloride for flushing lumens and connectors
- Ampoule of local anaesthetic for injection, as per local policy
- Large sharps bin
- Solution or applicator containing 2% chlorhexidine gluconate and 70% isopropyl alcohol
- Sterile gauze
- Sutures
- Transparent sterile semi-permeable polyurethane dressing, large enough for a CVC device
- Patient's healthcare record

Procedure

1. A trolley should be prepared, with all items required placed on its bottom shelf.
2. It is important to explain the procedure to the patient and to warn that his or her face may need to be covered by the drapes to protect against any infection risk during insertion.
3. The patient should be positioned ready for the procedure; it may be necessary to place a small rolled up towel or pillowcase under the shoulderblade on the side of insertion to allow the operator to clearly identify anatomical landmarks and to aid insertion. At this stage the patient can be sitting up if comfortable. The assistant should put on an apron and perform hand hygiene.
4. The gown pack should be opened onto a separate trolley, with care taken not to contaminate the inner layers. Once gowned, remove the outer layer of a sterile glove pack and drop the sterile glove inner pack onto the opened field.
5. The CVC pack should be carefully opened onto the main trolley, ready for the operator, who may have donned cap and mask if required by local policy, plus sterile gown and sterile gloves.[31]
6. Once the operator is ready and has opened the inner sterile field, the assistant should carefully open and drop the following sterile items onto the sterile field, ensuring that his or her hands do not touch any of the sterile items or inner packets:
   - Gauze
   - Syringes
   - Needles
   - Suture
   - Dressing
7. Use skin preparation solution containing 2% chlorhexidine gluconate and 70% isopropyl alcohol, either as a liquid or in a sterile applicator. If an applicator is chosen, the volume of liquid required is likely to exceed 3 ml, so a suitably sized applicator should be selected to allow for this.

8. Local anaesthetic should be offered to the operator, who should check the drug while the non-sterile vial is held by the assistant. The assistant should then open the vial and offer it to the operator to allow him or her to carefully draw up sufficient drug.
9. If the sterile drapes are packed separately, the assistant should open the outer wrapping and drop the drapes onto the sterile field.
10. The patient should be laid flat. If the patient is particularly hypovolaemic or shocked, it may be necessary to raise the foot of the bed to place the patient in the Trendelenberg position to aid the filling of the veins of the neck and chest.
11. The assistant should perform hand hygiene after assisting the patient into position.
12. The operator places the sterile drapes over the patient and secures them in place.
13. The assistant should then carefully position the trolley for the operator to use, without touching the sterile field.
14. The operator cleans the site thoroughly, using 2% chlorhexidine gluconate and 70% isopropyl alcohol.
15. Local anaesthetic is then administered, and the operator usually disposes of the syringe and needle once this is done. A sharps bin should be placed so that this can be done without breaching the sterile field or contaminating the hands. Some operators pass the syringe and needle out of the sterile field using a receiver. Sharps must never be passed hand to hand because of the risk of injury.
16. The operator should be offered the vials/pods of flush solution to check; these should be opened and offered for him or her to draw up.
17. The assistant should open the outer packaging of the CVC device and drop the sterile product carefully onto the sterile field. The operator then checks the device, makes any attachments necessary, primes all lumens using the flush solution and ensures that all clamps or on/off devices are closed. One lumen remains unconnected to allow the guidewire to be fed through at insertion.
18. Once ready, the operator first places the introducer through the skin and, once the vein has been located, feeds the guidewire through the introducer. This part of the process can be distressing for the patient because of the feeling of 'pushing' through the skin. The assistant should offer reassurance to the patient throughout.
19. Once the introducer has been removed, the CVC is inserted over the guidewire and the device slid to the correct position. The guidewire is then removed and the connector attached to the lumen.
20. The operator then cleans and dries the insertion site, removing any blood contamination of the skin.
21. The line is sutured into position, with care taken to place the suture needle either directly into a sharps bin once finished or into a receiver where it is clearly visibly to the assistant.
22. The drapes are removed and the dressing applied, with care taken to fully seal the insertion site.
23. The patient is then returned to a more comfortable position.
24. The position of the line needs to be confirmed radiographically, either by chest x-ray or by ultrasound.

25. Once the patient has been made comfortable, the trolley should be removed, hand hygiene performed, gloves and waste disposed of, with care to dispose of any used sharps safely.
26. The apron and gloves should be removed and hand hygiene performed.
27. Device insertion should be documented by the operator. The insertion may also need to be documented by the assistant in nursing notes or on observation charts.

Management of CVCs and Attachments

The principles of asepsis must be adhered to for all manipulations of the CVC and attachments. All lumens should always be attached to a closed connector. If used for pressure monitoring, one lumen is attached to a pressure transducer, and the bag of fluid, administration set and pressure transducer should be changed every 96 hours. The pressure transducer set should be labelled with the date and time of opening to facilitate this. The insertion site should be observed regularly—a minimum of twice every 24 hours—for signs of insertion site infection such as heat, redness, pain, discharge or cellulitis. The patient should be observed for signs of systemic infection such as persistent pyrexia above 38°C, rigors, hypotension and oliguria or anuria.

As soon as an infection is suspected, appropriate diagnostic and therapeutic measures must be performed.[39] When the site is observed, the condition of the dressing should also be inspected. If the dressing is not intact or is soiled, or if there is pooling of sweat or blood underneath the dressing, it should be replaced. This must be done as a sterile technique using a dressing pack, and the site should be cleaned using 2% chlorhexidine gluconate in 70% isopropyl alcohol before applying a new dressing. If the dressing remains clean and intact, it should be changed every seven days. This must be done as a sterile technique using a dressing pack, and the site should be cleaned using 2% chlorhexidine gluconate in 70% isopropyl alcohol before a new dressing is applied.

If the site is bleeding or oozing, or the patient is excessively sweaty, a gauze dressing should be used instead of a transparent dressing, and it must be changed daily or more often if it becomes damp, soiled or loose. It must be changed for a transparent semi-permeable polyurethane dressing as soon as possible.

If a tunnelled-line or implanted device has been inserted, the dressing should be changed every seven days until the site is completely healed. The site can then be left without a dressing, though careful observation must still be carried out for signs of infection each time the device is used. When a CVC is no longer required, it should be removed as soon as possible. After removal, the insertion site must be covered with a sterile dressing and observed daily until healed in order to detect any signs of infection.

Key Points

- The preceding section details an extended function which requires additional training and assessment.
- CVCs have their tip placed into the central circulation of the body, most usually via the internal jugular vein or, preferably, via the subclavian vein. These devices can be inserted for either short- or long-term use. In general devices inserted for long-term use are placed using methods that minimise infection in the longer term—for instance, by tunnelling through the skin before entering the vein or by inserting a specific type of device under the skin so that the skin covers or seals the device and, once healed, protects it from contamination with micro-organisms.
- Devices for short-term use are generally inserted directly through the skin into the central circulation. They provide very effective access for accurate cardiovascular monitoring, administration of inotropic and other vasoactive drugs, and rapid fluid resuscitation. They also provide ready access to the central bloodstream for micro-organisms unless managed robustly to prevent infection.
- Insertion of a CVC is a sterile procedure, and attention must be paid to maintaining full asepsis even in an emergency situation. If this is not done, a sick patient may subsequently succumb to overwhelming sepsis due to the CVC, and all of the emergency treatment delivered will have been for nothing. National guidance advises use of maximum sterile barrier precautions, though it is acknowledged that the evidence for use of caps and masks for CVC insertion is inconclusive. Healthcare workers should check local policy and wear caps and masks if required.
- The skill of the person assisting with the insertion can make a major difference in achieving correct asepsis by ensuring that all items required are available and prepared. This helps to prevent the person performing the procedure from being tempted to 'cut corners' and it can maintain maximum asepsis.

Key Points

- CVCs inserted for short-term use present a potentially major risk of infection.
- Unlike the risk with peripheral cannulae, the risk of infection per CVC device day remains constant and so there is no need to routinely replace it if it is still required and if there are no signs of sepsis or insertion site infection.
- Infection can best be prevented by strict attention to aseptic technique at insertion, followed by strict adherence to asepsis during all procedures while the line is in use.

NUTRITION

For many reasons, patients may be unable to take in adequate nutrition orally, including short-term problems and long-term or permanent conditions. Whilst intravenous fluids can provide hydration and basic electrolytes, this is not sufficient to maintain the body, which within a very short time will start to break down muscle and fat stores if it does not receive sufficient nutrients. Therefore additional nutrition is commenced early for many patients to stop the body from entering a catabolic state, leading to malnourishment and increasing the risk of infection further.

Key Point

- Nutritional status should be monitored regularly for all patients in order to predict or detect problems early and to initiate a nutritional plan to avoid malnutrition. Objective scoring systems, such as the MUST tool, are available for this.[81]

Parenteral Nutrition

Patients who cannot meet their nutritional requirements orally need nutritional support. If the patient has a functioning gastrointestinal tract, enteral feeding is the preferred choice of nutritional support owing to its safety of administration, reduced cost, enhanced nutritional use and maintenance of gut integrity. Selection of an access route for enteral nutrition depends primarily on the anticipated duration of feeding, as follows:

- Short-term enteral feeding can be administered through a fine-bore nasogastric tube.
- Long-term enteral feeding is more appropriately delivered through a percutaneous endoscopic gastrostomy.

Enteral feeds can be administered as a continuous infusion, an intermittent infusion, or bolus.[82]

Parenteral nutrition solutions are usually prepared by local pharmacy aseptic services, and additions should not be made at the clinical level because low numbers of micro-organisms introduced into parenteral nutrition solutions can rapidly multiply during use. The product must be stored in a working drug fridge prior to use. Whenever possible, parenteral nutrition should be administered via a dedicated single-lumen line. If a multi-lumen catheter is used, parenteral nutrition must be given using a dedicated lumen to prevent it from mixing with any other drugs. An unused lumen should be identified for parenteral nutrition at the time the CVC line is inserted. This lumen should be primed prior to insertion and flushed regularly using an aseptic technique if there is a delay in commencing parenteral nutrition.

Connecting a new bag of parenteral nutrition should be done as an aseptic technique, using the same principles as for the connection of any IV infusion. Connectors attached to lumens on CVCs should, whenever possible, be needle-less. They should be fully cleaned using wipes containing 2% chlorhexidine gluconate in 70% isopropyl alcohol, and allowed to dry, before connection of the administration set. The cleaned connector lumen and sterile end of the administration set must not be touched during the connection procedure. Parenteral nutrition must hang only for the prescribed time: a maximum of 24 hours. On removal of the completed infusion, the line must be flushed thoroughly using an aseptic technique if a new bag of nutrition is not connected immediately.

Key Points

- Parenteral nutrition provides patients with complete nutrition delivered via a central intravenous device.
- Parenteral nutrition solutions contain a mixture of lipid emulsions along with glucose and electrolytes. This makes it perfect for supporting rapid microbial growth, especially when the mixture is in use at room temperature.

Key Points

- Artificial feeding via the enteral route is preferred over the IV route. This is because it allows the gut to function in a relatively normal manner by absorbing nutrients. Maintaining normal gut function helps to protect against translocation of bacteria through the gut wall, reducing the risk of sepsis.
- Feed can be administered via artificial tubing into the stomach or small intestine.
- The feeding tube can be either placed through the nostril or inserted percutaneously through the abdominal wall into the intestine.

Enteral Feeding

Nasogastric (NG) feeding is usually the first route to be considered if the gastrointestinal tract is functioning but the oral route is contra-indicated or oral intake is inadequate.

Fine-bore Nasogastric Tube Insertion

Use of a fine-bore nasogastric tube is the most common delivery method and is suitable for feeding for 4 to 6 weeks.[82] Fine-bore tubes contain a metal guidewire that provides adequate stiffness to allow insertion. As the risk of accidental placement of fine-bore tubes into the lung is much greater, the tubes must be inserted only by, or under the close supervision of, someone skilled and competent in the procedure, as detailed in Exhibit 7.8. Additional training and assessment are required.

EXHIBIT 7.8 NG Tube Insertion

Equipment

- Alcohol-based handrub
- Disposable plastic apron
- Non-sterile gloves
- Clean injection tray or similar
- Fine-bore NG tube
- Tape to secure the NG tube
- Catheter-tipped syringe
- Lubricant and litmus paper

- Drainage bag, if required
- Spigot, if required
- Tissues
- Cup of water for the patient to sip from
- Clinical waste bag
- Patient's healthcare record

Procedure

1. Collect all equipment required.
2. Perform hand hygiene before contact with the patient.
3. Ensure that patient is comfortable in an upright or semi-upright position and knows how to signal for the healthcare worker to stop the procedure.
4. Put on the plastic apron and perform hand hygiene.
5. Open the nasogastric tube onto a clean tray and apply lubricant to the outside of the first 10–15 centimetres of the tube.
6. Don a pair of non-sterile gloves and begin inserting the NG tube, advancing it slowly.
7. Give the patient small sips of water to encourage him or her to swallow and assist with the passage of the tube until it reaches the correct length.
8. Gently remove the guidewire.
9. Stop at once and remove the tube if the patient becomes distressed, starts gasping or coughing or becomes cyanosed, or if the tube coils in the mouth.
10. Once the tube is in place, check its position by attaching the syringe and aspirating stomach contents, and affirming that the pH is less than 5.5. However, patients who are receiving acid-inhibiting drugs may have an altered stomach acid pH, and an x-ray may need to be performed to confirm the correct position of the tube.
11. Use pH indicator strips that have 0.5 gradations, or use paper with a range of 0 to 6 or 1 to 11.[85] It is important that the colour change on any indicator or paper be easily distinguishable, particularly between pH 5 and pH 6. Seek advice from an experienced colleague when there is any doubt about the pH level indicated.
12. Spigot or attach the tube to a drainage bag and secure it in place with tape.
13. Assist the patient to wipe away lubricant from the nose, without dislodging the tube.
14. Dispose of the waste, remove apron and gloves and perform hand hygiene.
15. Document the insertion of the NG tube.

Key Points

- Tubes can be inserted both for drainage of gastric contents and for the administration of enteral nutrition.
- Tubes range from semi-rigid drainage tubes to soft, pliable, fine-bore feeding tubes. Insertion of an NG tube is a clean procedure, as the tube is not being placed into a sterile body cavity but into the stomach. However, it is important to avoid introducing unnecessary micro-organisms as patients requiring enteral nutrition are likely to be more vulnerable to infection than are healthy individuals.
- The use of some drugs reduces the ability of gastric acid to kill bacteria that enter the stomach; this may increase vulnerability to infection.

Checking the Position of the NG Tube

Confirmation that the tip of the fine-bore NG tube is in the stomach is vital before the administration of any feed or medication. The position should also be checked at least once during continuous feeding; following evidence of vomiting, retching or coughing; or following evidence of tube misplacement.

The National Patient Safety Agency[83] has developed advice for healthcare workers on methods for checking position, as follows:

- Measuring the pH of the aspirate using pH indicator strips or paper
- Radiography (x-ray)

Methods that have traditionally been used, but should not be, are

- The 'whoosh' test, in which air is forced into the abdomen via a syringe and the healthcare worker uses a stethoscope placed on the patient's abdomen
- Testing the acidity/alkalinity of aspirate using blue litmus paper
- Interpreting the absence of respiratory distress as a sign of correct positioning
- Monitoring bubbling at the end of the tube
- Observing the appearance of feeding tube aspirate

The most accurate way of confirming tube position is by x-ray; however, this should not be used routinely as a method of checking[83] because of the following:

- It increases exposure of the patient to radiation.
- There is a loss of feeding time for the patient.
- The seriously ill patient requires more handling.
- There is potential for misinterpretation of x-rays if the healthcare worker is not trained in radiology.

The same methods should be followed for checking that a drainage NG tube is in the correct position. NG drainage tubes are used to drain the stomach of its contents and are not routinely used for feeding or drug administration because this can irritate the mucosal lining. The indications for drainage NG tubes are as follows[82]:

- Bowel obstruction
- Gastrointestinal surgery
- Excessive vomiting
- Pancreatitis
- Endotracheal intubation
- Positive pressure ventilation

Managing Percutaneous Endoscopic Gastrostomy Sites

A percutaneous endoscopic gastrostomy (PEG) medical device is placed endoscopically to deliver nutrients directly to the stomach through a tube passing through the abdominal wall. A PEG is used for both short-term and long-term feeding. The administration of enteral feed through a PEG tube can be as a bolus or as an intermittent or continuous infusion using a feeding pump.[82]

General Principles for Delivering Enteral Feeds

The administration system selected should be compatible with the feeding tube in use. Pre-prepared, sterile, ready-to-use feeds should always be used if possible, as they reduce the risk of accidental contamination during preparation and from exposure to air.[23] If pre-prepared feeds cannot be used, a high standard of food hygiene practice must be achieved, including effective hand hygiene, use of clean equipment and preparation in a clean area. Exhibit 7.9 lists necessary equipment and the detailed procedure.

EXHIBIT 7.9 Delivering Enteral Feeds

Equipment

- Alcohol-based handrub
- Disposable apron
- Non-sterile gloves
- pH indicator strips or test paper
- Catheter-tipped syringe
- Clean receiver
- Water for flushing (cooled boiled or sterile water)
- Clean spigot
- Feed
- Jug or disposable measuring bowl
- Administration set
- Patient's healthcare record

Procedure

To provide a break in feeding and ensure that an accurate pH reading is obtained, local protocol may require healthcare workers to switch off the feed for 1 to 6 hours prior to aspiration and connection of the new feed. This provides a break from feeding and allows gastric pH to return to a more normal level where it can suppress bacterial growth.[84]

1. Collect gloves, apron, syringe, water and spigot and take to the bedside.
2. Perform hand hygiene and put on gloves and apron.
3. Switch off feed.
4. Draw up 30 to 50 mls of water, disconnect the feeding tube and flush it slowly with water.
5. Spigot the tube using the clean spigot.
6. Remove the old feed, administration set, syringe, gloves and apron, and dispose of them as clinical waste.
7. Perform hand hygiene.
8. Document the removal of feed in the patient's healthcare record.

After the required time period has elapsed:

1. Collect all equipment and take it on a trolley to the bedside.
2. Place a strip of indicator paper in the clean receiver, ready to test the pH of stomach contents.

3. Perform hand hygiene and put on apron and gloves.
4. Take the syringe and withdraw the plunger so that the syringe is filled with air.
5. Remove the spigot, keeping it in the non-dominant hand, connect the syringe and inject the air into the tubing so that all water is removed.
6. Draw back on the syringe plunger to aspirate stomach contents. Check that the amount of undigested stomach contents is not excessive. The amount to be concerned about depends on the length of the feed break and local policy. Excessive quantities of aspirate should be ejected into the jug or disposable measuring bowl.
7. Replace the spigot in the feeding tube and drop a few mililitres of gastric aspirate onto the indicator paper to confirm that the pH is less than 5.5.
8. If there is any doubt about the pH, seek advice from an experienced colleague before commencing feeding.
9. If the indicator paper confirms a pH of 5.5 or less, recommence the feed.
10. Remove gloves and perform hand hygiene.
11. Always check the feed container to ensure that the packaging has not been damaged, as any damage breaches the sterility of the product. If there is any doubt about the condition of the packaging, the product must be rejected.
12. Open the administration set and connect it to the container of sterile feed.
13. Connect the bag to the sterile spike of the administration set without touching either the spike or the port on the feed bag. The spike must be pushed firmly into the port, ensuring that it pierces the internal membrane.
14. Prime and label the set with the date and time of opening.
15. Hang the feed next to the patient and place the patient end in a position where you can easily reach it.
16. Perform hand hygiene and apply non-sterile gloves.
17. Remove the sterile cap on the administration set and attach it to the NG tube using a non-touch technique. The feed can then be switched on.
18. When the feed is recommenced, confirm that the patient shows no sign of distress, coughing, gasping or cyanosis.
19. If excessive stomach contents require disposal, take the container to the dirty utility room and dispose of it into a slop hopper or directly into a macerator.
20. Remove gloves and apron, and dispose of waste materials in the clinical waste bag.
21. Perform hand hygiene and document commencement of the new feed.

Key Points

- Feeding tubes designed for long-term feeding are often inserted percutaneously through the abdominal wall. Once the insertion site has healed, usually 10 to 12 days after insertion, it forms a protective barrier, stopping ingress of bacteria and reducing the risk of infection. It is possible, however, for insertion sites to become infected. To minimise this risk the tube should be regularly rotated 360 degrees.[23]

- Until healed, new insertion sites should be protected by use of a non-adherent dressing. Once healed, the site does not require daily cleaning as part of normal patient hygiene procedures.
- PEG sites should be observed daily for redness, discharge, pain, cellulitis around the site or other signs of infection. If any of these are present, a medical opinion must be sought and antibiotics may need to be given because of the risk of infection tracking into the gut wall and deep tissue. If the site looks infected, a swab should be taken and sent to the laboratory for culture and sensitivity.
- If the site is infected, in addition to systemic treatment a suitable dressing should be selected to deliver topical agents to the wound to assist with healing. The dressing selected should ideally be able to mould to the shape of the wound and surround the PEG tube. Once the site is dry and healed, it should be treated as normal skin.

Locally prepared feeds must be mixed with cooled boiled water or sterile water, and must be stored in a refrigerator once prepared. They must be clearly labelled and stored for no longer than 24 hours. Maximum hanging time for pre-prepared feeds is 24 hours; for locally prepared feeds, the maximum hanging time is 4 hours. Administration sets and all equipment used to administer a feed must be disposed of per the manufacturer's instructions; generally they are single-use items.

Some manufacturers produce catheter-tipped syringes that are designed specifically for enteral feeding. Some are designated for single-patient use and may be ideal for enteral nutrition given in the patient's home. Healthcare workers should follow local policy on the use of these devices, as they pose a risk of infection if disassembled, washed and left wet in the clinical setting. Prior to administering a new feed, the position of the feeding tube should be checked and the tube should be flushed with cooled boiled water or sterile water. Local policy should be followed, as some recommend using tap water for flushing between feeds.

Key Points

- Enteral feeds provide effective nutrition for vulnerable individuals. However, because they hang at room temperature, they also provide an ideal growth medium for bacteria.
- Preparation, storage and delivery of enteral feeds must minimise the risk of microbial contamination in order to prevent rapid multiplication of bacteria once introduced into the feed tube.
- The delivery of enteral feeding in the United Kingdom is governed by the Food Safety Act.

CONCLUSION

As discussed in this chapter, it is very important that correct infection control procedures be employed when invasive devices are being used. Appropriate procedures and protocols of care help to reduce the risk of infection and cross-contamination concerns. Numerous guidelines and procedures are available to assist the healthcare worker when invasive devices are employed.[85] The use of skin antisepsis, in particular, is recommended before a device enters into the body. For example, agents, such as 2% chlorhexidine gluconate with 70% isopropyl alcohol, are often used, but numerous others are often reported to be effective.[86,87]

Another significant concern with the use of all invasive devices is the risk of colonisation by micro-organisms which quickly develop into a biofilm, thus increasing the patient's risk of infection and further distress.[40,88,89]

REFERENCES

1. Dohnt K, Sauer M, Müller M, Atallah K, Weidemann M, Gronemeyer P, et al. An *in vitro* urinary tract catheter system to investigate biofilm development in catheter-associated urinary tract infections. *J Microbiol Methods* 2011;**87**(3):302–8.
2. Collins AS. Preventing health care-associated infections. In: Hughes RG, editor. *Patient safety and quality: an evidence-based handbook for nurses*. Rockville, MD: Agency for Healthcare Research and Quality (US); 2008, Chapter 41.
3. Flodgren G, Conterno LO, Mayhew A, Omar O, Pereira CR, Shepperd S. Interventions to improve professional adherence to guidelines for prevention of device-related infections. *Cochrane Database Syst Rev* 2013;**28**:3.
4. *Bandolier. Urinary catheters*; Oxford Medical Knowledge Ltd. 1998; 58:3.
5. Wagenlehner FM, Vahlensieck W, Bauer HW, Weidner W, Piechota HJ, Naber KG. Prevention of recurrent urinary tract infections. *Minerva Urol Nefrol* 2013;**65**(1):9–20.
6. Plowman R, Graves N, Griffin M, Roberts JA, Swan AV, Cookson B, Taylor L. *The socio-economic burden of hospital acquired infection*, vols. I, II, III and executive summary. London: Public Health Laboratory Service; 1999.
7. Pratt RJ, Pellowe CM, Loveday HP, et al. The epic project: developing national evidence based guidelines for preventing healthcare associated infections, phase 1: guidelines for preventing hospital-acquired infections. *J Hosp Infect* 2001;**47**(Suppl):S1–82.
8. Bond P, Harris C. Best practice in urinary catheterisation and catheter care. *Nurs Times* 2005; **101**:54–8.
9. Pottinger PS. Our lights are on for safety: comment on preventing catheter-associated urinary tract infection in the United States. *JAMA Intern Med* 2013;**173**(10):879–80.
10. Saint S, Greene MT, Kowalski CP, Watson SR, Hofer TP, Krein SL. Preventing catheter-associated urinary tract infection in the United States: a national comparative study. *JAMA Intern Med* 2013; **173**(10):874–9.
11. Tambyah PA, Oon J. Catheter-associated urinary tract infection. *Curr Opin Infect Dis* 2012; **25**(4):365–70.
12. Getliffe K, Newton T. Catheter-associated urinary tract infection in primary and community health care. *Age Ageing* 2006;**35**:477–81.
13. Getliffe K. The characteristics and management of patients with recurrent blockage of long-term catheters. *J Adv Nurs* 1994;**20**:140–9.
14. Sorbye L, Finne-Soveri H, Ljunggren G, Topinkova E, Bernabei R. Indwelling catheter use in home care: elderly, aged 65+ in 11 different countries in Europe. *Age Ageing* 2005;**34**:377–81.
15. Landi F, Cesari M, Onder G, et al. Indwelling urethral catheter and mortality in frail elderly women living in the community. *Neurol Urodyn* 2004;**23**:697–701.
16. Dingwall L, McLafferty E. Nurses' perceptions of indwelling urinary catheters in older people. *Nurs Stand* 2006;**21**:35–42.
17. Hart S. Urinary catheterisation. *Nurs Stand* 2008;**22**(27):44–8.
18. Stamm W. Urinary tract infections. In: Root R, Waldvogel F, Corey L, Stamm W, editors. *Clinical infectious diseases. A practical approach*. Oxford University Press; 1999. p. 649–56.

19. Nazarko L. Preventing catheter-related urinary tract infection: focus on *Proteus mirabilis*. *Br J Community Nurs* 2011;**16**(11) 528, 530–3.
20. Růžička F, Holá V, Mahelová M, Procházková A. Yeast colonization of urinary catheters and the significance of biofilm formation. *Klin Mikrobiol Infekc Lek* 2012;**18**(4):115–9.
21. Chenoweth C, Saint S. Preventing catheter-associated urinary tract infections in the intensive care unit. *Crit Care Clin* 2013;**29**(1):19–32.
22. Maki D, Tambyah P. Engineering out the risk of infection with urinary catheters. *Emerg Infect Dis* 2001;**7**:2.
23. National Institute for Clinical Excellence. *Infection control guidelines in primary and community care.* London: NICE; 2005.
24. Janzen J, Geerlings S. Appropriate and inappropriate use of indwelling urinary catheters. *Ned Tijdschr Geneeskd* 2012;**156**(37) A5052.
25. Jansen IA, Hopmans TE, Wille JC, van den Broek PJ, van der Kooi TI, van Benthem BH. Appropriate use of indwelling urethra catheters in hospitalized patients: results of a multicentre prevalence study. *BMC Urol* 2012;**6**(12):25.
26. Andreessen L, Wilde MH, Herendeen P. Preventing catheter-associated urinary tract infections in acute care: the bundle approach. *J Nurs Care Qual* 2012;**27**(3):209–17.
27. Conway LJ, Larson EL. Guidelines to prevent catheter-associated urinary tract infection: 1980–2010. *Heart Lung* 2012;**41**(3):271–83.
28. Rebmann T, Greene LR. Preventing catheter-associated urinary tract infections: an executive summary of the Association for Professionals in Infection Control and Epidemiology, Inc., Elimination Guide. *Am J Infect Control* 2010;**38**(8):644–6.
29. Gokula M, Smolen D, Gaspar PM, Hensley SJ, Benninghoff MC, Smith M. Designing a protocol to reduce catheter-associated urinary tract infections among hospitalized patients. *Am J Infect Control* 2012; **40**(10):1002–4.
30. Department of Health. *HSC 2001/023—Good practice in consent: achieving the NHS Plan commitment to patient-centred consent practice.* London: Department of Health; 2001.
31. Pratt RJ, Pellowe C, Wilson J, Loveday HP, Harper P, Jones SR, et al. Epic2: national evidence-based guidelines for preventing healthcare-associated infections in NHS hospitals in England. *J Hosp Infect* 2007;**65**(Suppl)
32. Brosnahan J, Jull A, Tracy C. Types of urethral catheters for management of short-term voiding problems in hospitalised adults (Cochrane Review) *The Cochrane Library, Issue 1.* John Wiley and Sons, Ltd; 2004.
33. Doherty W. Male urinary catheterisation. *Nurs Stand* 2006;**20**(35):57–63.
34. Department of Health. *The Health Act (2006): Code of Practice for the prevention and control of healthcare-associated infection*: updated February 2008. London: Department of Health; 2006.
35. Rees J, Sharpe A. The use of bowel management systems in the high-dependency setting. *Br J Nurs* 2009;**18**(7) S19–20, S22, S24.
36. Lewis SJ, Heaton KW. Stool form scale as a useful guide to intestinal transit time. *Scand J Gastroenterol* 1997;**32**(9):920–4.
37. Padmanabhan A, Stern M, Wishin J, Mangino M, Richey K, DeSane M. Clinical evaluation of a flexible faecal incontinence management system. *Am J Crit Care* 2007;**16**(4):384–93.
38. Ayliffe GAJ, English MP. *Hospital infection: from Miasmas to MRSA*. Cambridge University Press; 2003.
39. Lew D, Schrenzel J. Intravenous catheter-related infections, suppurative thrombophlebitis and mycotic aneurysms. In: Root R, Waldvogel F, Corey L, Stamm W, editors. *Clinical infectious diseases. A practical approach.* Oxford University Press; 1999. p. 637–42.
40. Percival SL, Kite P. Intravascular catheters and biofilm control. *J Vasc Access* 2007;**8**(2):69–80.
41. Crnich C, Maki D. The promise of novel technology for the prevention of intravascular device-related bloodstream infection. Pathogenesis and short-term devices. *Clin Infect Dis* 2002;**34**:1232–42.
42. RCN IV Therapy Forum. *Standards for infusion therapy*. London: Royal College of Nursing; 2007.
43. Centers for Disease Control and Prevention. Guidelines for the prevention of intravascular catheter-related infections. *Morb Mortal Wkly Rep* 2002;**51**:(RR-10):1–29.
44. Zhang L, Marsh N, McGrail MR, Webster J, Playford EG, Rickard CM. Assessing microbial colonization of peripheral intravascular devices. *J Infect* 2013;**67**(4):353–5.

45. Caguioa J, Pilpil F, Greensitt C, Carnan D. HANDS: standardised intravascular practice based on evidence. *Br J Nurs* 2012;**21**(14) S4, S6, S8–11.
46. Singhai M, Malik A, Shahid M, Malik A, Rawat V. Colonization of peripheral intravascular catheters with biofilm producing microbes: evaluation of risk factors. *Niger Med J* 2012;**53**(1):37–41.
47. Sochor M, Pelikánová Z, Sercl M, Mellanová V, Lazarov PP, Fáčková D. Bloodstream infections of the intravascular access devices—case reports and review of the literature. *Klin Onkol* 2012; **25**(5):375–81.
48. Schwaiger K, Christ M, Battegay M, Widmer A. Prevention of catheter-related infections. *Anaesthesist* 2012;**61**(10):915–24.
49. Department of Health. *Winning Ways: working together to reduce healthcare infection in England*. London: The Stationary Office; 2003.
50. Infection Control Nurses Association. *Audit tools for monitoring infection control standards*; 2004. <www.inicc.org/guias/audit_tools_acute.pdf>.
51. Lai K. Safety of prolonging peripheral cannula and IV tubing from 72 to 96 hours. *Am J Infect Control* 1998;**26**:66–70.
52. Joanna Briggs Institute. *Management of peripheral intravascular devices.* Evidence-based practice information sheets for health professionals, 1998;**2**:1.
53. Curran ET, Coia JE, Gilmour H, McNamee S, Hood J. Multi-centre research surveillance project to reduce infections/phlebitis associated with peripheral vascular catheters. *J Hosp Infect* 2000; **46**:194–202.
54. Parliamentary Office of Science and Technology. *Infection control in healthcare settings*. Postnote; 2005.
55. Maki DG, Ringer M. Risk factors for infusion related phlebitis with small peripheral venous catheters. A randomised controlled trial. *Ann Int Med* 1991;**114**:845–54.
56. Centers for Disease Control and Prevention, and O'Grady NP, Alexander M, Dellinger EP, Gerberding JL, Heard SO, et al. Hospital Infection Control Practices Advisory Committee—guidelines for the prevention of intravascular catheter-related infections. *Morb Mortal Wkly Rep* 2002;**1**(RR10):1–26.
57. Cadorna EA, Watanakunakorn C. Septicemic shock from urinary tract infection caused by *Staphylococcus epidermidis*. *South Med J* 1995;**88**:879–80.
58. Kennedy HF, Morrison D, Kaufmann ME, Jackson MS, Bagg J, Gibson BE, et al. Origins of *Staphylococcus epidermidis* and *Streptococcus oralis* causing bacteraemia in a bone marrow transplant patient. *J Med Microbiol* 2000;**49**:367–70.
59. Public Health Laboratory Service. *Surveillance of hospital-acquired bacteraemia in English hospitals 1997–1999*. London: PHLS; 2000.
60. Jones ID, Case AM, Stevens KB, Boag A, Rycroft AN. *In vitro* comparison of bacterial contamination of peripheral intravenous catheter connectors. *Vet Rec* 2009;**164**(18):556–7.
61. Meers P, Jacobsen W, McPherson M. *Hospital infection control for nurses*. Chapman and Hall; 1992.
62. Wilson J. *Infection control in clinical practice*, 3rd ed. Bailliere Tindall; 2006.
63. Horton R, Parker L. *Informed infection control practice*, 2nd ed. Churchill Livingstone; 2002.
64. RCN. *Standards for infusion therapy*. Becton, Dickinson and Company; 2003.
65. Messner R, Pinker M. Preventing a peripheral IV infection. *Nursing* 1992;**22**(6):34–41.
66. Catney M, Hillis S, Wakefield B, Simpson L, Domino L, Keller S, et al. Relationship between peripheral intravenous catheter dwell time and the development of phlebitis and infiltration. *J Infus Nurs* 2001;**24**(5):332–41.
67. Tomford J, Hershey CO, McLaren CE, Porter DK, Cohen DI. Intravenous therapy team and peripheral venous catheter-associated complications. *Arch Int Med* 1984;**144**:1191–4.
68. Miller J, Goetz A, Squier C, Muder R. Reduction in nosocomial intravenous device-related bacteremias after institution of an intravenous therapy team. *J Intraven Nurs* 1996;**19**(2):103–6.
69. Nicole M, Bavin C, Bedford-Turner S, Cronin P, Rawlings-Anderson K. *Essential nursing skills*, 2nd ed. Mosby; 2004.
70. Jackson A. A battle in the vein: infusion phlebitis. *Nurs Times* 1998;**94**(4):68–71.
71. Webster J, Lloyd S, Hopkins T, Osborne S, Yaxley M. Developing a research base for intravenous peripheral cannula re-sites (DRIP trail). A randomised controlled trial of hospital in-patients. *Int J Nurs Stud* 2008;**44**(5):664–71.

72. Bregenzer T. Is routine replacement necessary? *Arch Int Med* 1998;**158**:151–6.
73. Holmes KR. Comparison or push-pull versus discard method from central venous catheters for blood testing. *J Intraven Nurs* 1998;**21**(2):282–5.
74. Carlson K, Perdue MB, Hankins J. Infection control. In: Hankins J, Lonsway RW, Hedrick C, Perdue MB (Eds). *Infusion therapy in clinical practice*, 2nd ed. WB Saunders; 2001. p. 126–40.
75. Lok CE, Mokrzycki MH. Prevention and management of catheter-related infection in hemodialysis patients. *Kidney Int* 2011;**79**(6):587–98.
76. Bouza E, Guembe M, Muñoz P. Selection of the vascular catheter: can it minimise the risk of infection? *Int J Antimicrob Agents* 2010;**36**(Suppl. 2):S22–5.
77. Crowley L, Wilson J, Guy R, Pitcher D, Fluck R. Chapter 12, Epidemiology of *Staphylococcus aureus* bacteraemia amongst patients receiving dialysis for established renal failure in England in 2009–2011: a joint report from the Health Protection Agency and the UK Renal Registry. *Nephron Clin Pract* 2012;**120**(Suppl. 1):c233–45.
78. Zingg W, Cartier-Fässler V, Walder B. Central venous catheter-associated infections. *Best Pract Res Clin Anaesthesiol* 2008;**22**(3):407–21.
79. Alexandrou E, Murgo M, Calabria E, Spencer TR, Carpen H, Brennan K, et al. Nurse-led central venous catheter insertion-procedural characteristics and outcomes of three intensive care based catheter placement services. *Int J Nurs Stud* 2012;**49**(2):162–8.
80. Bianco A, Coscarelli P, Nobile CG, Pileggi C, Pavia M. The reduction of risk in central line-associated bloodstream infections: knowledge, attitudes, and evidence-based practices in health care workers. *Am J Infect Control* 2013;**41**(2):107–12.
81. British Association for Parenteral and Enteral Nutrition. *Malnutrition universal screening tool*. Redditch: BAPEN; 2003.
82. Stayt L, Randle J. Gastrointestinal system. In: Randle J, Coffey F, Bradbury M, editors. *Clinical skills in nursing*. Oxford University Press; 2009.
83. National Patient Safety Agency. *Reducing the harm caused by misplaced nasogastric feeding tubes. Interim advice for healthcare workers*. London: NPSA; 2005.
84. Stroud M, Duncan H, Nightingale J. Guidelines for enteral feeding in adult hospital patients. *Gut* 2003;**52**(Suppl. VII):vii,1–12.
85. O'Grady NP, Alexander M, Burns LA, Dellinger EP, Garland J—Healthcare Infection Control Practices Advisory Committee (HICPAC). Guidelines for the prevention of intravascular catheter-related infections. *Clin Infect Dis* 2011;**52**(9):e162–93.
86. Inwood S. Skin antisepsis: using 2% chlorhexidine gluconate in 70% isopropyl alcohol. *Br J Nurs* 2007;**16**(22):1390, 1392–4.
87. Goudet V, Timsit JF, Lucet JC, Lepape A, Balayn D, Seguin S, et al. Comparison of four skin preparation strategies to prevent catheter-related infection in intensive care unit (CLEAN trial): a study protocol for a randomized controlled trial. *Trials* 2013;**14**:114.
88. Singhai M, Malik A, Shahid M, Malik MA, Goyal R. A study on device-related infections with special reference to biofilm production and antibiotic resistance. *J Glob Infect Dis* 2012;**4**(4):193–8.
89. Wang X, Lünsdorf H, Ehrén I, Brauner A, Römling U. Characteristics of biofilms from urinary tract catheters and presence of biofilm-related components in *Escherichia coli*. *Curr Microbiol* 2010; **60**(6):446–53.

CHAPTER EIGHT

Wounds and Infection

Steven L. Percival

INTRODUCTION

The skin is the most important line of defence in the human body, protecting tissue from external factors such as harmful chemicals and micro-organisms.[1] Once this barrier has been breached by trauma or surgical interventions, the body's main goal is the re-establishment of this barrier as quickly as possible. Re-establishing this barrier limits the loss of tissue components and fluid and therefore reduces invasion by micro-organisms. However, it is inevitable that all wounds, be they acute or chronic, will eventually become contaminated and then colonised with micro-organisms.[2] Consequently, wounds will harbour micro-organisms derived from a variety of sources, and this complex micro-organism community within the wound constitutes an infection risk.

A wound environment is therefore considered to support a complex of micro-organisms or 'biofilm' which can cause a significant effect on 'normal' wound healing.[3,4] The community of micro-organisms residing in a wound will contain ones which are aerobic, anaerobic, micro-aerophilic and pathogenic often resistant to an array of antimicrobials.[5]

Key Points

- The definition of a wound is a loss of continuity of skin or tissue.[6]
- The community of micro-organisms within a wound also constitutes a cross-contamination risk to patients, healthcare workers and the healthcare environment as a whole.

It is widely accepted that as the number and diversity of micro-organisms increase then the probability of a wound becoming infected also increases. However, the risk of infection depends on an array of endogenous (derived internally) and exogenous (derived externally) factors.

Biofilms in Infection Prevention and Control
Doi: http://dx.doi.org/10.1016/B978-0-12-397043-5.00008-6

Micro-organisms proliferate in wounds because this habitat contains a surface and environment for microbial attachment and therefore a safe haven, moisture and nutrition.[1] Such favourable conditions promote microbial growth and enhance virulence and pathogenicity, which all help to resist clearance by the body's immune system.[7]

In a large number of patients with wounds, the 'normal' healing process can be severely affected and colonising bacteria will proliferate to an extent where microbial virulence is enhanced and the risk of infection is subsequently heightened. Therefore, from a microbiological and clinical perspective, the challenge regarding infection control and wounds remains that in order to reduce the risk of infection at a wound site, the quantity and pathogenicity of micro-organisms needs to be reduced to a level to allow the patient's own defensive and repair mechanisms to prevail and thus heal the wound.[8] However, when wounds do become infected, the healing process is significantly affected. Additionally, the evidence of an infection can lead to significant distress to a patient, which may delay recovery. Wound infections also increase a patient's length of stay in hospital and require the use of expensive wound dressings that contain antimicrobials.

The burden to both the patient's quality of life and healthcare resources are significant to wound management. Factors that impact the patient's quality of life include cosmetically unacceptable scars, persistent pain and itching, restriction of movement and a noticeable impact on emotional well-being.[9] If healthcare workers provide support to the patient in relation to wound management, it can positively affect a person's perceived quality of life, irrespective of whether the wound heals or not[10]; therefore, it is vital that healthcare workers are aware of their own limitations and seek specialist support whenever necessary.[11]

Because of the vast array of concerns with wound infections, appropriate infection control procedures are fundamental in helping to reduce infection incidence and prevalence. Prevention of wound infections should be part of an ongoing programme of education in healthcare settings and are currently being proactively addressed in many public health facilities and hospitals. Although regular support programmes have helped to decrease wound infections, this is still a concern and warrants greater support and resources.

To fully implement infection control procedures for wound care, it is necessary to understand both the role micro-organisms play and the different types of wounds known to be vulnerable to infection, and the patients' underlying pathophysiology. Additionally, a sound comprehension of the role of anti-infectives is also significant in the management of infections in wounds.[12]

WOUNDS AND THE HEALING PROCESS

All wounds contain micro-organisms, but this in itself does not mean that specific wound management interventions need to be initiated[13,14] as it is often possible for wounds to heal successfully without intervention. Wounds that are closed surgically heal by primary intention and wounds that are left open to heal do so by secondary intention.[15] Wound healing by primary intention joins the edges of the wound together and minimises the need for new tissue formation thus minimising scar formation. Wound healing by secondary intention takes longer because it relies on the granulation of tissue arising from the base of the wound to fill the space created by the wound itself.[16] Some wounds may be closed by delayed primary closure where non-viable tissue is removed, the wound is left open for four to six days and then surgically closed. This is necessary for wounds with edges that cannot be fully opposed or after trauma or bowel surgery.[17]

Key Points

- When a wound is infected, it takes longer to heal because the microbial bioburden prevents healthy granulation tissue from developing.
- Infected wounds that are closed by primary intention tend to break down quickly.[18]

A chronic wound is any wound that fails to heal as anticipated or one that has been stuck in one phase of the healing process for a period of six weeks or more.[19,20] The wound-healing process involves the following four major phases[21,22]:

- Inflammation
- Destructive
- Proliferation
- Maturation

Normal wound healing is complex and involves various processes which occur at different rates and at different speeds,[23,24] as follows:

- The *inflammation phase* lasts up to 96 hours. The function of the inflammation phase is to ensure that the wound bed is free from micro-organisms and other contaminants so as to create the optimum environment for the production and multiplication of the various layers of the skin.[25,26] Phagocytosis also occurs during this phase,[27] as well as other processes involving enzymes, growth factors, vasodilatory agents and matrix metalloproteases.[28,29]
- The *destructive phase* can last for up to 2 to 5 days and involves macrophages and fibroblasts removing dead tissue.
- The *proliferation phase* involves fibroblasts and epithelial cells proliferating; granulation tissue forms.
- The *maturation phase* can take up to two years and involves fibroblasts and proteases. Fibroplasts pull the wound edges together and the granulation tissue matures into scar tissue, which then pales, shrinks and thins.[30]

Key Points

- Phagocytosis is the engulfing and destruction of micro-organisms and foreign bodies by phagocytes in the blood.
- Patients whose wound history extends more than two years may be more prone to recurrent wound infections.

Burton[31] identifies the following factors that delay the healing process:

- Infection
- Haematoma, foreign body or necrosis
- Low albumin level
- Poor vascular supply, anaemia
- Poor nutritional intake
- Chronic medical conditions such as diabetes
- Mechanical stress on the wound

WOUNDS AND MICRO-ORGANISMS

Bacterial colonisation of a wound allows the establishment of populations of micro-organisms at the wound site.[32] Micro-organisms produce specific protein-degrading enzymes called 'proteinases', and if these are within a contaminated wound site, they lead to enhanced tissue breakdown.[33] If the community of micro-organisms is protected, then the infected wound will not heal. This becomes an important aspect of the patient's care and destruction of the community of micro-organisms is part of the treatment of the non-healing wound.

Overall, wound healing and infection are influenced by the relationship between the ability of micro-organisms to create a stable, prosperous community within a wound environment and the ability of the host to control the bacterial community. A prolonged exposure to chronic wound micro-organisms leads to an ongoing inflammatory response—complex processes which have a potentially detrimental effect on cellular processes involved in wound healing.

For instance, bacteria, such as *Pseudomonas aeruginosa,* release proteinases known to affect growth factors and many other tissue proteins necessary for the wound-healing process.[29] During the healing process, exudate is produced and chronic wound exudate affects cell proliferation and wound healing.

Key Points

- Contamination of a wound is when micro-organisms are present but they are transient and wound healing is not delayed.

- Colonisation of a wound refers to the multiplication of micro-organisms, but they do not cause damage or initiate wound infection.
- Wound infection is when microbial growth, multiplication and invasion into the host's tissue leads to cellular injury and initiates the host's immunological reactions. Healing is interrupted.[15]

Risk Factors

There are various reasons why wounds become infected. The risks or factors, which are known to enhance the likelihood of an infection, need to be addressed to evaluate any infection prevention and control strategy. The risk of wound infection of both acute and chronic wounds is related to a number of risk factors, including the environment, the patient, the type of surgical category and procedure, the appropriateness of the correct care and healthcare worker training and education.

Environment

For post-operative infections, the patient environment is very important as a potential conduit for infection. For example, in operating theatres there should be adequate supplies of disinfectants and handwashing facilities outside of the main sterile operating rooms to ensure effective hand hygiene takes place during the immediate pre- and post-operative care of the patient.

In addition to this, filtered air ventilation into the operating theatre plus frequent, regular changes of air is important both to remove bacteria that may be brought in from outside and to remove bacteria shed on skin scales by staff present in the theatre during the surgical procedure. Appropriate theatre clothing should be worn because this reduces shedding of material fibres and skin scales. Also, access into the 'sterile' environments should be restricted, which helps to keep infection risk to a minimum.

Patient

Individuals vary widely in susceptibility to infection, and this depends on many factors, including age, general health, underlying illness, state of nutrition, previous exposure to infection and vaccination, occurrence of invasive procedures, acuity level and undisciplined use of antibiotics.[34] The patient constitutes a high risk in respect of endogenous infection from the individual's own microbial flora and contamination of any wound. Additional factors (e.g., skin condition) also play a role in increasing the risk of a wound infection, with damaged skin being at increased risk of colonisation with resistant organisms; therefore it poses an increased risk to wounds.

Types of Operation and Surgical Procedures

The types of operation and surgical procedures, which may have induced the wound initially, will add an enhanced risk to the generation of infection in a wound. For example, risks of deep wound infection are increased for procedures, such as transplants and implantation of orthopaedic devices, because necessary items left inside the body can harbour very low numbers of organisms unless sterility is maintained absolutely. Once implanted these micro-organisms can multiply, sometimes over many months, and then cause symptoms of infection.

For this reason surveillance definitions for surgical site infection generally include deep infection in implant surgery up to one year after the procedure. Surgery involving a site with pre-existing infection, or where necrosed tissue is present, also presents a high risk of post-operative infection.[35] In addition to this, the length of time a surgical procedure takes can increase risk as well as the use of medical devices (e.g., catheters, cannula and drains), which are well-known risk factors.

Appropriate Usage of Prophylaxis

As with many surgical procedures, antibiotic and antiseptic prophylaxis is often administered. This has been shown to be very beneficial at reducing the risk of wound infection in particular types of surgical intervention. However, there is little sound clinical evidence to support the indiscriminate use of prophylaxis.[36] For any incision, appropriate skin antisepsis should take place to prevent skin/wound contamination.[37]

Other Problems

Additional problems associated with both acute and chronic wounds exist. Solutions to manage such risks have included careful design of operating theatres; antimicrobial washes; wearing of correct clothing; and the appropriateness of training that has been provided to medical staff, in particular to nurses, in relation to wound management. In respect of wound care, the correct use of appropriate dressings, in addition to technically correct application and removal from the wound site, all represent opportunities for lowering the risk of a wound infection occurring. It should also be noted that wound cleansing should be performed to remove pus, exudate, debris and old dressing material.[38]

RECOGNISING AND PREVENTING WOUND INFECTION

Traditional criteria for recognising wound infections have been considered as limiting,[39,40] so additional criteria are used (see Table 8.1). Traditional criteria include:

- Pain
- Presence of abscess, cellulitis or discharge
- Elevated temperature

Additional criteria include:

- Abnormal smell
- Wound breakdown
- Pocketing at the base of the wound
- Friable tissue that bleeds easily

Table 8.1 Criteria for Recognising Wound Infection

Primary wounds	Secondary wounds
Cellulitis	Cellulitis
Pus/abscess	Pus/abscess
Delayed healing	Delayed healing
Erythema with/without induration	Erythema with/without induration
Haemopurulent exudate	Haemopurulent exudate
Seropurulent exudate	Seropurulent exudate
Oedema	Oedema
Increase in local skin temperature	Increase in local skin temperature
Unexpected pain or tenderness	Unexpected pain or tenderness
Wound breakdown or enlargement	Wound breakdown or enlargement
Serous exudate with erythema	Increase in exudate volume
Malodour	Malodour
Swelling	Pocketing
	Friable granulation tissue that bleeds easily
	Discolouration

Cutting and colleagues[41] have also documented that different types of wounds have different infection criteria.

Preventing Wound Infections

The chance of a wound infection occurring can be minimised and the following subsections describe some strategies that can be implemented.

Delay Non-urgent Procedures

If a patient is at increased risk from pending surgery, such as for multiple concurrent diseases and ailments, it may be wise to delay non-urgent procedures if measures can be taken to reduce the risk of infection during the pre-operative period. This may assist in reducing the increased risk of cross-contamination from various reservoirs of infection; however, it is only helpful if measures can be taken to reduce risk.

Prophylaxis Use

The use of antibiotic/antiseptic prophylaxis may be necessary if the site of a surgical procedure is an area contaminated with an array of potential or pathogenic micro-organisms. The use of prophylaxis remains an area of controversy and debate, which is beyond the scope of this chapter. However, it has been demonstrated that pre- and peri-operative prophylaxis is highly effective for some types of procedures.

Local guidelines should be available, generally written by clinical staff, pharmacists and microbiologists, in order to ensure that they are appropriate and effective. Prophylactic antimicrobial efficacy is dependent on the timing of its administration[42] and the Department of Health high-impact intervention for prevention of surgical site infection requires that the pre-operative dose be given within one hour of surgery.[43]

It should be noted though that adequate time must elapse for perfusion of the antibiotic to the surgical area (e.g., administration immediately prior to applying a tourniquet to a limb will not provide adequate time for perfusion to the affected area) before the blood supply is reduced. There is evidence that if the correct antibiotics are given, at the correct time and dosage over the correct time period, positive clinical outcomes will be achieved together with a low risk of infection at high-risk areas.

Drainage

In the case of wounds, open drains are considered very high risk and should, if at all possible, not be used; however, with many wounds a drainage device is required and it usually is connected to a bottle or bag in theatre. Such drains present opportunities for bacteria to access and contaminate deep tissue—in particular, there is a risk of cross-contamination from healthcare workers who are involved with changing such devices unless they use an effective aseptic technique.

Skin Preparation

Skin preparation for surgery is an important area of potential risk. Shaving is known to bruise and abrade the skin, which increases the risk of infection. This is because many *staphylococci* in particular are known to reside around the sebaceous glands adjacent to hair follicles. By removing the hair follicles, the *staphylococci* disperse onto the patient's skin surface and increase the risk of cross-contamination to other individuals or to wounds. In these situations the use of skin antiseptics is often warranted before an incision is made.

Alcohol solutions of chlorhexidine or iodine are the most commonly used skin antiseptics.[44] After an area has been coated in the skin preparation solution and before the incision is made, the solution should be allowed to dry completely to reduce the risk of infection and to avoid the risk of igniting alcohol vapour if diathermy is to be used.

Surgical Environment

The surgical environment (usually an operating theatre) is very important in the prevention of a wound infection. In addition, the movement of healthcare workers through a surgical facility should be substantially reduced as countless pathogenic bacteria are airborne on skin scales shed by everyone, which can lead to increased risk of cross-contamination in the wound vicinity. The use of appropriate theatre clothing is very important—these should be such that they reduce the transfer of pathogens. Therefore clothing used in the operating theatre should be different and separate from clothing used in everyday life, and there should be adequate areas for showering and handwashing.

Ventilation

Adequate ventilation in operating theatres is required. Mechanical ventilation is the preferred option and here the air should be changed 20 to 24 times per hour and any air entering the theatre should be filtered. In addition to this standard, for very high-risk procedures involving implant surgery, the use of ultraclean theatres is now the standard. In ultraclean theatres there is an area under a canopy that has an extremely high rate of air exchange and the levels of contamination under the canopy are kept extremely low ($<10\,cfu/m^3$). This reduces the risk of airborne contamination into a deep wound.

Although theatre ventilation and the environment is monitored and strictly controlled, staff working in theatres should check on a daily basis that they can hear the ventilation system working and that they can see the balancing vents opening and closing when external theatre doors are opened and closed. Plus, they should be able to feel airflow between the main theatre and lay-up and scrub-up rooms. This will ensure that any failure is detected and can be reported immediately.

Wound Dressing

Appropriate choice of a wound dressing that helps to manage the wound bioburden, remove corrosive enzymes, to limit cross-contamination to other wound sites and to promote wound healing is of paramount importance to infection control in wound care, and it will facilitate wound healing.

To reduce the risk of wound infection, the procedures in the preceding subsections need to be strictly adhered to with clinicians guided by the various protocols of care that are in place and discussed throughout this book. Essential requirements that help to reduce wound infection include the following:

- Use of sterile gloves
- Correct surgical procedures that reduce the potential for cross-contamination
- Use of appropriate disinfectants and antiseptics
- Correct environment for surgical interventions
- Appropriate use of prophylactic antibiotics

After a surgical procedure, patients should be kept in hospital for the shortest possible time to limit the risk of infection.

SURGICAL SITE INFECTIONS

Surgical site infections (SSIs) are wound infections that can occur after an invasive procedure and have been shown to be the cause of up to 20% of all healthcare-associated infections (HCAI). The majority of SSIs are preventable, but many infections are caused by contamination of an incision with micro-organisms from the patient's own body (endogenous spread) during surgery.[45] However, as already noted in this chapter, SSIs can result from exogenous spread; this occurs when micro-organisms from instruments or the theatre environment contaminate a surgical wound, or post-operatively as a result of cross-infection, according to the National Institute for Health and Clinical Excellence and the National Collaborating Centre for Women's and Children's Health.[46]

CATEGORIES OF SURGICAL WOUNDS AND INFECTION RISK

Surgical wounds are classified according to the likelihood and degree of wound contamination at the time of surgery into the following four categories[47]:

Clean—Uninfected wounds that do not enter body systems naturally colonised by micro-organisms.

Clean-contaminated—Surgical wounds that enter body systems naturally colonised with micro-organisms, including the respiratory, alimentary, genital and urinary tracts. This includes when they are entered under controlled conditions and without unusual contamination, providing that there is no evidence of infection or a major break in aseptic technique.

Contaminated—Operations on new, open traumatic wounds; operations where there is a major break in aseptic technique; and operations in which there is gross spillage from the gastrointestinal tract or acute inflammation without pus being encountered.

Dirty or infected wounds—Operations in which acute inflammation with pus is encountered, or in which perforated viscera are found; this includes operations on traumatic wounds that have retained dead and severely damaged tissue, foreign bodies or faecal contamination. Operations include those where the micro-organisms causing post-operative infection are likely to have been present before surgery.

Surgery on body sites that are considered 'clean' (e.g., in prosthetic surgery) have relatively low SSI rates whereas operations at dirty body sites (e.g., the bowel) can exceed 10%.[48] The Centers for Disease Control and Prevention (CDC) identifies the following three levels of SSI:

1. *Superficial incisional*—This affects the skin and subcutaneous tissue. Signs and symptoms of infection are redness, pain, heat or swelling at the site of incision or by the drainage of pus.
2. *Deep incisional*—This affects the fascial and muscle layers. Signs and symptoms are the presence of pus or an abscess, fever, wound site tenderness or separation of the edges of the incision, which exposes the deeper tissues.
3. *Organ/space infection*—This affects any part of the anatomy other than the actual wound incision. Signs and symptoms are drainage of pus or formation of an abscess.

Key Points

- Obtained cultures may provide microbiological evidence of SSI; however, skin is normally colonised by a range of micro-organisms that could cause infection. Consequently, signs and symptoms of infection are required before it is classified as a SSI.[46]
- *Staphylococcus aureus* is the most commonly cultured micro-organism in SSIs.
- The majority of SSIs become apparent within 30 days of a surgical procedure and most frequently within 5 to 10 post-operative days.
- When a prosthetic implant is used, SSIs affecting the deeper tissues can occur several months post-operatively.[46]

During the post-operative period, it is good practice to cover the surgical wound with a dressing for the first 48 hours during the initial period of healing by primary intention unless it is contraindicated—for example, if there is excessive wound leakage or haemorrhage.[46] For those wounds healing by secondary intention, contact should be made with a healthcare worker who has tissue viability expertise—a Tissue Viability Nurse. Many organisations have local policies or local wound formularies listing approved dressings and provide guidance on wound dressing selection; these should always be consulted to ensure that the most suitable approved wound dressing is selected.

There are a plethora of wound dressings available to healthcare workers.[49] The purposes of surgical wound dressings are to:

- Allow appropriate visual assessment
- Absorb exudate
- Ease pain
- Provide protection for newly formed tissue
- Provide an optimal moist wound environment[50]
- Allow early bathing or showering and therefore promote patient rehabilitation[46]

A robust health technological assessment[46] concluded that there is no significant evidence to support the use of one dressing over another, although the use of gauze should be avoided as it has the potential to cause pain when it is removed; it can also disrupt the healing process because it is likely to stick to the wound. Instead, it is generally accepted that a dressing which has a semi-permeable film membrane with or without an absorbent part can be used in the majority of clinical situations.

CONCLUSION

Non-healing and infected wounds are microbiologically diverse. This diversity has a number of advantages to micro-organisms because it is likely to help microbial interactions and enhance their survival. However, these interactions are likely to have detrimental effects on the host. Wound infections as a result of the colonising bacteria are diagnosed on the basis of clinical signs.[47] The objective of infection control and, in particular, wound care is to minimise the risk of transmission of micro-organisms and to prevent patients from acquiring an infection.

Despite an array of infection control procedures, however, this is often not achievable. Failing this, to help reduce the risk of the wound becoming infected, the microbial load in the wound bed should be significantly reduced to allow the host immune system to prevail. By reducing the number of micro-organisms, their related interactions and pathogenicity will also be reduced. Through the correct use of antimicrobials, the bioburden can be decreased significantly and therefore aid in wound healing.

REFERENCES

1. Percival SL. Skin anatomy and microbiology. In: Percival SL, editor. *Microbiology of aging*. Humana Press Inc./Springer; 2009.
2. Percival SL, Rogers AA. The significance and role of biofilms in chronic wounds. In: McBain A, Allison D, Pratten J, Spratt D, Upton M, Verran J, editors. *Biofilms, persistence and ubiquity*. Bioline; 2005.
3. Percival SL, Thomas J, Williams D. Biofilms and bacterial imbalances in chronic wounds: anti-Koch. *Int Wound J* 2010;**7**(3):169–75.
4. Cutting K, Wolcott R, Percival SL. Biofilms and significance to wound healing. In: Percival SL, Cutting K, editors. *Microbiology of wounds*. CRC Press; 2010. p. 233–48.
5. Percival SL, Dowd S. The microbiology of wounds. In: Percival SL, Cutting K, editors. *Chronic wounds*. CRC Press; 2010. p. 187–218.
6. Ayto M. Wounds that won't heal. *Nurs Times* 1985;**81**:16–19.
7. Woods E, Davis P, Barnett J, Percival SL. Wound healing, immunology and biofilms. In: Percival SL, Cutting K, editors. *Microbiology of wounds*. CRC Press; 2010. p. 271–92.
8. Percival SL, Cooper R, Lipsky B. Antimicrobial interventions for wounds. In: Percival SL, Cutting K, editors. *Microbiology of wounds*. CRC Press; 2010. p. 293–328.
9. Bayat A, McGrouther D, Ferguson M. Skin scarring. *Br Med J* 2003;**36**:88–92.
10. Franks PJ, Collier MC. Quality of life: the cost to the individual. In: Morison MJ, editor. *The prevention and treatment of pressure ulcers*. Mosby; 2001.
11. Collier M. Wound bed preparation: theory to practice. *Nurs Stand* 2003;**17**:45–52.
12. Percival SL, Thomas JT, et al. The efficacy of silver alginate against a broad spectrum of wound isolates. *Vet Microbiol* 2011;**150**(1–2):152–9.
13. Gilchrist B. Should iodine be reconsidered in wound management? *J Wound Care* 1997;**6**(3):148–50.
14. Thomas JG, Motlagh H, Percival SL. The role of microorganisms and biofilms in dysfunctional wound healing: wound care. In: Farrar D, editor. *Advanced wound repair therapies*. Woodhead Publishing; 2011.
15. Vuolo J. Assessment and management of surgical wounds in clinical practice. *Nurs Stand* 2006; **20**:46–56.
16. Nguyen DT, Orgill DP, Murphy GF. The pathophysiologic basis for wound healing and cutaneous regeneration. *Biomaterials for treating skin loss*. Woodhead Publishing and CRC Press; 2009. p. 25–57.
17. Gottrup F, Melling A, Hollander D. *An overview of surgical site infections, aetiology, incidence and risk factors* <www.worldwidewounds.com/2005/september/Gottrup/Surgical-Site-Infections-Overview.html>; 2005.
18. Wolcott R, Cutting K, Dowd S, Percival SL. Types of wounds and infection. In: Percival SL, Cutting K, editors. *Microbiology of wounds*. CRC Press; 2010. p. 219–32.
19. Collier M. Limited resources in wound management: a reality for some. *J Commun Nurs* 2002; **16**:38–42.
20. Stadelmann WK, Digenis AG, Tobin GR. Physiology and healing dynamics of chronic cutaneous wounds. *Am J Surg* 1998;**176**(2A Suppl):26S–38S.

21. Wilson J. *Infection control in clinical practice*, 3rd ed. Balliere Tindall; 2006.
22. Gurtner GC, Werner S, Barrandon Y, Longaker MT. Wound repair and regeneration. *Nature* 2008; **453**(7193):314–21.
23. Bale S, Harding K, Leaper D. *An introduction to wounds*. Emap Healthcare Ltd; 2000.
24. Enoch S, Price P. *Cellular, molecular and biochemical differences in the pathophysiology of healing between acute wounds, chronic wounds and wounds in the elderly* <www.worldwidewounds.com>; 2004.
25. Midwood KS, Williams LV, Schwarzbauer JE. Tissue repair and the dynamics of the extracellular matrix. *Int J Biochem Cell Biol* 2004;**36**(6):1031–7.
26. Harding KG, Morris HL, Patel GK. Science, medicine and the future: healing chronic wounds. *BMJ* 2002;**324**(7330):160–3.
27. Calvin M. Cutaneous wound repair. *Wounds* 1998;**10**:12–32.
28. McCarty S, Clegg PD, Cochrane CA, Percival SL. Equine wound-isolated *Staphyloccous aureus* isolates up-regulated fibroblast MMP expression *in vitro*: a focus on the stimulatory effect of bacterial proteases. *Wound Rep Reg* 2011;**19**:A8–A62.
29. McCarty SM, Cochrane CA, Clegg PD, Percival SL. The role of endogenous and exogenous enzymes in chronic wounds: a focus on the implications of aberrant levels of both host and bacterial proteases in wound healing. *Wound Repair Regen* 2012;**20**(2):125–36.
30. National Institute for Health and Clinical Excellence. *Prevention of surgical site infection*. London: NICE. Available online at <www.nice.org.uk>; 2008.
31. Burton F. Best practice overview: surgical and trauma wounds. *Wound Essent* 2006;**1**:98–107.
32. Percival SL, Bowler P. Biofilms and their potential role in wound healing. *Wounds* 2004;**16**(7):234–40.
33. Cochrane C, Percival SL. Wounds, enzymes and proteases. In: Percival SL, Cutting K, editors. *Microbiology of wounds*. CRC Press; 2010. p. 249–70.
34. Duffy J. Nosocomial infections: important acute care nursing-sensitive outcome indicators. *Am Assoc Crit Care Nurs* 2002;**13**:358–66.
35. Wolcott R, Cutting K, Dowd S, Percival SL. Surgical site infections: biofilms, dehiscence and wound healing. *US Dermatol* 2008:56–9.
36. Timmons J, Bell A. Wound infection. *Prim Health Care* 2000;**10**(2):31–8.
37. Moore P, Foster L. Acute surgical wound care. 1: An overview of treatment. In: White R, editor. *Trends in wound care*. Mark Allen; 2002.
38. Miller M, Dyson M. *Principles of wound care*. London: Professional Nurse/Emap Healthcare; 1996.
39. Cutting KF, Harding KG. Criteria for identifying wound infection. *J Clin Nurs* 1994;**7**:539–46.
40. Cutting KF, Harding KG. Criteria for identifying wound infection. *J Wound Care* 1994;**3**(4):198–201.
41. Cutting KF, White RJ, Mahoney P, Harding K. Clinical identification of wound infection: a Delphi approach *Identifying criteria for wound infection, EWMA position document*. London: MEP; 2005.
42. Holzheimer RG, Haupt W, Thriede A, et al. The challenge of post-operative infections. Does the surgeon make a difference? *Infect Control Hosp Epidemiol* 1997;**18**:449–56.
43. Department of Health. *Saving lives—high impact intervention. 4: Care bundle to prevent surgical site infection*. From Clean, Safe Care website at <www.clean-safecare.nhs.uk/toolfiles/22_SL_HII_4_v2.pdf>; 2007 [accessed 30.12.08].
44. Maiwald M, Chan ES. The forgotten role of alcohol: a systematic review and meta-analysis of the clinical efficacy and perceived role of chlorhexidine in skin antisepsis. *PLoS One* 2012;**7**(9):e44277.
45. Hospital Healthcare Europe. <www.hospitalhealthcare.com/default.asp?page=article.display&title=GuidelinesaimtopreventSSIs&article.id=14209>; 2008 [accessed 04.12.08].
46. National Institute for Health and Clinical Excellence and the National Collaborating Centre for Women's and Children's Health. *Surgical site infection: prevention and treatment of surgical site infection. clinical guideline*. London; 2008.
47. Health Protection Agency. *Protocol for surveillance of surgical site infection, version 4 (July)*. From HPA website <www.hpa.org.uk/web/HPAwebFile/HPAweb_C/1194947388966>; 2008 [accessed 30.12.08].
48. Health Protection Agency. Surveillance of surgical site infection in England October 1997-September 2005. London: Health Protection Agency; 2005.
49. Jones V, Grey JE, Harding KG. Wound dressings. *BMJ* 2006;**332**(7544):777–80.
50. Winter GD. Formation of the scab and the rate of epithelisation of superficial wounds in the skin of the young domestic pig. *J Wound Care* 1962;**4**(8):366–7, discussion 368–71.

PART 2

Biofilms and Infection Control

CHAPTER NINE

Biofilms

From Concept to Reality

Sara McCarty, Emma Woods and Steven L. Percival

INTRODUCTION

It is now widely accepted that 99% of all micro-organisms attach to a surface and grow as a biofilm. The biofilm mode of growth is an important survival strategy for micro-organisms in the healthcare environment.[1,2] Biofilms consist of microbial cells encased in a self-generated extracellular polymeric substance (EPS).[3] The first evidence of their existence was reported by Anton van Leeuwenhoek in the seventeenth century following the analysis of plaque scraped from his own teeth. In 1933, Henrici observed micro-colonies forming on glass slides placed in an aquarium which, he reported, steadily grew in size. These bacterial micro-colonies were firmly adherent to the surface and could not be removed through washing.[4]

In 1936 Zobell and Anderson observed the enhanced growth rate of bacteria in the interstices of small inert particles of sand and glass beads placed in seawater when compared with growth in seawater alone; they attributed this to the proximity of the water to solid surfaces[5] and noted that bacterial growth is enhanced in smaller containers, given the greater amount of solid surface per unit volume of stored water. They concluded that solid surfaces are beneficial to bacteria in dilute nutrient solutions.[5,6]

In 1940, Heukelekian and Heller[7] observed substantially enhanced bacteria growth when microbes were attached to a surface; *Escherichia coli* grown with limited nutrients grew better with the addition of glass beads, demonstrating that solid surfaces, which act as a source of concentrated nutrients, enable the bacteria to survive in substrates which would otherwise be too dilute for growth.[7]

Zobell,[6] in 1943, provided extensive evidence of biofilms by observing higher numbers of bacteria on surfaces than in the surrounding seawater. He proposed that the adhesion of bacteria occurred in two phases: one reversible, the other irreversible. Furthermore, he noted that nutrients were concentrated on solid inert surfaces, enhancing bacterial activity, with many of the bacteria found to be sessile in seawater because they were attached to a solid surface.

Biofilms in Infection Prevention and Control
Doi: http://dx.doi.org/10.1016/B978-0-12-397043-5.00009-8

It was not until the 1970s that research on biofilms truly began despite early evidence of their existence. Marshall and colleagues in 1971 noted that a motile strain of *Pseudomonas* produced polymeric fibrils in arteficial seawater, which they thought might be involved in the irreversible sorption of the bacteria to surfaces.[8] Late in the 1970s a clearer definition emerged,[9,10] with Costerton et al.[10] proposing that biofilms consist of communities of bacteria attached to surfaces encased in a "glycocalyx" matrix. In 1987, Costerton and colleagues[11] proposed that biofilms consisted of single cells and micro-colonies which become embedded in a hydrated polyanionic exopolymer matrix. Soon after, Characklis and Marshall[12] described other characteristics, such as spatial and temporal heterogeneity and substances holding the matrix of the biofilm together.

In 1995, Costerton and colleagues reported that biofilms can adhere to surfaces and interfaces and to each other.[13] In their definition of a biofilm, they included microbial aggregates and floccules as well as the adherent populations in the pore spaces of porous media.[13] They also clarified that biofilms constitute a distinct growth phase of bacteria which is in contrast to that of their planktonic counterparts.[13] Such pioneering research helped to introduce the world to biofilmology—'the study of biofilms', through which we better understand biofilm structure. Biofilm architecture has been and continues to be extensively studied, using such techniques as optical sectioning,[14] confocal laser scanning microscopy (CLSM),[15,16] scanning electron microscopy (SEM)[17] and three-dimentional imaging.[18] These techniques in particular have helped researchers to clearly and simply define a *biofilm* as a community of bacteria and/or other microbes and their extracellular polymers attached to a surface.[19,20]

Biofilms exhibit a number of fundamental characteristics which the reader may find useful[19,21]:

- Biofilms form a three-dimensional structure.
- Biofilms contain one or more microbial species.
- Biofilms form at interfaces. These can be solid/liquid, liquid/air, liquid/liquid or solid/air.
- Biofilms exhibit spatial heterogeneity. This is because of the physicochemical and chemical gradients which develop in them and to the adaptation inflicted on them by outside perturbations.
- Bioflms are often permeated by water channels which are sometimes referred to as the 'circulatory system'.
- The micro-organisms within biofilms, compared to their planktonic counterparts, exhibit a marked decrease in susceptibility to antimicrobial agents and host defence systems.

Although the precise structure, chemistry and physiology of a biofilm varies depending on the resident microbes and the surrounding environment, its structural integrity strongly depends on the extracellular matrix produced by the resident cells.[22]

Initial studies by Costerton and Lappin-Scott[23] found that, in addition to its other common characteristics, a biofilm's cells differ from their planktonic counterparts. This has been demonstrated in altered gene expression in particular, which has been shown to aid adhesion and biofilm formation. An early biofilm study demonstrated the variation in gene expression between bacterial cells residing in the biofilm and their planktonic counterparts.[24]

In 2002, Donlan and Costerton[25] proposed a new biofilm definition to include characteristics such as altered growth rate and altered gene transcription:

> *A microbially derived sessile community characterized by cells that are irreversibly attached to a substratum or interface or to each other, are embedded in a matrix of extracellular polymeric substances that they have produced, and exhibit an altered phenotype with respect to growth rate and gene transcription in page 1.*

However, the environment in which it is formed must also be taken into consideration when defining a biofilm. As noted by Harding and colleagues,[26] one widely accepted definition is 'a community of microbial cells permeated by water channels allowing efficient biomass exchange between the population and the environment', which considers only biofilms in an aqueous environment.

PREVALENCE OF BIOFILMS

Biofilms are observed on an variety of surfaces and materials, both biotic and abiotic; furthermore, they can be benign or pathogenic. The ability of micro-organisms to produce biofilms on surfaces is believed to contribute substantially to the pathogenesis of infection.[1] In fact, it is reported that biofilm 'infections' constitute, or are associated with, 65% of human infections reported in the developed world.[27,28] Indeed, biofilms have commonly been identified in both medical device-related and non-device-related infections. These may be caused by a single species or a mixture of species of micro-organisms.[29,30]

There are many advantages to the biofilm mode of growth for microbial survival and spread which are not observed for free-floating planktonic cells. In particular, it offers residing cells protection from an often hostile environment; a broader habitat range for growth, increased metabolic diversity and efficiency; enhanced resistance to environmental stress, antimicrobial agents and host defences.[28,31]

In the medical setting, biofilms have been implicated in the rejection of many medical devices, including artificial dental and medical prostheses. They are well known to be prevalent on teeth,[32] inside the gut and in the lungs of cystic fibrosis (CF) patients, as well as on and in endotracheal tubes, intravascular catheters,[33] heart valves, the bladder, the middle ear,[34] prosthetic joints[35] and chronic wounds.[36]

A number of studies have emphasised the high incidence of biofilms among samples from root canals associated with periodontitis.[28] Ricucci and Siqueira[28] evaluated the presence of biofilms in both treated and untreated root canals associated with apical periodontitis; they found them in 77% of the root canals analysed, with untreated canals having more than treated canals. Comparatively, Noiri and colleagues[37] found that of 11 samples consisting of both extracted teeth and root-filling gutta-percha points associated with refractory periapical periodontitis, 9 showed evidence of bacterial biofilm as observed via SEM. Similarly, Carr and colleagues[38] found that a comlex, variable, multi-species biofilm was associated with a resected, failing endodontically re-treated molar root tip.

Pseudomonas aeruginosa biofilms are a major factor in the persistence of infection in CF patients; the polysaccharide alginate is the main constituent of the biofilm matrix in these patients' lungs.[39] Mucoid *P. aeruginosa* biofilms are commonly observed in CF lung tissue.[39] For example, Bjarnsholt and colleagues[40] investigated *P. aeruginosa* biofilms in samples of CF lung tissue and sputum. They found that, in preserved tissues taken from CF patients who had not undergone intensive antiobiotic therapy, mucoid (alginate) bacteria in aggregating structures surrounded by polymorphonuclear leukocyte (PMN) inflammation in the respiratory zone correlated with *P. aeruginosa* infection and destruction of the lung.[40] Detached biofilm-containing alveoles can also be found in the sputum of CF patients.[41] Biofilms in chronically infected CF patients also act as a bacterial reservoir from which planktonic cells can shed, colonising new sites and thus causing an inflammatory response and clinical symptoms.[42]

Singh and colleagues[43] also found evidence of *P. aeruginosa* biofilm in sputum samples obtained from CF patients via morphological analysis and through the identification of quorum-sensing (QS) signals. In an *in vitro* study, Matsui and team[44] highlighted how the physicochemical properties of mucus in the PMN of CF patients influence *P. aeruginosa* biofilm formation—through the promotion of macro-colony formation and restricted bacterial motility by CF-like concentrated mucus.[44] Furthermore, the concentrated, dehydrated mucus gel acts to inhibit host antimicrobial defence by diminishing lactoferrin activity.[44]

Chronic otitis media, a common paediatric infectious disease, is also associated with biofilm formation.[45] Using CLSM with generic and species-specific bacterial probes, Hall-Stoodley and colleagues[45] found mucosal biofilms in 46 of 50 middle-ear mucosa biopsy specimens from children with otitis media with effusion and with recurrent otitis media.

MODELS TO STUDY BIOFILMS

Biofilms in different settings form a variety of structures composed of different microbial consortia dictated by biological and environmental parameters, which can quickly respond and adapt. Indeed, a significant problem associated with the investigation of environmental systems is the inherent complexity within them.[46]

Although it is difficult to accurately mimic the *in vivo* environment in an experimental model, a number of *in vitro* biofilm models have proven useful in the study of biofilm processes and antimicrobial efficacy against biofilm-forming organisms.[25] However, some key parameters may affect the rate and extent of biofilm formation in a model system, including the medium used, the inoculum, the system's hydrodynamics and the substratum.[25] Indeed, biofilm architecture is influenced by various environmental conditions, including nutrient status, the presence or absence of heavy metals, material composition, roughness of the substratum and hydrodynamic shear force.[47] As noted by Trappetti and colleagues,[48] biofilm models are less standardised than the classic mid-log growth phase, in which most microbiological research has been done; thus it is important to identify the most appropriate experimental biofilm model to obtain significant and comparable results.[48]

Laboratory-grown biofilms are designed to mimic the *in vivo* milieu, and through the modification of various parameters, it is possible to grow them with differing structures and functions.[49] One commonly used experimental model is the Calgary Biofilm Device (CBD), otherwise known as the Minimum Biofilm Eliminating Concentration (MBEC) biofilm assay (MBEC Innovatech, Calgary, Alberta). This model consists of two parts which are designed to fit together: a polystyrene lid with 96 pegs forms the top component; a standard 96-well plate forms the bottom component.[50]

The CBD provides a high-throughput method often employed for testing antimicrobial susceptibility in an easy-to-use single experimental system[51]; it produces 96 equivalent biofilms which generate reproducible results, and it has the potential for automation given its 96-well plate design.[51] The CBD was originally developed as a reproducible method of assessing the antimicrobial efficacy of antibiotics against biofilms associated with indwelling medical devices.[51,52]

Harrison and colleagues[47] described the use of the CBD to evaluate biofilm structure under multivariate growth conditions and exposure. They noted that it allows the user to test single- or multi-species biofilms against a range of antimicrobial treatments and controlled variables in a single system; also, it permits study of variations in growth medium formulation, exposure times and antimicrobial concentration alone or in combination with this biofilm model.[47] The authors used SEM and CLSM to qualitatively evaluate biofilms formed using the CBD.[47]

The CBD facilitates various techniques for studying biofilm structure and antimicrobial susceptibility in the same system. For example, some pegs may be removed for microscopy, others may be used for viable cell counts, while others may be used for susceptibility testing.[47]

One of the CBD limitations is that it does not feature controlled fluid dynamic conditions and so consideration must be given when interpreting data from various biofilm systems.[47] In addition, growth does not evenly cover the surface of a CBD peg and therefore examination of pegs should be made systematically to reflect this.[47] The height of the biofilm formed in the CBD should also be considered; biofilms grown in it may be less 'mature' than multi-layered biofilms grown in a continuous-flow cell system.[53] In spite of these drawbacks, the CBD is simple to use, provides a method for rapid screening and avoids the use of complex flow systems and tubes which may be a source for contamination.[50]

The Modified Robbin's device (MRD), originally designed to study biofilms in tubular devices (e.g., catheters and pipelines),[54] facilitates the study of biofilms under flowing conditions with controlled hydrodynamics.[46] Like the CBD, the MRD has the advantage of multiple sampling points so that several samples can be obtained simultaneously or at various time points.[46] However, it can be used in both recirculating and flow-through systems and can be connected to a chemostat for controlled growth conditions.[46]

The limitations of the MRD include the inability to visualise the biofilm *in situ*, possible nutrient gradients along the length of the device and possible deviations in hydrodynamics.[46] However, it provides a more accurate representation of the *in vivo* environment, given the presence of shear forces which cause the detachment of weakly adhered cells. Thus, the MRD represents true sessile micro-organisms.[55]

Flow cell systems allow study of mature biofilms within a flow of liquid to mimic the *in vivo* environment, and they permit microscopic observation and imaging of the biofilm.[56] Flow cells permit *in situ* visualisation, which can overcome the drawbacks associated with fixation for visualisation in other systems.[46] However, the setup of these devices can be difficult and researchers may be challenged with air bubble formation.[56] Channel depth can also be a limitation.[46]

The CDC bioreactor is another system for investigating biofilm formation, structure and response to antimicrobial treatment in flow phase or batch phase. It facilitates biofilm growth under moderate to high fluid shear stress.[57,58] The CDC bioreactor consists of removable rods which each hold three removable coupons; thus, the bioreactor enables repeatable bacteria growth on 24 removable coupons.[59] Ngo and colleagues[60] conducted a study using the CDC bioreactor to generate biofilms which were then used in an *in vitro* model of a biofilm-infected chronic wound that was in constant contact with moisture and nutrients.[60] However, it is essential to choose the appropriate reactor and growth conditions since these affect laboratory biofilm.[57]

FORMATION OF BIOFILMS

Biofilm formation is a complex process, but generally it can be recognised as consisting of a series of stages. O'Toole et al.[61] described biofilm formation as including initiation, maturation, maintenance and dissolution. However, numerous others have classified these stages differently to reflect the environment in which they are studying biofilms.

The development of a biofilm is influenced by different processes, including adhesion, detachment, mass transport, quorum sensing, cell death and active dispersal.[62]

Development of the Conditioning Film

The characteristics of the substratum and the presence of a conditioning film are important factors in microbial attachment to a solid surface, and the physicochemical properties of the surface may influence the rate and extent of attachment.[25] It has long been noted that micro-organisms attach more readily to hydrophobic surfaces (e.g., plastic) compared with hydrophillic surfaces such as glass and metals.[63–66]. However, depending on the conditioning film and ionic strength conditions, coatings can either enhance or reduce initial adhesion of bacteria.[66,67]

A 'conditioning film' was initially reported by Loeb and Neihof in 1975.[68] They found that it influences the interaction of the bacteria with the interface. Neu found that the organic molecules of the conditioning film altered the wettability and surface charge of the original surface.[69] Furthermore, Neu and others found that the properties of the substratum affect the composition and orientation of the molecules, which make up the conditioning film, during in the first hour of exposure.[69,70] They note too that surface active compounds (SACs) produced by the bacteria interact with the interface, affecting adhesion and de-adhesion.[69] It is now widely accepted that the composition of the conditioning film affects the rate and extent of microbial attachment.[70]

Depending on the environment, the conditioning film has been shown to consist of polysaccharides, glycoproteins and humic compounds. In the human host, the conditioning film is determined by the site being conditioned. For example in the oral cavity, particularly the teeth, the film is referred to as a proteinaceous 'pellicle' composed of albumin, glycoproteins, lipids, lysozyme, phosphoproteins and gingival crevice fluids. Other types of conditioning films have also been reported on biomaterials and found to be composed of blood, tears, urine, saliva, intervascular fluid and respiratory secretions. Interestingly, alterations in the chemistry of the conditioning film have been shown to have an affect on bacterial adhesion.[71]

Mechanisms Involved in Adhesion of Micro-organisms

For non-motile organisms, transport to a surface occurs via a number of routes. Brownian motion has been shown to play a part, or an organism can be carried by the flow of the bulk fluid. Motile organisms are thought to search for a surface by phototactic, chemotactic or even aerotactic responses. The transport of microbial cells and nutrients to a surface is generally achieved by a number of well-established fluid dynamic processes. These include mass transport and thermal and gravity effects.

The velocity characteristics of the liquid affect the rate of microbial attachment to a surface, and association with the surface is affected by cell size and cell motility.[25] Cell surface hydrophobicity, the presence of fimbriae and flagella and EPS production collectively influence the extent and rate of microbial attachment.[25] While various cell surface structures are involved in attachment, each structure may be specific to an attachment surface and the expression of these structures may vary as determined by the surrounding environment.[61]

Adhesion

Zobell[6] proposed that adhesion consists of two steps: one reversible and the other irreversible. By definition, *reversible* adhesion is an initial weak attachment of microbial cells to a surface whereas *irreversible* adhesion is a permanent bonding. Surface interactions, including Van Der Waals, hydrogen bonds and other electrostatic forces, play an important role in initial attachment.[66] This adhesion stage seems to be related to the bacteria's distance from the surface. For example, Van Der Waals forces and electrostatic interactions exist at a distance from the surface of 10 to 20 nm. Adhesion has been studied using an array of techniques, including atomic force microscopy,[72,73] which has shown that bacterial adhesion can be affected by factors such as surface roughness.[74–76] However, the host itself has also been shown to have an affect.[77]

Once at a surface, sessile bacteria can be redistributed. O'Toole and Kolter[78] found that while both wild type and type IV pilli mutant strains of *Pseudomonas aeruginosa* form a dispersed monolayer of cells on polyvinylchloride (PVC) plastic, only the wild type is able to form micro-colonies and very few cells of a non-motile strain attach to the PVC surface even after eight hours. This suggests that these cell structures are important in micro-colony formation.[78] Furthermore, type IV pilli-mediated twitching motility was found to be involved in the migration of cells along the surface during the formation of multi-cell aggregates[78]; indeed, O'Toole and Kolter found that the wild type strain does move along the surface, forming cell aggregates via the recruitment of cells from the adjacent monolayer.[78] Through observation of *P. aeruginosa*, O'Toole and colleagues[61] found that the organism swims along the surface prior to initial attachment and, once a monolayer is formed, it continues to move across the surface using twitching motility.

The bacterial cell's surface has a major impact on adhesion to a substratum.[79–81] Surface hydrophobicity, fimbriae and flagella, and particularly the extent and composition of EPS influence the rate and extent of microbial cell attachment. Bacterial cell surface structures, including fimbriae, other proteins, lipopolysaccharides (LPS), EPS, and flagella all play an important role in the attachment process. Cell surface polymers with non-polar sites such as fimbriae, other proteins, and components of certain Gram-positive bacteria (mycolic acids) appear to dominate attachment to hydrophobic substrata, while EPS and LPS are more important in attachment to hydrophilic materials. Furthermore, bacterial attachment appears to occur most readily on surfaces that are rougher, more hydrophobic and coated by surface conditioning films. This is influenced, of course, by flow velocity, water temperature or nutrient concentration.

Phenotypic changes that are known to take place once a cell becomes sessile are numerous. These incude changes in the rate of oxygen uptake, respiration rate, synthesis of extracellular polymers, substrate uptake, rate of substrate breakdown, heat production and changes in growth rate, to name but a few. Molecular changes are documented to occur when bacteria are attached to a surface. The up- and down-regulation of a number of genes during substratum adhesion suggests that adhesion mechanisms involve many genes.

Numerous strategies have been investigated to control bacterial adhesion. This remains an area of intensve research in the quest to control biofilm development.[82,83]

Extracellular Polymeric Substances

If cells reside at a surface for a certain time, irreversible adhesion forms through the mediation of a cementing substance which is extracellular in origin. This substance has been referred to as 'slime', which suggests that the biofilm is not rigid; however, mechanical stability is an important feature of biofilms, with cohesive and adhesive forces acting within them.[84] The colonisation stage involves the synthesis of large amounts of extracellular material, multiplication of the attached organisms and/or attachment of other bacteria to the already-adhered cells; this is known as *co-adhesion*.

Biofilms are composed primarily of microbial cells and EPS, which may account for 50–90% of their total organic carbon.[85] Extracellular polymeric substances are primarily polysaccharides which are neutral or polyanionic in the case of Gram-negative bacteria. Uronic acids (e.g., D-glucuronic D-galacturonic and mannuronic) or ketal-linked pryruvates constitute part of the EPS matrix and are known to provide the anionic property of the biofilm, allowing the cross-linking of divalent cations such as calcium and magnesium.[86]

EPS is also highly hydrated, both hydrophilic and hydrophobic, with varying degrees of solubility. Its polysaccharide content has a marked affect on the biofilm. EPS absorbs nutrients, both organic and inorganic, absorbs microbial products; and other microbes and aids in the protection of immobilised cells. It influences the physical properties of the biofilm, including diffusivity, thermal conductivity and rheology. The predominantly polyanionic, highly hydrated nature of EPS also means that it can act as an ion exchange matrix, serving to increase local concentrations of ionic species.

Extracellular DNA, which is evident in the biofilm matrix, is known to affect microbial attachment[87] and the development of the mushroom-shaped structure of the biofilm.[88]

Micro-colony and Biofilm Formation

Adsorption of macro-molecules followed by the attachment of microbial cells to a substratum is the reported and proposed sequence of events during the first stages of biofilm development. Next is the growth of bacteria, development of micro-colonies, recruitment of additional attaching bacteria and often colonisation of other organisms. As the bacteria attach, they begin to grow; extracellular polymers are produced and accumulated so that the bacteria are eventually embedded in a hydrated polymeric matrix. Biofilm communities tend to be complex, both taxonomically and functionally. This provides the potential for synergistic interaction among constituent organisms. Biofilm architecture is heterogenous and is constantly changing as a result of both external and internal processes.[25]

Microbial succession is a common feature of biofilms. During adhesion the pioneering or primary coloniser to any surface has defined requirements dictated by the conditioning film. The succession of the biofilm community is then governed by a number of physiological and biological events initiated by this pioneering species. Stress is known to increase the bacteria's propensity to form a biofilm.[89]

It has been estimated that only 10% or less of a 'mature' biofilm's dry weight is in the form of microbial cells which are considered to be diverse in taxonomy.[90] The less 'mature' biofilms generally contain very few species[91]; however, over time microbial diversity increases to form a stable climax community. The complexity of the microbial community is often underestimated because of the selectivity and inadequacy of pure-culture isolation techniques and the large presence of viable but non-culturable micro-organisms.[92]

DETACHMENT AND DISSEMINATION OF BIOFILMS

It is well known that biofilms are in dynamic flux and thus detachment of micro-organisms from the matrix occurs on a regular basis.[93] This is generally referred to as the *release of cells* (either individually or in clumps) from a biofilm or surface. Detachment can occur actively[94–96] or passively. Active detachment or dispersal is considered to be highly regulated as opposed to passive detachment, which is determined by external forces such as abrasion of shear.[96] However, environmental conditions can also result in microbial detachment and dispersal because of factors such as oxygen or nutrition changes. This effect has been observed with *P. aeruginosa.*[97,98] Dispersal is thought to be important for the survival of microbial species because it expands the bacterial population and allows for the transmission of bacteria between hosts and the spread of infection in a host.[96]

Detachment of biofilm cell clusters facilitates transmission of pathogens which can disseminate to new surfaces both in the form of detached biofilm clumps and via the fluid-driven dispersal of biofilm clusters along the surface.[99] Biofilm cells may be dispersed by the shedding of daughter cells from actively growing cells, detachment as a result of nutrient levels or quorum sensing. The detachment and dispersal of planktonic bacteria from the biofilm enables bacteria within a sessile biofilm community to colonise new areas, forming a new sessile population.[29]

In addition to the mechanical detachment of biofilm segments via the flow of the bulk fluid, it has been found that programmed cell detachment involves the hydrolysis of the extracellular polymeric matrix.[29] Hall-Stoodley and Stoodley[99] proposed three dispersal methods for biofilm-residing bacteria: swarming (using twitching motility), clumping (using rolling motility) and surface/clumping (using sliding motility).

Many parameters known to affect biofilm detachment have been studied, including pH, temperature and the presence of organic macro-molecules.[100] Nutrient starvation is another environmental factor which has been shown to have an impact on the release of microbes from a biofilm.[101] The effects these conditions have on bacterial detachment are generally species-specific.

Substratum surface roughness may also be a significant factor in detachment, with early events in biofilm formation being controlled by hydrodynamic forces. Macro- and micro-roughness may significantly influence detachment rates because of a sheltering effect from hydrodynamic shear. Indeed, cell detachment is affected by variable hydrodynamic shear when changes occur in the flow rate.[99] QS systems have also been shown to be involved in the detachment and dispersal of biofilm bacteria.[102]

Detachment is important in biofilm development and survival, and it has implications for infection, contamination and public health. Despite the lack of research in this area, there is emerging evidence that micro-organisms eminating from biofilms can overcome the host immune system and cause infection.[103]

The mode of dispersal apparently affects the phenotypic characteristics of organisms. Eroded or sloughed aggregates from the biofilm are likely to retain certain biofilm characteristics, such as antimicrobial resistance, whereas cells that have been shed as a result of growth may revert quickly to the planktonic phenotype. Biofilm cell dispersal/detachment has very important implications for public health medicine. In fact, Raad et al.[104] determined a relationship between biofilm formation and catheter-related septicemia. Also, the detachment of clumps from biofilms from native heart valves has been implicated in infective endocarditis. These clumps also may contain platelets or erthyrocytes that lead to the production of emboli which may cause serious complications in the host.

Detachment from surfaces is also important in biofilm diversification. For example, biofilms with greater detachment rates have been found to have larger fractions of active bacteria. Detachment is important not just for promoting genetic diversity but also for escaping unfavorable habitats, aiding in the development of new niches.[105,106]

BIOFILM STRUCTURE

The structure and architecture of a biofilm have been studied extensively within both mixed and pure culture systems evident in many different environments.[107–109]

The present conceptual model of a biofilm is described as cell clusters or stacks separated by interstitial voids. Depending on the site of biofilm formation, it is accepted that the structure is complex, with layers of organisms of different types.[109]

The thickness of a biofilm is very difficult to define. If we consider an oral biofilm, it has been shown that the thickness may be up to 1 mm. Biofilms in CAPD catheters are approximately 30 μm thick. Those studied in silicone Foley catheters have been shown to be 200 to 500 μm thick,with plate counts showing bacterial load as high as 10^8 cells per cm^2.[110]

Biofilm growth allows sessile organisms to create their own environment. Thus, it influences the micro-environment, which in turn influences the physiology of its component cells. Phenotypic variations in physiology and biochemistry occur in response to growth in a diffusion gradient system with continuously varying concentrations of substrates, products and other solute molecules. Such phenotypic changes may, for example, be due to the deployment of antimicrobials and, as a general rule, the biofilm mode of growth leads to a lowered sensitivity to antimicrobial agents compared to their freely dispersed counterparts.[111]

An understanding of the physical and chemical characteristics of the biofilm matrix and its relationship to resident organisms will provide a greater understanding of biofilm structure and functioning. A common feature of biofilms is their high EPS content (i.e., 50–60%).[112] The vast majority of bacterial EPS are polysaccharides. Common sugars, such as glucose, galactose, mannose, fructose, rhamnose, N-acetylglucosamine, glucuronic acid, galacturonic acid, mannuronic acid and guluronic acid, are typical constituents. Biofilms are very heterogeneous, containing micro-colonies of bacterial cells encased in a three-dimensional EPS matrix.

It has been shown that within the EPS matrix there are bywater channels or voids.[113] It is possible that these voids act as pools of genes that permit genetic acquisition and exchange. Complex biofilms can develop to form a three-dimensional sructure with cell aggregates, interstital pores and conduit channels. This structure allows nutrients to reach the biomass, and the channels possibly are shaped by the protozoa that graze the bacteria.

The structure of medical biofilms is substantially influenced by the interaction of particles of non-microbial components from the host or environment. In the human body, biofilms on native heart valves provide a clear example of this type of interaction, in which the biofilm's bacterial micro-colonies develop in a matrix of platelets, fibrin, and EPS.[114] Bacterial endocarditis lesions are composed of aggregates of bacterial cells,platelets and fibrin that adhere to the damaged epithelium of cardiac valves.[46] Damaged endothelium exposes the underlying basement membrane, providing a substratum to which bacteria can adhere.[46] The inflammatory process then stimulates the deposition of fibrin, creating an insoluble clot composed of fibrin and platelets.[46] The fibrin capsule that develops protects the organisms from the host's leukocytes, leading to infective endocarditis.

Biofilms on urinary catheters may contain organisms that have the ability to hydrolyse urea to form free ammonia through the action of urease. The ammonia may then raise the pH at the biofilm-liquid interface, resulting in the precipitation of minerals such as calcium phosphate (hydroxyapatite) and magnesium ammonium phosphate (struvite).[115] These minerals can then become entrapped in the biofilm and cause encrustation of the catheter[116]; cases have been described in which the catheter becomes recurrently blocked by this mineral build-up.[116–118]

Overall, a biofilm is heterogeneous in both space and time, constantly changing because of external and internal processes.[109]

THE BIOFILM COMMUNITY

Once the basic micro-colony of the biofilm is formed, a complex environment is established. It is at this point that the biofilm becomes a functioning ecosystem with the development of a mono- and/or mixed-species community of bacteria and the cycling of nutrients. Developing in the biofilm are nutrient gradients and areas where gene exchange is prevelant and where, evidence shows, there is possible species domination by quorum sensing. These are discussed in turn.

Gene Transfer

Biofilms provide an ideal environment for the exchange of extrachromasomal DNA (plasmids),[25,119] given their high cell density, increased genetic competance and accumulation of mobile genetic elements.[120] Indeed, *conjugation*—the major mechanism involved in horizontal gene transfer between donor and recipient cells—has been linked to the exchange of antimicrobial resistance genes between bacteria; it has also been shown to occur in biofilms at a high frequency.[121,122] Enhanced conjugation is likely a consequence of the minimal shear and closer cell-to-cell contact in the biofilm environment.

Interestingly, it has also been found that conjugative plasmids influence bacteria to form biofilms, suggesting that medically relevant plasmid-bearing strains may preferentially form a biofilm and thus may increase the chance of biofilm infection and the risk of the spread of virulence factors.[123] Gene transfer has also been shown to induce an enhanced stabilisation of the biofilm structure.[124]

Gene transfer by plasmid conjugation and DNA transformation has been found to occur frequently and effectively in many bacterial biofilms, both in natural environments and in more artificial settings.[124] Evidence suggests that efficient gene transfer influences both biofilm development and stability, and therefore that the process of gene transfer is both a consequence and a cause of biofilm development and performance.[124]

Quorum Sensing

As discussed previously, it is evident that micro-organisms in biofilms interact with each other. While there may be physical interaction, it is known that, to enable interaction, bacteria secrete small molecules. known as quorum-sensing molecules.[125] Quorum sensing is the intercellular signalling in bacteria when signals build extracellularly because of cell density and then become internalised to co-ordinate gene expression.[126] Because QS is important to biofilm formation, interfering with it helps to control biofilm formation.

Quorum sensing has been shown to influence biofilm development, with studies suggesting its involvement in the attachment, maturation and dispersal phases. For instance, the cyclic-peptide-dependent accessory gene regulator (agr) QS system in *Staphylococcus aureus* acts, in part, to decrease the expression of adhesion factors and increase potential dispersal factors, thus affecting the attachment of cells to a surface.[127]

The N-acylhomoserine lactone-mediated QS system of *Serratia liquefaciens*, on the other hand, has been linked with biofilm maturation and has been shown to be crucial for normal biofilm development and differentiation.[128] *P. aeruginosa* is a prime example of how quorum sensing and biofilm formation are interconnected; QS systems in *P. aeruginosa* control differentiation as well as virulence in a cell-density-dependent manner.[43] At a critical cell density, quorum-sensing signals can accumulate and trigger the expression of specific sets of genes.[29] Drugs that target QS systems in *P. aeruginosa* have been shown to inhibit virulence factor expression.[129]

Induction of genetic competence has also been shown to be mediated by QS systems; for instance, Li et al.[130] found that a QS peptide pheromone–signalling system in *Steptococcus mutans*—a bacterium found in dental plaque biofilms—controls genetic competence and functions optimally when the cells are in the biofilm mode of growth.[130]

Interactions in the Biofilm

Many strict and opportunistic bacterial pathogens have been shown to associate with, and grow, in biofilms, including *P. aeruginosa*,[131] *Staphylococcus aureus*, *Acinetobacter baumannii*, *Legionella pneumophila*, *S. aureus*,[132] *Listeria monocytogenes*,[133] *Campylobacter* spp.,[134] *E. coli* O157:H7, *Salmonella typhimurium*, *Vibrio cholerae*[135] and *Helicobacter pylori*.[136] Although all these organisms have the ability to attach to surfaces and existing biofilms, most if not all appear incapable of extensive growth. This may be because of their fastidious growth requirements or because of their inability to compete with indigenous organisms. The mechanisms of interaction and growth apparently vary with the micro-organism.[137] The association and metabolic interactions with indigenous organisms might also enhance survival and growth of pathogenic organisms in biofilms.

Micro-organism interactions have been extensively studied in dental plaque, which has been used in the study of micro-organism interactions in other envionments.[138]

Biofilm Recalcitrance

Microbial cells in biofilms have been reported to be 10 to 100 times more resistant to antibiotics when compared to their planktonic counterparts.[138] For an antimicrobial to have an affect on a biofilm, it must diffuse through the protective matrix which encases the bacterial cells. This matrix limits the transport of agents through the biofilm structure, thus reducing antimicrobial penetration. Antimicrobials may be bound within the matrix and thus are unable to act on encased cells. Furthermore, enzymes trapped in the matrix structure may inactivate it. The mucoid exopolysaccharide produced by *P. aeruginosa* was shown to protect biofilm cells from tobramycin. Dispersed cells were 15 times more susceptible to the agent than the cells in the biofilm matrix.[139]

The reduced growth rate of biofilm bacteria has been shown to render them less susceptible to an antimicrobial agent. This reduced growth rate results in a diminished uptake of actives by biofilm cells. A study examining the effects of the quartenary ammonium compound cetrimide on *E. coli* biofilms showed that the bacteria are more resistant to the agent at the slowest growth rates.[140]

The altered micro-environment in biofilms (e.g., pH, oxygen content) can reduce antimicrobial activity. Tresse et al.[141] used agar-entrapped *E. coli* to study the effects of antibiotics under different oxygen tensions. They concluded that there is a reduced uptake of the amonoglycoside antibiotic when the bacteria are grown in oxygen-deficient conditions. This has particular significance in the recalcitrance of thicker biofilms where conditions are more oxygen-limited.

In 1991 Dagostino et al.[142] speculated that altered gene expression by biofilm organisms can result in a phenotype with reduced susceptibility to an antimicrobial agent. They demonstrated that the association of bacteria with a surface results in physiologic changes brought about by the induction or repression of genes; these changes do not happen when cells are free-floating in liquid media.

Because of the recalcitrance of biofilms, as previously described, it is often challenging to eradicate them. To enhance antimicrobial efficacy, ultasound or electric currents in the presence of antibiotics can be used to produce a synergistic effect that can kill the biofilms' organisms. Methods to help control biofilms are numerous[143–146] and are discussed in later chapters.

CONCLUSION

Biofilms are responsible for a vast array of conditions in healthcare settings. According to the CDC, they are associated with 65% of all HCAI, including those associated with medical devices, chronic wounds and surgical sites, and this is an issue of growing concern.[147,148] Biofilms as they relate to medical device infections are discussed in further detail throughout this book.

REFERENCES

1. Kwon AS, Park GC, Ryu SY, Lim DH, Choi CH, Park Y, et al. Higher biofilm formation in multidrug-resistant clinical isolates of *Staphylococcus aureus*. *Int J Antimicrob Agents* 2008;**32**:68–72.
2. Nadell CD, Xavier JB, Foster KR. Review: the sociobiology of biofilms. *FEMS Microbiol Rev* 2009; **33**(1):206–24.
3. Flemming HC, Neu TR, Wozniak DJ. The EPS matrix: the house of biofilm cells. *J Bacteriol* 2007; **189**:7945–7.
4. Henrici AT. Studies of freshwater bacteria: I. A direct microscopic technique. *J Bacteriol* 1933; **25**(3):277.
5. Zobell CE, Anderson DQ. Vertical distribution of bacteria in marine sediments. *Bull Am Assoc Petrol Geol* 1936;**20**:258–69.
6. Zobell CE. The effect of solid surfaces upon bacterial activity. *J Bacteriol* 1943;**46**:39–56.
7. Heukelekian H, Heller A. Relation between food concentration and surface for bacterial growth. *J Bacteriol* 1940;**40**:547–58.
8. Marshall KC, Stout R, Mitchell R. Mechanism of the initial events in the sorption of marine bacteria to surfaces. *J Gen Microbiol* 1971;**68**:337–48.
9. Marshall KC. *Interfaces in microbial ecology*. Harvard University Press; 1976. p. 44–47.
10. Costerton JW, Geesey GG, Cheng K-J. How bacteria stick. *Sci Am* 1978;**238**:86–95.
11. Costerton JW, Cheng KJ, Geesey GG, Ladd TIM, Nickel JC, Dasgupta M, et al. Bacterial biofilms in nature and disease. *Annu Rev Microbiol* 1987;**41**:435–64.
12. Characklis WG, Marshall KC. Biofilms: a basis for an interdisciplinary approach. In: Characklis WG, Marshall KC, editors. *Biofilms*. Wiley; 1990. p. 3–15.
13. Costerton JW, Lewandowski Z, Caldwell DE, Korber DR, Lappin-Scott HM. Microbial biofilms. *Annu Rev Microbiol* 1995;**49**:711–45.
14. Lawrence JR, Korber DR, Hoyle BD, Costerton JW, Caldwell DE. Optical sectioning of microbial biofilms. *J Bacteriol* 1991;**173**(20):6558–67.
15. Lawrence JR, Wolfaardt GM, Korber DR. Determination of diffusion coefficients in biofilms by confocal laser microscopy. *Appl Environ Microbiol* 1994;**60**(4):1166–73.
16. Inaba T, Ichihara T, Yawata Y, Toyofuku M, Uchiyama H, Nomura N. Three-dimensional visualization of mixed species biofilm formation together with its substratum. *Microbiol Immunol* 2013:3.
17. Głowacki R, Tomaszewski KA, Stręk P, Tomaszewska IM, Zgórska-Świerzy K, Markiewicz B, et al. The influence of bacterial biofilm on the clinical outcome of chronic rhinosinusitis: a prospective, double-blind, scanning electron microscopy study. *Eur Arch Otorhinolaryngol* 2013; 18 July.
18. Li J, Busscher HJ, van der Mei HC, Sjollema J. Surface enhanced bacterial fluorescence and enumeration of bacterial adhesion. *Biofouling* 2013;**29**(1):11–19.
19. Wilson M. Bacterial biofilms and human disease. *Sci Prog* 2001;**84**:235–54.
20. Maianskiĭ AN, Chebotar' IV. Staphylococcal biofilms: structure, regulation, rejection. *Zh Mikrobiol Epidemiol Immunobiol* 2011;**1**:101–8.
21. Palmer Jr. R, White DC. Developmental biology of biofilms: implications for treatment and control. *Trends Microbiol* 1997;**5**:435–40.
22. Branda SS, Vik A, Friedman L, Kolter R. Biofilms: the matrix revisited. *Trends Microbiol* 2005; **13**:20–6.
23. Costerton JW, Lappin-Scott HM. Behaviour of bacterial biofilms. *Am Soc Microbiol News* 1989; **55**:650–4.
24. Davies DG, Geesey. GG. Regulation of the alginate biosynthesis gene algC in *Pseudomonas aeruginosa* during biofilm development in continuous culture. *Appl Environ Microbiol* 1995;**61**:860–7.
25. Donlan RM, Costerton JW. Biofilms: survival mechanisms of clinically relevant microorganisms. *Clin Microbiol Rev* 2002;**15**(2):167–93.
26. Harding MW, Marques LLR, Howard RJ, Olson ME. Can filamentous fungi form biofims? *Trends Microbiol* 2009;**17**(11):475–80.
27. Potera C. Forging a link between biofilms and disease. *Science* 1999;**283**:1837–8.
28. Ricucci D, Siqueira JF. Biofilms and apical periodontitis: study of prevalence and association with clinical and histopathologic findings. *JOE* 2010;**36**(8):1277–88.

29. Costerton JW, Stewart PS, Greenberg EP. Bacterial biofilms: a common cause of persistent infections. *Microbes, Immun Dis* 1999;**284**:1318–22.
30. Sauer K, et al. Biofilms and biocomplexity. *Microbe* 2007;**2**:347–53.
31. Marsh PD. Dental plaque: biological significance of a biofilm and community life-style. *J Clin Periodontol* 2005;**32**(Suppl. s6):7–15.
32. Palmer RJ, Darveau R, Lamont RJ, Nyvad B, Teles RP. Human oral bacterial biofilms: composition, dynamics, and pathogenesis. In: Bjarnsholt T, Moser C, Jensen PØ, Høiby N, editors. *Biofilm infections*. Springer; 2010. p. 35–68.
33. Percival SL, Kite P. Intravascular catheters and biofilm control. *J Vasc Access* 2007;**8**(2):69–80.
34. Homøe P, Bjarnsholt T, Wessman M, Sørensen HC, Johansen HK. Morphological evidence of biofilm formation in Greenlanders with chronic suppurative otitis media. *Eur Arch Otorhinolaryngol* 2009; **266**(10):1533–8.
35. Pozo JL, Patel R. Infection associated with prosthetic joints. *N Engl J Med* 2009;**361**:787–94.
36. Percival SL, Rogers AA. The significance and role of biofilms in chronic wounds. In: McBain A, Allison D, Pratten J, Spratt D, Upton M, Verran J, editors. *Biofilms, persistence and ubiquity*. Bioline; 2005. p. 171–8.
37. Noiri Y, Ehara A, Kawahara T, Takemura N, Ebisu S. Participation of bacterial biofilms in refractory and chronic periapical periodontitis. *J Endod* 2002;**28**(10):679–83.
38. Carr GB, Schwartz RS, Schaudinn C, Gorur A, Costerton JW. Ultrastructural examination of failed molar retreatment with secondary apical periodontitis: an examination of endodontic biofilms in an endodontic retreatment failure. *JOE* 2009;**35**(9):1303–9.
39. Høiby N, Ciofu O, Bjarnsholt T. *Pseudomonas aeruginosa* biofilms in cystic fibrosis. *Future Microbiol* 2010; **5**(11):1663–74.
40. Bjarnsholt T, Jensen PØ, Fiandaca MJ, Pedersen J, Hansen CR, Andersen CB, et al. *Pseudomonas aeruginosa* biofilms in the respiratory tract of cystic fibrosis patients. *Pediatr Pulmonol* 2009;**44**:547–58.
41. Høiby N. *P. aeruginosa* in cystic fibrosis patients resists host defences, antibiotics. *Microbe* 2006; **1**(12):571–7.
42. VanDevanter DR, Van Dalfsen JM. How much do pseudomonas biofilms contribute to symptoms of pulmonary exacerbation in cystic fibrosis? *Pediatr Pulmonol* 2005;**39**:504–6.
43. Singh PK, Schaefer AL, Parsek MR, Moninger TO, Welsh MJ, Greenberg EP. Quorum-sensing signals indicate that cystic fibrosis lungs are infected with bacterial biofilms. *Nature* 2000;**407**(6805): 762–4.
44. Matsui H, Wagner VE, Hill DB, Schwab UE, Rogers TD, Button B, et al. A physical linkage between cystic fibrosis airway surface dehydration and *Pseudomonas aeruginosa* biofilms. *PNAS* 2006; **103**(48):18131–6.
45. Hall-Stoodley L, Ze Hu,F, Gieseke A, Nistico L, Nguyen D, Hayes J, et al. Direct detection of bacterial biofilms on the middle-ear mucosa of children with chronic otitis media. *JAMA* 2006; **296**(2):202–11.
46. Hall-Stoodley L, Costerton JW, Stoodley P. Bacterial biofilms: from the natural environment to infectious diseases. *Nat Rev Microbiol* 2004;**2**:95–108.
47. Harrison JJ, Ceri H, Yerly J, Stremick CA, Hu Y, Martinuzzi R, et al. The use of microscopy and three-dimensional visualization to evaluate the structure of microbial biofilms cultivated in the Calgary biofilm device. *Biol Proced Online* 2006;**8**:194–215.
48. Trappetti C, Gualdi L, Di Meola L, Jain P, Korir CC, Edmonds P, et al. The impact of the competence quorum sensing system on *streptococcus pneumoniae* biofilms varies depending on the experimental model. *BMC Microbiol* 2011;**11**:75.
49. Goeres DM, Loetterle LR, Hamilton MA, Murga R, Kirby DW, Donlan RM. Statistical assessment of a laboratory method for growing biofilms. *Microbiol* 2005;**151**(3):757–62.
50. Parahitiyawa NB, Samaranayake YH, Samaranayake LP, Ye J, Tsang PWK, Cheung BPK, et al. Interspecies variation in *Candida* biofilm formation studied using the Calgary biofilm device. *Apmis* 2006;**114**(4):298–306.
51. Ceri H, Olson ME, Stremick C, Read RR, Morck D, Buret A. The Calgary Biofilm Device: new technology for rapid determination of antibiotic susceptibilities of bacterial biofilms. *J Clin Microbiol* 1999;**37**(6):1771–6.

52. Ali L, Khambaty F, Diachenko G. Investigating the suitability of the Calgary biofilm device for assessing the antimicrobial efficacy of new agents. *Bioresour Technol* 2006;**97**(15):1887–93.
53. Aaron SD, Ferris W, Ramotar K, Vandemheen K, Chan F, Saginur R. Single and combination antibiotic susceptibilities of planktonic, adherent, and biofilm-grown *Pseudomonas aeruginosa* isolates cultured from sputa of adults with cystic fibrosis. *J Clin Microbiol* 2002;**40**(11):4172–9.
54. Jass J, Costerton JW, Lappin-Scott HM. Assessment of a chemostat-coupled modified Robbins device to study biofilms. *J Ind Microbiol* 1995;**15**:283–9.
55. Honraet K, Nelis HJ. Use of the modified Robbins device and fluorescent staining to screen plant extracts for the inhibition of s. mutans biofilm formation. *J Microbiol Methods* 2006;**64**(2):217–24.
56. Crusz SA, Popat R, Rybtke MT, Cámara M, Givskov M, Tolker-Nielsen T, et al. Bursting the bubble on bacterial biofilms: a flow cell methodology. *Biofouling* 2012;**28**(8):835–42.
57. Goeres DM, Loetterle LR, Hamilton MA, Murga R, Kirby DW, Donlan RM. Statistical assessment of a laboratory method for growing biofilms. *Microbiol* 2005;**151**(3):757–62.
58. McBain AJ. Chapter 4: *In vitro* biofilm models: an overview. *Adv Appl Microbiol* 2009;**69**:99–132.
59. Hadi R, Vickery K, Deva A, Charlton T. Biofilm removal by medical device cleaners: comparison of two bioreactor detection assays. *J Hosp Infect* 2010;**74**(2):160–7.
60. Ngo QD, Vickery K, Deva AK. The effect of topical negative pressure on wound biofilms using an in vitro wound model. *Wound Repair Regen* 2012;**20**:83–90.
61. O'Toole G, Kaplan HB, Kolter R. Biofilm formation as microbial development. *Annu Rev Microbiol* 2000;**54**:49–79.
62. Fagerlind MG, Webb JS, Barraud N, McDougald D, Jansson A, Nilsson P, et al. Dynamic modelling of cell death during biofilm development. *J Theor Biol* 2012;**295**:23–36.
64. Bartley SN, Tzeng YL, Heel K, Lee CW, Mowlaboccus S, Seemann T, et al. Attachment and invasion of Neisseria meningitidis to host cells is related to surface hydrophobicity, bacterial cell size and capsule. *PLoS One* 2013;**8**(2):e55798.
63. Fletcher M, Loeb GI. The influence of substratum characteristics on the attachment of a marine Pseudomonas to solid surfaces. *Appl Environ Microbiol* 1979;**37**:67–72.
65. Pringle JH, Fletcher M. Influence of substratum wettability on attachment of freshwater bacteria to solid surfaces. *Appl Environ Microbiol* 1983;**45**:811–7.
66. Hwang G, Kang S, El-Din MG, Liu Y. Impact of conditioning films on the initial adhesion of *Burkholderia cepacia. Colloids Surf B Biointer* 2012;**91**:181–8.
67. van der Mei HC, Rustema-Abbing M, de Vries J, Busscher HJ. Bond strengthening in oral bacterial adhesion to salivary conditioning films. *Appl Environ Microbiol* 2008;**74**(17):5511–5.
68. Loeb GI, Neihof RA. Marine conditioning films. *Adv Chem Series* 1975;**145**:319–35.
69. Neu TR. Significance of bacterial surface-active compounds in interaction of bacteria with interfaces. *Microbiol Rev* 1996;**60**:151–66.
70. Lorite GS, Rodrigues CM, de Souza AA, Kranz C, Mizaikoff B, Cotta MA. The role of conditioning film formation and surface chemical changes on *Xylella fastidiosa* adhesion and biofilm evolution. *J Colloid Interf Sci* 2011;**359**(1):289–95.
71. Gabi M, Hefermehl L, Lukic D, Zahn R, Vörös J, Eberli D. Electrical microcurrent to prevent conditioning film and bacterial adhesion to urological stents. *Urol Res* 2011;**39**(2):81–8.
72. Chen Y, Busscher HJ, van der Mei HC, Norde W. Statistical analysis of long- and short-range forces involved in bacterial adhesion to substratum surfaces as measured using atomic force microscopy. *Appl Environ Microbiol* 2011;**77**(15):5065–70.
73. Ribeiro M, Monteiro FJ, Ferraz MP. Infection of orthopedic implants with emphasis on bacterial adhesion process and techniques used in studying bacterial-material interactions. *Biomatter* 2012;**2**(4):176–94.
74. Crawford RJ, Webb HK, Truong VK, Hasan J, Ivanova EP. Surface topographical factors influencing bacterial attachment. *Adv Colloid Interface Sci* 2012;**179–182**:142–9.
75. Al-Ahmad A, Wiedmann-Al-Ahmad M, Fackler A, Follo M, Hellwig E, Bächle M, et al. *In vivo* study of the initial bacterial adhesion on different implant materials. *Arch Oral Biol* 2013
76. Mei L, Busscher HJ, van der Mei HC, Ren Y. Influence of surface roughness on streptococcal adhesion forces to composite resins. *Dent Mater* 2011;**27**(8):770–8.
77. Słotwińska SM. Host and bacterial adhesion. *Pol J Vet Sci* 2013;**16**(1):153–6.

78. O'Toole GA, Kolter R. Flagellar and twitching motility are necessary for *Pseudomonas aeruginosa* biofilm development. *Mol Microbiol* 1998;**30**:295–304.
79. Takahashi Y, Urano-Tashiro Y, Konishi K. Adhesins of oral streptococci. *Nihon Saikingaku Zasshi* 2013;**68**(2):283–93.
80. Heilmann C. Adhesion mechanisms of staphylococci. *Adv Exp Med Biol* 2011;**715**:105–23.
81. Moschioni M, Pansegrau W, Barocchi MA. Adhesion determinants of the *Streptococcus* species. *Microb Biotechnol* 2010;**3**(4):370–88.
82. Poncin-Epaillard F, Herry JM, Marmey P, Legeay G, Debarnot D, Bellon-Fontaine MN. Elaboration of highly hydrophobic polymeric surface—a potential strategy to reduce the adhesion of pathogenic bacteria?. *Mater Sci Eng C Mater Biol Appl* 2013;**33**(3):1152–61.
83. Krachler AM, Orth K. Targeting the bacteria-host interface: strategies in anti-adhesion therapy. *Virulence* 2013;**4**(4):284–94.
84. Flemming HC, Wingender J. The biofilm matrix. *Nat Rev Microbiol* 2010;**8**(9):623–33.
85. Flemming H-C, Wingender J, Griegbe C, Mayer C. Physico-chemical properties of biofilms. In: Evans LV, editor. *Biofilms: Recent advances in their study and control.* Harwood Academic Publishers; 2000. p. 19–34.
86. Vu B, Chen M, Crawford RJ, Ivanova EP. Bacterial extracellular polysaccharides involved in biofilm formation. *Molecules* 2009;**14**(7):2535–54.
87. Rice KC, Mann EE, Endres JL, Weiss EC, Cassat JE, Smeltzer MS, et al. The cidA murein hydrolase regulator contributes to DNA release and biofilm development in *Staphylococcus aureus*. *Proc Natl Acad Sci* 2007;**104**(19):8113–8.
88. Allesen-Holm M, Barken KB, Yang L, Klausen M, Webb JS, Kjelleberg S, et al. A characterization of DNA release in *Pseudomonas aeruginosa* cultures and biofilms. *Mol Microbiol* 2006;**59**(4):1114–28.
89. Landini P. Cross-talk mechanisms in biofilm formation and responses to environmental and physiological stress in *Escherichia coli*. *Res Microbiol* 2009;**160**(4):259–66.
90. Hamilton WA. Sulphate-reducing bacteria and anaerobic corrosion. *Annu Rev Microbiol* 1985; **39**:195–217.
91. Atlas RM. Diversity of microbial communities. *Adv Microb Ecol* 1984;**7**:1–47.
92. Hegarty JP, Pickup R, Percival SL. Detection of viable but non-culturable bacterial pathogens. In: Gilbert PG, Allison D, Walker JT, Brading M, editors. *Biofilm community interactions: Chance or necessity species consortia.* Bioline; 2001, p. 39–51.
93. Kaplan JB. Biofilm dispersal: mechanisms, clinical implications, and potential therapeutic uses. *J Dent Res* 2010;**89**(3):205–18.
94. Stoodley P, Wlson S, Hall-Stoodley L, Boyle JD, Lappin-Scott HM, Costerton JW. Growth and detachment of cell clusters from mature mixed species biofilms. *Appl Environ Microbiol* 2001;**67**:5608–13.
95. Davies DG. Regulation of matrix polymer in biofilm formation and dispersion. In: Wingender J, Neu TT, Flemming H-C, editors. *Microbial extracellular polymeric substances.* Springer; 1999. p. 93–112.
96. Derlon N, Massé A, Escudié R, Bernet N, Paul E. Stratification in the cohesion of biofilms grown under various environmental conditions. *Water Res* 2008;**42**(8–9):2102–10.
97. Hunt SM, Werner EM, Huang B, Hamilton MA, Stewart PS. Hypothesis for the role of nutrient starvation in biofilm detachment. *Appl Environ Microbiol* 2004;**70**(12):7418–25.
98. Mirani Zulfiqar Ali, Mubashir Aziz, Mohammad Naseem Khan, et al. Biofilm formation and dispersal of *Staphylococcus aureus* under the influence of oxacillin. *Microb Pathog* 2013;**61–62**:66–72.
99. Hall-Stoodley L, Stoodley P. Biofilm formation and dispersal and the transmission of human pathogens. *Trends Microbiol* 2005;**1**:7.
100. McEldowney S, Fletcher M. Effect of pH, temperature, and growth conditions on the adhesion of a gliding bacterium and three nongliding bacteria to polystyrene. *Microb Ecol* 1988;**16**:183–95.
101. Hunt SM, Werner EM, Huang B, Hamilton MA, Stewart PS. Hypothesis for the role of nutrient starvation in biofilm detachment. *Appl Environ Microbiol* 2004;**70**(12):7418–25.
102. Boles BR, Horswill AR. Staphylococcal biofilm disassembly. *Trends Microbiol* 2011;**19**(9):449–55.
103. Ward KH, Olson ME, Lam K, Costerton JW. Mechanisms of persistent infection associated with peritoneal implants. *J Med Microbiol* 1992;**36**:406–13.
104. Raad II, Sabbagh MF, Rand KH, Sherertz RJ. Quantitative tip culture methods and the diagnosis of central venous catheter-related infections. *Diagn Microbiol Infect Dis* 1992;**15**:13–20.

105. McDougald D, Rice SA, Barraud N, Steinberg PD, Kjelleberg S. Should we stay or should we go: mechanisms and ecological consequences for biofilm dispersal. *Nat Rev Microbiol* 2011;**10**(1):39–50.
106. Abee T, Kovács AT, Kuipers OP, van der Veen S. Biofilm formation and dispersal in Gram-positive bacteria. *Curr Opin Biotechnol* 2011;**22**(2):172–9.
107. Yang L, Liu Y, Wu H, et al. Current understanding of multi-species biofilms. *Int J Oral Sci* 2011;**3**:74–81.
108. Lazăr V, Chifiriuc MC. Architecture and physiology of microbial biofilms. *Roum Arch Microbiol Immunol* 2010;**69**(2):95–107.
109. Tolker-Nielsen T, Molin S. Spatial organization of microbial biofilm communities. *Microb Ecol* 2000; **40**:75–84.
110. Ganderton L, Chawla J, Winters C, Wimpenny J, Stickler D. Scanning electron microscopy of bacterial biofilms on indwelling bladder catheters. *Eur J Clin Microbiol Infect Dis* 1992;**11**:789–96.
111. Nichols WW, Evans MJ, Slack MPE, Walmsley HL. The penetration of antibiotics into aggregates of mucoid and non-mucoid *Pseudomonas aeruginosa. J Gen Microbiol* 1989;**135**:1291–303.
112. Characklis WG, Cooksey KE. Biofilms and microbial fouling. *Adv Appl Microbiol* 1983;**29**:93–138.
113. Lewandowski Z. Structure and function of biofilms. In: Evans LV, editor. *Biofilms: Recent advances in their study and control.* Harwood Academic Publishers; 2000. p. 1–18.
114. Durack DT. Experimental bacterial endocarditis. IV: structure and evolution of very early lesions. *J Pathol* 1975;**115**:81–9.
115. Tunney MM, Jones DS, Gorman SP. Biofilm and biofilm-related encrustations of urinary tract devices. In: Doyle RJ, editor. *Methods in enzymology, vol. 310: Biofilms*. Academic Press; 1999. p. 558–66.
116. Stickler DJ. Bacterial biofilms in patients with indwelling urinary catheters. *Nat Clin Pract Urol* 2008;**5**(11):598–608.
117. Getliffe KA. The characteristics and management of patients with recurrent blockage of long-term urinary catheters. *J Adv Nurs* 1994;**20**:140–9.
118. Macleod SM, Stickler DJ. Species interactions in mixed-community crystalline biofilms on urinary catheters. *J Med Microbiol* 2007;**56**(11):1549–57.
119. Reisner A, Holler BM, Molin S, Zechner EL. Synergistic effects in mixed *Escherichia coli* biofilms: conjugative plasmid transfer drives biofilm expansion. *J Bacteriol* 2006;**188**:3582–8.
120. Fux CA, Costerton JW, Stewart PS, Stoodley P. Survival strategies of infectious biofilms. *Trends Microbiol* 2005;**13**:34–40.
121. Ehlers LJ, Bouwer EJ. Rp4 plasmid transfer among species of pseudomonas in a biofilm reactor. *Water Sci Technol* 1999;**39**(7):163–71.
122. Hausner M, Wuertz S. High rates of conjugation in bacterial biofilms as determined by quantitative in situ analysis. *Appl Environ Microbiol* 1999;**65**(8):3710–3.
123. Ghigo J-M. Natural conjugative plasmids induce bacterial biofilm development. *Lett Nature* 2001; **412**:442–5.
124. Molin S, Tolker-Nielsen T. Gene transfer occurs with enhanced efficiency in biofilms and induces enhanced stabilisation of the biofilm structure. *Curr Opin Biotechnol* 2003;**14**:255–61.
125. Miller MB, Bassler BL. Review: quorum sensing in bacteria. *Annu Rev Microbiol* 2001;**55**:165–99.
126. Jayaraman A, Wood TK. Review: Bacterial quorum sensing: signals, circuits, and implications for biofilms and disease. *Annu Rev Biomed Eng* 2008;**10**:145–67.
127. Yarwood JM, Schlievert PM. Quorum sensing in staphylococcus infections. *J Clin Investig* 2003; **112**(11):1620–5.
128. Labbate M, Queck SY, Koh KS, Rice SA, Givskov M, Kjelleberg S. Quorum sensing-controlled biofilm development in serratia liquefaciens MG1. *J Bacteriol* 2004;**186**(3):692–8.
129. Hentzer M, et al. Attenuation of *Pseudomonas aeruginosa* virulence by quorum sensing inhibitors. *EMBO J* 2003;**22**:3803–15.
130. Li YH, et al. A quorum-sensing signaling system essential for genetic competence in *Streptococcus mutans* is involved in biofilm formation. *J Bacteriol* 2001;**184**:2699–708.
131. Bjarnsholt T, Jensen PØ, Fiandaca MJ, et al. *Pseudomonas aeruginosa* biofilms in the respiratory tract of cystic fibrosis patients. *Pediatr Pulmonol* 2009;**44**:547–58.

132. Madeo J, Frieri M. Bacterial biofilms and chronic rhinosinusitis. *Allergy Asthma Proc* 2013; **34**(4):335–41.
133. Wirtanen G, Alanko T, Mattila-Sandholm T. Evaluation of epifluorescence image analysis of biofilm growth on stainless steel surfaces. *Colloids and Surfaces B: Biointer* 1996;**5**:319–26.
134. Buswell CM, Herlihy YM, Lawrence LM, McGuiggan JTM, Marsh PD, Keevil CW, et al. Extended survival and persistence of Campylobacter spp. in water and aquatic biofilms and their detection by immunofluorescent-antibody and -rRNA staining. *Appl Environ Microbiol* 1998;**64**:733–41.
135. Watnick PI, Kolter R. Steps in the development of a *Vibrio cholerae* El Tor biofilm. *Mol Microbiol* 1999;**34**:586–95.
136. Stark RM, Gerwig GJ, Pitman RS, Potts LF, Williams NA, Greenman J, et al. Biofilm formation by *Helicobacter pylori*. *Lett Appl Microbiol* 1999;**28**:121–6.
137. Hansen SK, Rainey PB, Haagensen JA, Molin S. Evolution of species interactions in a biofilm community. *Nature* 2007;**445**:533–6.
138. Hoiby N, Bjarnsholt T, Givskov M, Molin S, Ciofu O. Antibiotic resistance of bacterial biofilms. *Int J Antimicrob Agents* 2010;**35**:322–32.
139. Hoyle BD, Costerton JW. Bacterial resistance to antibiotics: the role of biofilms. *Prog Drug Res* 1991;**37**:91–105.
140. Evans DJ, et al. Susceptibility of *Pseudomonas aeruginosa* and *Escherichia coli* biofilms towards ciprofloxacin: effect of specific growth rate. *J Antimicrob Chemother* 1991;**27**:177–84.
141. Tresse O, Jouenne T, Junter GA. The role of oxygen limitation in the resistance of agar-entrapped, sessile-like *Escherichia coli* to aminoglycoside and beta-lactam antibiotics. *J Antimicrob Chemother* 1995;**36**(3):521–6.
142. Dagostino L, Goodman AE, Marshall KC. Physiological responses induced in bacteria adhering to surfaces. *Biofouling* 1991;**4**:113–9.
143. Rogers GB, Carroll MP, Bruce KD. Enhancing the utility of existing antibiotics by targeting bacterial behaviour? *Br J Pharmacol* 2012;**165**(4):845–57.
144. Donlan RM. Biofilm elimination on intravascular catheters: important considerations for the infectious disease practitioner. *Clin Infect Dis* 2011;**52**(8):1038–45.
145. Sintim HO, Smith JA, Wang J, Nakayama S, Yan L. Paradigm shift in discovering next-generation anti-infective agents: targeting quorum sensing, c-di-GMP signaling and biofilm formation in bacteria with small molecules. *Future Med Chem* 2010;**2**(6):1005–35.
146. Seil JT, Webster TJ. Antimicrobial applications of nanotechnology: methods and literature. *Int J Nanomedicine* 2012;**7**:2767–81.
147. Leaper D, McBain AJ, Kramer A, Assadian O, Sanchez JL, Lumio J, et al. Healthcare associated infection: novel strategies and antimicrobial implants to prevent surgical site infection. *Ann R Coll Surg Engl* 2010;**92**(6):453–8.
148. Høiby N, Ciofu O, Johansen HK, Song ZJ, Moser C, Jensen PØ, et al. The clinical impact of bacterial biofilms. *Int J Oral Sci* 2011;**3**(2):55–65.

CHAPTER TEN

Healthcare-Associated Infections and Biofilms

Louise Suleman, Debra Archer, Christine A. Cochrane and Steven L. Percival

INTRODUCTION

Nosocomial infections, more commonly known as healthcare-associated infections (HCAI), develop as a result of direct medical care or treatment in hospitals, care homes or the patient's own home, or as a result of contact with a healthcare setting.[1] The prevalence of HCAI in England is 6.4%, according to figures produced by the Health Protection Agency (HPA).[2] There are more than 100,000 cases of HCAI every year, causing fatalities in both healthy and immuno-compromised patients. The economic implications are primarily prolonged patient stay, reduced in-hospital bed availability and increased diagnostic testing and infection control.[3,4]

Infection can arise through the use of indwelling medical devices, including ventilators, catheters, shunts and endoscopes, but can also occur following surgical procedures as a result of contact with contaminated surfaces or infection with air-borne fungal spores.[5–8] The following are the most common HCAI:

- Ventilator-associated pneumonia (VAP)
- Central-line-associated septicaemia
- Catheter-associated urinary tract infection
- *Clostridium difficile* infection
- Surgical site infection

Several species of bacteria and fungi have been associated with HCAI, the most widely publicised being the hospital 'superbug' known as meticillin-resistant *Staphylococcus aureus* (MRSA), which is a common cause of bacteraemia and septicaemia.[2] The emergence of these antibiotic-resistant strains of bacteria has proven difficult to overcome because of the lack of treatment options presently available.[9] Furthermore, bacteria and fungi are known to colonise surgical and chronic wounds and to grow as a biofilm.[10–12]

Biofilms are complex communities of micro-organisms that are highly resistant to antibiotic and antiseptic treatment, partly because they produce extracellular polymeric substances (EPSs) which act as inherent barriers that help to neutralise, sequester and breakdown these agents.[13]

Biofilms in Infection Prevention and Control
Doi: http://dx.doi.org/10.1016/B978-0-12-397043-5.00010-4

This chapter discusses what is known about biofilm-associated infections in a clinical setting. Particular focus is given to what defines HCAI, how biofilms are formed, host responses to microbial biofilms, how these infections are detected, evidence of biofilms in HCAI and potential preventative anti-biofilm strategies.

Healthcare-Associated Infections

The most common types of HCAI in England include pneumonia and lower respiratory tract infections (22.8% of cases) followed by urinary tract infections (UTIs) (17.2% of cases) and surgical site infections (SSIs) (15.7% of cases).[2] SSI is a type of HCAI in which wound infection occurs after a surgical procedure.[14] It primarily develops from contamination of the wound site from the patient's own endogenous micro-flora located on the skin.[15–17]

The Centers for Disease Control and Prevention (CDC), a US government agency, has published definitions of nosocomial infection which are used by UK healthcare professionals:

Pneumonia: Must

1. Be present with dullness to percussion during physical chest examination AND either present with change in sputum, show organisms in blood culture or show organisms in bronchial brushings or biopsies.
2. Be present with progressive infiltrate or pleural effusion upon radiographic examination and either present with sputum change, show organisms in blood culture, show organisms in tracheal brushing or biopsies, present with viral antigens in respiratory secretions, or show histopathology evidence of pneumonia.
3. Be a patient of ≤ 12 months of age and have two of the following: apnoea, bradycardia, wheezing, rhonchi, tachypnoea or cough; and either show increased respiratory secretions, show sputum change, show organisms in blood culture, show organisms in tracheal brushings or biopsies, present viral antigens in respiratory secretions, or show histopathology evidence for pneumonia. OR
4. Be a patient of ≤ 12 months of age and show new or progressive infiltrate and pleural effusion in chest radiology examination; and either show increased respiratory secretions; show sputum change; show organisms in blood culture; show pathogens in transtracheal aspirate or bronchial brushings/biopsy; present with viral antigens in respiratory secretions; or show histopathology evidence of pneumonia.

Primary bloodstream infections: Must either[1] show a recognised pathogen in blood culture not related to a secondary site of infection[2]; be present with fever $>38\,^{\circ}C$; or[3] be ≤ 12 months of age and present with fever $>38\,^{\circ}C$.

Urinary tract infection: Must either[1] be present with fever >38 °C and show $>10^5$ CFU/ml; or, according to the Health Protection Agency (HPA),[2] be present with fever >38 °C and show positive results for leukocyte esterase and/or nitrate, Pyuria, Gram-positive stains or positive pathogens of $\geq 10^2$ CFU/ml in two separate urine samples.

Surgical wound infection: Must occur within 30 days after surgery and involve skin, subcutaneous tissue and muscle.[17–19]

In addition to these guidelines, the HPA, which now forms part of Public Health England, provides UK healthcare professionals with advice on how to deal with HCAI; it also monitors antibiotic resistance across the United Kingdom. (See Exhibit 10.1.)

How Do HCAI Develop?

HCAI occur as a result of infection by a number of micro-organisms, the most common being bacteria, but they can also arise from infection by viruses, fungi, parasites or prions.

To infect a new host, micro-organisms must exit from 'reservoirs' such as human skin, water or food sources. Transmission to a new host may occur through several kinds of exit portals such as faeces, urine or blood. Micro-organisms reach the new host either directly, by contact with the infected person, or indirectly, by contact with contaminated surfaces. This can occur when surfaces have been touched by unwashed hands or droplets of body fluid. Alternative modes of transmission include airborne contamination or the consumption of contaminated food. The new host can contract the micro-organisms through inhalation; ingestion; breaks in the skin barrier from surgery or intravenous lines; or mucous membranes including eyes, mouth and nose (Figure 10.1). The newly infected host acts as a reservoir for the micro-organisms, potentially infecting a new host, and thus the cycle continues, as shown in Figure 10.2.

EXHIBIT 10.1 Risk Factors for Healthcare-Associated Infections

- Long hospital stays
- Residence in a nursing home
- Overcrowded community of patients treated together
- Invasive surgery
- Immuno-compromised patients due to medical treatments (e.g., chemotherapy) or illnesses including cancer, heart disease and diabetes
- Home wound management

Source: As outlined by HPA, Public Health England, from www.hpa.org.uk/Topics/InfectiousDiseases/InfectionsAZ/HCAI/GeneralInformationOnHCAI/.

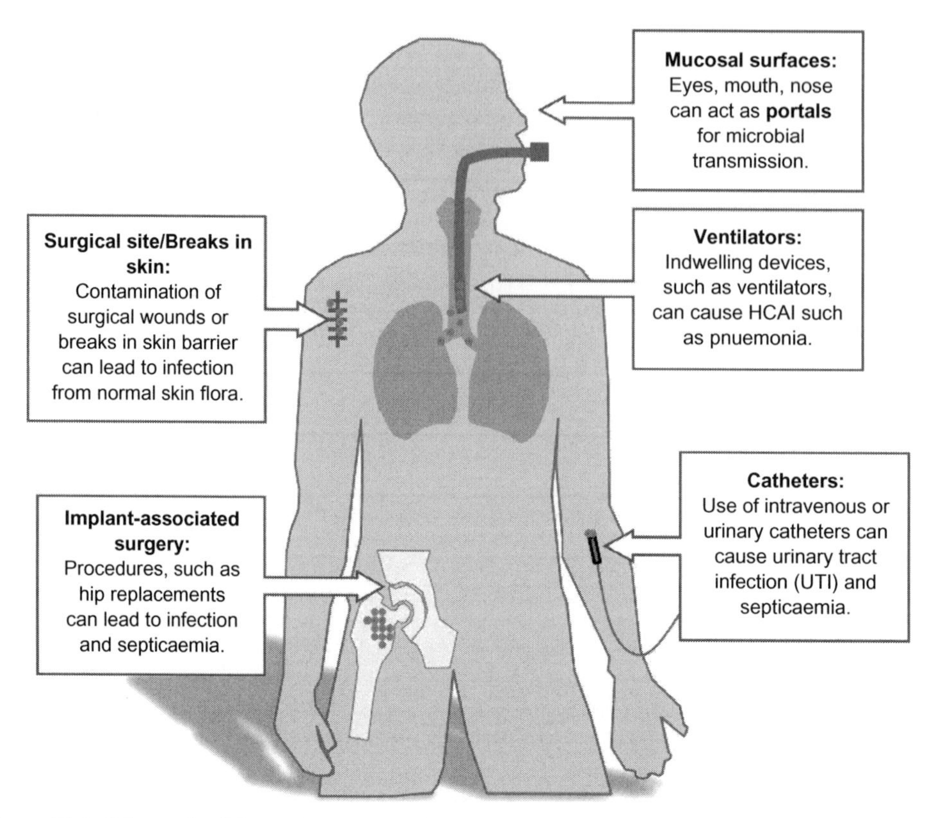

Figure 10.1 *Schematic of the various routes of microbial transmission in HCAI.*

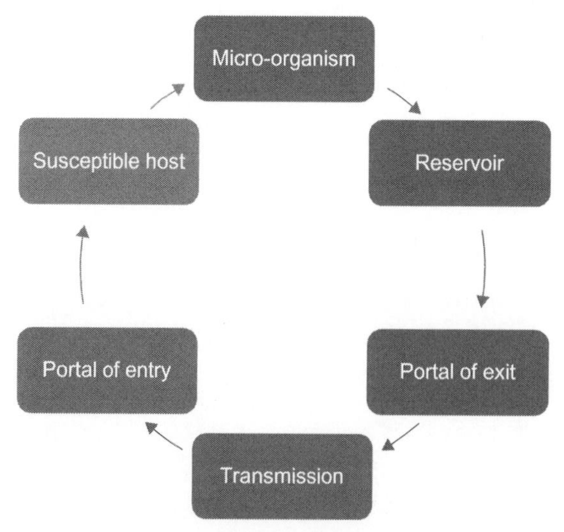

Figure 10.2 ***Cycle of infection.*** *Source: Adapted from Health Protection Agency website at www.hpa.org .uk/Topics/InfectiousDiseases/InfectionsAZ/HCAI/GeneralInformationOnHCAI/.*

BIOFILM FORMATION: A STRATEGY FOR SURVIVAL

Upon contamination of a new host, there is increased risk of biofilm formation by micro-organisms. Biofilms do not grow in a uniform manner, but are complex communities of multi-species micro-organisms that irreversibly attach to a surface, are surrounded by a self-produced matrix of EPS and are capable of nutrient–waste exchange through open water channels.[20] Dower and Turner[21] investigated biofilm formation on closed-suction drains used in many surgical procedures as a means of preventing haematoma formation or a build-up of fluid. In this study, closed-suction drains from patients undergoing a variety of procedures were removed at several time points between 2 and 42 hours and colonisation was found to occur as soon as two hours following placement.[21]

Biofilm formation (Figure 10.3) comprises several stages, as detailed in the following subsections. Stage I involves reversible attachment of planktonic micro-organisms on a medium-conditioned surface. Stage II represents irreversible attachment, the formation of an EPS matrix and the growth and division of micro-organisms with a biofilm. Stage III illustrates the dispersal of parts of the biofilm and planktonic organisms.

Surface Conditioning

It was originally thought that biofilms could not form on smooth surfaces. However, they can form equally well on smooth and rough surfaces, and the physical characteristics of a surface have been shown to have only minor effects on biofilm formation. It is the external mediators, such as high shear force, that strongly influence the formation of robust biofilms that are resistant to mechanical disturbance, as seen in a variety clinical cases.[22]

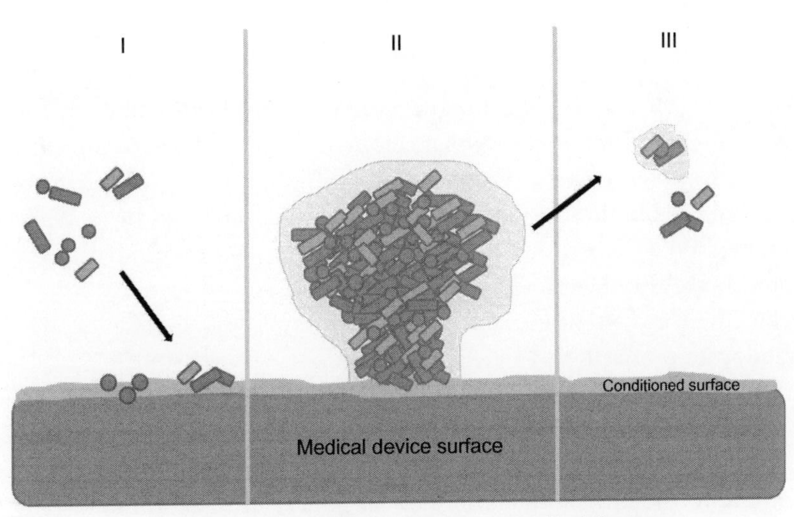

Figure 10.3 ***Schematic of polymicrobial biofilm formation on medical devices.***

In addition, it has been shown that micro-organisms are more likely to attach to a surface that has been conditioned; more specifically, when a surface is exposed to surrounding media, such as tears or blood, it undergoes biochemical modification.[23] For example, clinical bloodstream isolates of the fungus *Candida albicans* have been shown to increase *in vitro* biofilm formation in the presence of serum on metallic and non-metallic surfaces, with varying degrees of surface texture and hydrophobicity.[24]

Reversible Attachment

Planktonic micro-organisms under increased shear are 'forced' to reversibly attach to a surface. This phase is thought to be the weakest link in the chain of biofilm formation events because of an equilibrium of attached and planktonic cells.[25]

Irreversible Attachment

The attachment of a cell to a surface is very much dependent on cell properties such as the presence of fimbriae and flagella, hydrophobicity and EPS production.[23] For example, *Klebsiella pneumonia*, a Gram-negative opportunistic pathogen, has been reported to express type 3 fimbriae that act as appendages for the cellular attachment to a surface.[26] As the cells become more firmly attached to the surface, they differentiate and produce EPS, which acts as a protective matrix to external disruptors.

Colonisation

When micro-organisms become irreversibly attached, they begin to divide and grow, forming micro-niches within an EPS matrix. It is at this point that biofilms become distinctly different from their planktonic counterparts and are able to resist external factors such as mechanical force and antimicrobials.[27]

Dispersion

Micro-organisms can detach from a biofilm as a result of external disturbances, such as fluid shear force, and internal disturbances, such as EPS release or enzymatic breakdown. Although it seems like a relatively unavoidable process, given the diverse environments in which biofilms reside, dispersion is also thought to be a mechanism for further colonisation at new sites. Also referred to as 'detachment' and 'dispersal', dispersion is a process that can be categorised into three phases:

1. Detachment of cells from the biofilm
2. Translocation of cells to a new site
3. Attachment of cells to a surface at the new site

In terms of HCAI, the dispersion of pathogenic micro-organisms from a biofilm plays a key role in aiding transmission from external reservoirs, such as water supplies or human skin, to the new host, and it aids the spread of infection in already infected hosts.[28]

Quorum Sensing

Bacteria in biofilms have shown the ability to communicate with each other through biochemical signalling, also known as 'quorum sensing'. In signalling, the bacteria in a biofilm can sense the density and numbers of micro-organisms in a biofilm.[29] Not all biofilms release the same chemical signals. For example, Gram-negative bacteria release molecules called acyl-homoserine lactones, whereas Gram-positive bacteria release peptide molecules.[29] Some of the well-studied quorum-sensing (QS) molecules are associated with *P. aeruginosa* such as N-3–oxo–dodecanoyl-L-homoserine. In addition, the N-3–oxo–dodecanoyl-L-homoserine QS molecule has been reported to increase *P. aeruginosa* biofilm virulence and repress host immune responses.[30] Given that *P. aeruginosa* is one of the most common micro-organisms associated with HCAI, these QS molecules have become the target for drug development using quorum-sensing inhibitors.[31]

HOST RESPONSES TO BIOFILMS

In the past, a great deal of research was dedicated to understanding host immune responses towards both commensal and pathogenic micro-organisms. However, a large portion of this work was focussed on interactions with planktonic bacteria. Now that it is appreciated that micro-organisms more commonly live in biofilms, particularly in a clinical setting, research into host immune responses against biofilms has become a priority.

Biofilms have been reported to evade host immune responses, and it was initially thought that immune cells could not penetrate them. However, Leid and colleagues[32] found that, in the case of *Pseudomonas aeruginosa* biofilms, the secreted matrix of EPS, known as the exopolysaccharide alginate, effectively protected against interferon-γ (IFN-γ)-mediated macrophage killing.[32] In addition, micro-organisms in biofilms have been shown to evade host immune cell responses by using polymorphonuclear leukocytes (PMLs) to enhance biofilm formation. In patients with cystic fibrosis (CF), there is a characteristic chronic influx of PMLs and persistent *P. aeruginosa* infection.

Walker and colleagues[33] devised an *in vitro* experiment using PMLs isolated from healthy human volunteers and *P. aeruginosa* PA01. The presence of these cells resulted in improved biofilm formation and development through deposition of neutrophil-derived polymers, actin and DNA, which *P. aeruginosa* uses as a biological scaffold to form a biofilm.[33] Although this research used only PMLs from healthy donors and not from CF patients, it is indicative of how a *P. aeruginosa* biofilm can utilise components of the innate immune system to form biofilms in healthy individuals.

In the context of HCAI due to medical devices, a more recent study developed a novel *in vivo* model demonstrating the relationship between *P. aeruginosa* biofilms grown on silicone implants and PMLs. In this study, the researchers observed the death of PMLs after wild-type (WT) biofilm contact; however, mutant biofilms, defective for the biofilm virulence factor rhamnolipids, were effectively cleared by PMLs, indicating an important role for this virulence factor in biofilm survival.[34]

However, bacterial biofilms (e.g., *P. aeruginosa*) are not the only micro-organism to successfully form biofilms and evade host immune responses. The fungus *C. albicans* has been reported to show enhanced biofilm formation in the presence of enriched peripheral blood mononuclear cells (PBMCs) because of the soluble factors that are released into the co-culture medium. Increased levels of pro- and anti-inflammatory cytokines were also detected.[35]

Biofilms and Drug Resistance

Bacteria and fungi that grow in biofilms have shown increased resistance to antibiotics compared to their planktonic counterparts. In addition, plasmid exchange (the transfer of genetic material between bacterial cells) occurs at a higher rate, increasing the chances of developing naturally occurring and antimicrobial-induced resistance.[36] Resistance mechanisms that allow bacteria and fungi in biofilms to protect themselves against antmicrobial treatments have been proposed, including

Incomplete or slow penetration of antimicrobial agents through the surface layers of the biofilm.[37]

An altered micro-environment such as nutrient depletion or low levels of oxygen that may affect antibiotic action.[37]

Resistant phenotypes in a portion of the biofilm population, known as 'dormant cells' or 'persister cells', which have a much slower growth rate than other micro-organisms in the biofilm and have been found to be more resistant to the action of antimicrobials.

Kim and colleagues[38] tested the effects of non-antibiotic antimicrobial agents on active and dormant cell populations of *P. aeruginosa* PA01 biofilms *in vitro*. Although dormant cells were shown to have decreased resistance to chlorine, they showed increased resistance to other non-antibiotic antimicrobials such as silver ions and hydrogen peroxide.[38]

Multi-species Biofilms—Impact on Infection

There are numerous studies on biofilm formation and their relation to HCAI in the literature; however, few focus on the impact of polymicrobial biofilms in a clinical setting, despite the fact that in a natural environment and in a majority of clinical cases, biofilm-related infections are rarely single-species. The lack of polymicrobial biofilm studies may be due to the complications of increased variables and unknowns in *in vitro* models. Although it is beneficial that we investigate how micro-organisms work in single-species biofilms, the introduction of other species may alter cell interactions and virulence.

For instance, Vandecandelaere and colleagues[39] set out to identify micro-organisms in biofilms from endotracheal tubes (ETTs) using traditional culture techniques and 16S rRNA sequencing. The results revealed the presence of a diverse range of micro-organisms, from the common oral-associated micro-flora to the more clinically relevant isolates including *P. aeruginosa*, *Staphylococcus aureus* and *Staphylococcus epidermidis*.[39] This study highlighted not only the importance of culture-dependent and -independent techniques but also the existence of multi-species biofilms in the investigation of HCAIs.

Harriott and Noverr[40] performed an *in vitro* study of polymicrobial biofilms consisting of *S. aureus* and *C. albicans*. In this study, *S. aureus* was shown to develop substantial biofilms in the presence of *C. albicans* when compared to monoculture biofilms, with *C. albicans* acting as a scaffold for *S. aureus* formation. Furthermore, this polymicrobial biofilm showed a phenotypic change in *S. aureus* as well as increased resistance to the antibiotic vancomycin.[40] A more recent study further developed this bacterial–fungal interaction by investigating the mechanisms behind *C. albicans* and *S. aureus* binding in a biofilm. Using both *in vitro* and *in vivo* methods, Peters and colleagues[41] revealed that *S. aureus* is able to bind to a receptor on *C. albicans* referred to as Als3p.[41]

Increased biofilm stability and virulence due to interactions between different species confirms the importance of studying physiologically relevant polymicrobial biofilms.

BIOFILM DETECTION

There are relatively few methods available for identifying biofilms and few guidelines on their use in a clinical setting. The most common microbiological method of detecting bacterial growth is conventional plate counting, whereby swabs or scrapings of a surface are quantified on agar and expressed as colony-forming units (CFU). However, this traditional technique, although cost effective, does not indicate whether the micro-organism has been taken from a biofilm, nor does it reveal the biofilm's current developmental stage. It is more common now to visualise biofilms using microscopy, where fluorescent markers can identify the presence of bacteria. Furthermore, morphology, bacterial cell-surface attachment and *in situ* cell–cell interaction can be analysed using such techniques as confocal scanning laser microscopy (SCLM), scanning electron microscopy (SEM) and atomic force microscopy (AFM).[29]

Although microscopy is an effective way to visualise a biofilm in an *in vitro* setting, biofilms are not easily identifiable by healthcare professionals in a clinical environment. Clinical signs of a biofilm-associated infection include fever and persistent inflammation at the site. It is only when these signs appear that patients undergo blood testing for infection involving microbiological plate counting. However, the results of such tests may take up to 72 hours and are dependent on it being possible to culture the micro-organism.[42]

With this in mind, it is essential that rapid and accurate diagnostic tests be undertaken to detect biofilms in HCAI. More sophisticated techniques use molecular methods for biofilm identification such as fluorescence *in-situ* hybridisation, whereby fluorescent probes specific to a nucleotide sequence with bacterial RNA or DNA identify live and dead bacteria in complex samples.[43] Other techniques, such as matrix-assisted laser desorption ionisation coupled with time of flight analysis mass spectrometry (MALDI-TOF/MS), have been used to identify bacteria. MALDI-TOF/MS uses laser ionisation of bacteria to detect peptides and peptide ions on the cell surface. Laser ionisation can be compared to an extensive database enabling the species of bacteria to be identified.[44]

THE MICRO-ORGANISMS

In addition to the already potentially compromised immune system of some patients entering a hospital environment, surfaces on indwelling medical devices are at high risk of bacterial and fungal biofilm development.[29] This section focusses on the micro-organisms commonly associated with HCAI.

The most typical HCAI have been reported to be associated with Gram-negative bacteria. A report on the most commonly isolated bacteria from a range of HCAI showed that Gram-negative bacteria (e.g., *Pseudomonas aeruginosa*, *Klebsiella pneumonia* and *Escherichia coli*) are among the more prevalent.[45] Even in a planktonic state, Gram-negative bacteria can evade antibiotic treatment through mechanisms such as transmembrane efflux pumps, which export the antibiotic out of the cell, and antibiotic-modifying enzymes, which prevent antibiotics from binding to their target.[46]

P. aeruginosa

P. aeruginosa is a Gram-negative bacillus most commonly found in rivers and lakes because of its high tolerance to a wide range of environmental conditions. In a clinical setting, it is an opportunistic pathogen that affects mainly immuno-compromised patients but also has the potential to cause mild illness in relatively healthy individuals. *P. aeruginosa* can be found in hospital water systems, and patient exposure can be due to contaminated tap water through bathing, showering, drinking or tap water-contaminated medical equipment. All of these routes have been implicated in the development of HCAIs associated with this organism.

A recent study highlighted just how prevalent these bacteria are in hospital water systems. A total of 11 hospitals were investigated, and 44 water samples were tested for *Pseudomonas sp* and *P. aeruginosa* using a nested PCR method. Samples from 9 of the 11 hospitals tested positive for *P. aeruginosa*; 32% of the samples also tested positive.[47] *P. aeruginosa* causes a variety of HCAI, including urinary tract and wound infections; however, it is most commonly associated with pulmonary HCAI such as VAP.

Enterobacteriaceae

A part of the *Enterobacteriaceae* family, *Klebsiella pneumoniae* is a Gram-negative opportunistic pathogen typically found in water and soil environments; however, in a clinical setting humans are the primary reservoir. *K. pneumoniae* has primarily been implicated in urinary and respiratory tract infections, but there is also evidence for the presence of this organism in septicaemia and SSIs. Biofilm formation of *K. pneumoniae* has been implicated in its pathogenesis in HCAI. More specifically, *K. pneumoniae* are able to form biofilms through the use of fimbriae which act as appendages.[48] Furthermore, it has been shown that type 1 and 3 appendages play an important role in the colonisation of *K. pneumoniae* on silicone urinary catheter tubing in an *in vivo* mouse model.[48]

Another member of the *Enterobacteriaceae* family is *E. coli*, a Gram-negative pathogen that is more commonly associated with urinary tract infections and septicaemia. The number of *E. coli* septicaemia cases has risen in the United Kingdom over recent years, with extended surveillance warranted according recent HPA reports.[2] A study of the biofilm formation of symptomatic and asymptomatic strains of urinary tract-associated *E. coli* showed that asymptomatic *E.coli* strains are more superior in biofilm formation than are symptomatic strains. In this case the study authors suggested that biofilm formation by *E. coli* is a survival strategy rather than a mechanism for virulence.[49]

Staphylococcus aureus

S. aureus is a Gram-positive bacterium that can reside asymptomatically on the skin and in the nostrils of approximately one-third of human beings.[50] Staphylococcal infection on indwelling medical devices such as catheters is a common problem and is thought to be the result of poor vascularisation at implantation sites which impedes host defences against colonisation.[51]

In addition, *S. aureus* has an astonishing ability to develop resistance against multiple antibiotics, including penicillin and meticillin, with meticillin-resistant strains emerging just two years after the introduction of this antibiotic in response to penicillin resistance.[52] It is important to note that although recent reports published by the HPA have shown that MRSA-related bacteraemia is decreasing in the United Kingdom, MRSA is a prime example of the misuse of antibiotics. It is thus crucial to focus on antimicrobial strategies that do not involve antibiotics.

Clostridium difficile

C. difficile is a Gram-positive, anaerobic bacillus that is widely known for its role in hospital-acquired infectious diarrhoea, or *C. difficile* infection (CDI). *C. difficile* colonises the gastrointestinal tract through faecal–oral routes by which transmission can occur through direct contact of the infected person, via a vector such as the healthcare worker, or through *C. difficile* spores.[53] Although CDI is one of the most predominant HCAI worldwide and was highly publicised for high infection rates in the past, the number of *C. difficile* infections has been continuing to fall in the United Kingdom, according to the data gathered from 2008 to 2011.[2]

CDI is thought to occur when the normal micro-flora of the gut is disturbed through use of antibiotics. In terms of biofilm formation, two *C. difficile* strains taken from non-epidemic (strain 630) and epidemic (strain R20291) outbreaks in 2004 and 2005 were tested for their ability to form biofilms in an *in vitro* model. Results from this study revealed that *C. difficile* forms more biofilms in the epidemic strain which are encapsulated in a matrix resistant to the antibiotic vancomycin, and produces more toxins and virulence factors. The authors of this study hypothesised that *C. difficile* may colonise the gut and protect itself from antibiotic treatment by forming biofilms and secreting toxins.[54] More research into biofilm formation of these strains in an *in vivo* model is essential to understanding the pathogenesis of CDI.

C. albicans

Candida species are the most common cause of hospital-acquired fungal infections. While *C. albicans* is isolated most frequently, other sub-species are being recognised as key players in fungal infection, especially in biofilm formation and HCAI (see Table 10.1). A study of the biofilm-forming capabilities of *Candida spp* isolated from clinical cases of septicaemia discovered that biofilms are most commonly observed in non-*albicans Candida* species compared to *C. albicans*. These findings demonstrated that biofilm formation of non-*Candida* species may be more crucial than *C. albicans* biofilms in septicaemia. In addition, given that *C. albicans* is more pathogenic than other *Candida* species, the authors proposed alternative mechanisms behind *C. albicans* pathogenicity in septicaemia.[55]

Evidence of Polymicrobial Biofilms in HCAI

Advances in microbiological techniques have allowed the identification of isolates that would not have been detected using traditional culture techniques, and they have made it much easier to detect polymicrobial biofilms in clinical settings.

Catheters can be used to supply a part of the body with fluids or can be used to drain waste fluids. Either way, they can be the source of a variety of HCAI, including central-line-associated septicaemia and UTIs.[60] Larsen and colleagues[58] assessed microbial activity in intravascular catheters taken from 18 patients and observed polymicrobial biofilms using culture-dependent and -independent techniques. The most abundant isolates included *Staphylococcus epidermidis* and coagulase-negative *Staphylococci* (CNS); bacteria, such as *P. aeruginosa* and *K. pneumonia*, were recorded in only a few cases.[58]

Table 10.1 Most Commonly Isolated Micro-organisms in Biofilm-Associated HCAI

HCAI	Micro-organism	References
Medical device–related		
Catheter-associated urinary tract infection (source: urinary catheters)	Coagulase-negative *Staphylococci* (CNS) *Escherichia coli* *Klebsiella pneumonia* *Pseudomonas aeruginosa* *Staphylococcus aureus* *Staphylococcus epidermidis*	56, 57
Central-line-associated septicaemia (source: intravascular catheters	CNS *Candida albicans* *Klebsiella pneumonia* *Pseudomonas aeruginosa* *Staphylococcus aureus* *Staphylococcus epidermidis*	55, 57, 58
Ventilator-associated pneumonia (source: endotracheal tubes)	*Candida* *Klebsiella pneumonia, Pseudomonas aeruginosa* *Staphylococcus aureus* *Staphylococcus epidermidis*	39, 57
Surgical site infection		
Surgical wound, prosthesis-related infection	*Meticillin-resistant Staphylococcus aureus* (MRSA) *Staphylococcus aureus,* *Staphylococcus epidermidis*	59

In a separate study, the colonisation of urinary tract catheters from 45 patients revealed mixed Gram-negative and Gram-positive bacterial species. CNS was found on 52% of samples after short-term catheterisation (less than one month); however, it was found on only 16% of catheters after long-term catheterisation. Furthermore, *P. aeruginosa*, *K. pneumoniae* and *E. coli* were identified in biofilm form, with *P. aeruginosa* being the strongest biofilm former when tested *in vitro*.[56] In addition, a study into the microbial activity of several devices, including ETTs, intravascular catheters and urinary catheters, revealed polymicrobial biofilms involving micro-organisms (e.g., *K. pneumoniae* and *Staphylococcus*). Interestingly, this study assessed biofilm formation of clinical isolates using the tube method, which correlated with microscopy evidence of biofilms and thus provided a technique for healthcare facilities that do not have access to microscopes.[57]

Rohacek and others[61] have investigated subclinical infection in implanted cardiac medical devices, including pacemakers and defibrillators. Their study revealed that 38% of cardiac medical devices that were not explanted because of signs of infection were positive for multi-species bacteria, mainly including *Propionibacterium acnes*,

CNS and Gram-negative rods.[61] This research showed that bacterial growth can occur on implanted medical devices without signs of clinical infection. Another study by Stoodley and colleagues[59] revealed the presence of a mixed meticillin-resistant *S. aureus* and *S. epidermidis* in an explanted ankle arthroplasty. The micro-organisms were detected using methods such as coupled PCR-mass spectrometry as well as traditional culture methods. However, the presence of biofilm was determined using CLSM.[59]

BIOFILM PREVENTION

SSIs are a type of healthcare-associated infection that may be prevented if medical professionals adhere to NICE guidelines regarding hygiene procedures. With regard to indwelling medical device–related infections, whilst ensuring that high standards of hygiene are maintained, it is also imperative to explore ways in which we can prevent microbial growth on the materials used to manufacture such devices (see Figure 10.4). If it is more difficult for bacteria to attach to a surface without using antibiotics, the risk of HCAI due to indwelling devices may be significantly reduced.

Equipment Sterilisation

Biel and colleagues[62] investigated the antimicrobial role of photodynamic therapy on antibiotic-resistant *P. aeruginosa* and MRSA (clinical isolates) biofilms grown in ETTs. This involved spraying a methylene blue–based photosensitiser into the lumen of the ETT and then exposing the ETT to light from a fibre-optic diffuser at 644 nm wavelength. A reduction in polymicrobial biofilm growth of more than 99.9% was achieved. Ren and colleagues[63] assessed various detergents with the aim of removing *Escherichia coli* from flexible endoscopes using an artificial biofilm model. Their study revealed that more bacterial biofilm was found with enzymatic detergent when compared with non-enzymatic detergent.[63]

Surface Modification and Coating of Medical Devices

Once a medical device (e.g., implant material) is infected, it needs to be removed promptly. Although the use of pre-operative antibiotic treatment has been proposed, the potential coating of medical devices in antiseptics such as polyhexamethylene biguanide (PHMB), octenidine and triclosan, offers a broader spectrum of antimicrobial killing and a much lower risk of antimicrobial resistance.[64]

Raad and colleagues tested the efficiency of antimicrobial gardine- and gendine-coated ETTs against silver-coated ETTs *in vitro*. They subsequently showed that MRSA, *P. aeruginosa*, *C. albicans* and *K. pneumonia* biofilm growth could be completely inhibited with the gardine- and gendine-coated ETTs for up to two weeks when compared to the silver-coated ETTs, which still displayed growth of up to 10^7 CFU/cm.[65]

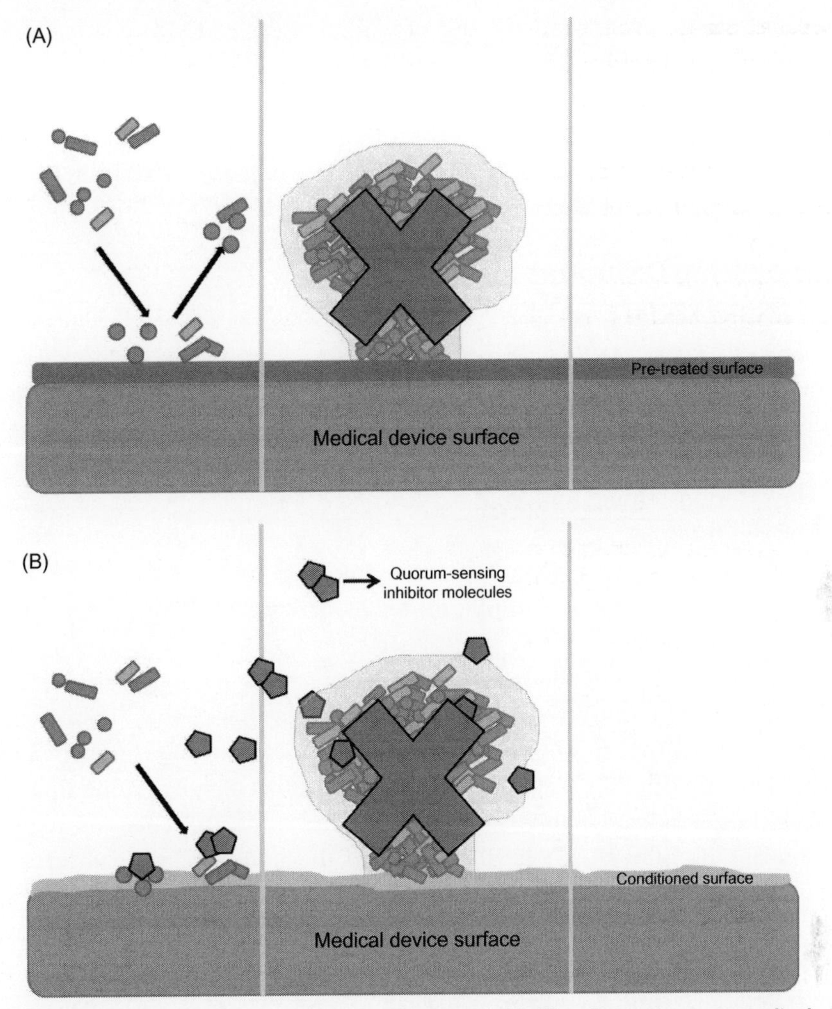

Figure 10.4 ***Potential strategies to prevent or disrupt biofilm formation on medical devices.*** A. Pretreating the surface of medical devices to reduce microbial attachment and prevent biofilm formation. B. Using quorum-sensing inhibitors to disrupt the development of a biofilm.

A recent study also explored the effect of nanomodified ETT on *S. aureus* biofilm formation.[66] This involved the creation of a textured, nanomodified surface using *Rhizopus arrhisus*, a fungal lipase which enzymatically degrades the ETT's PVC material. The nanomodified ETT was then exposed to a constant flow of *S. aureus* medium and incorporated in an airway model. The results showed significantly reduced CFU/ml of bacteria in the nanomodified ETT when compared to the untreated control. In addition, there was an increase in protein absorption by the nanomodified ETT which, the study authors hypothesised, may prevent the colonisation and formation of biofilms.

A separate study investigated the use of trimethylsilane (TMS) coating to prevent the formation of *S. epidermidis* biofilms on 316L stainless steel and grade 5 titanium alloy, which are commonly used materials in indwelling medical devices. The results showed a significant reduction in crystal violet staining of biofilms on TMS-coated surfaces when compared to the untreated control. In addition, the bacteria on these surfaces were more susceptible to the antibiotic ciprofloxacin.[67]

Quorum-Sensing Disruptors

Quorum sensing has been associated with biofilm development and increased microbial virulence.[68] Therefore, it is logical to target aspects of it to reduce biofilm virulence and disrupt development. Three main targets of quorum sensing have been proposed:

1. The signal generator
2. The quorum-sensing molecule
3. The signal receptor

In this case, it is the signal receptor that is more often the target of pharmacological action.[69] It is thought that QS inhibitors reduce bacterial pathogenicity rather than disturb microbial growth.

One mechanism by which these inhibitors affect the pathogenicity of a biofilm is reduction of the biofilm's resistance to antimicrobial treatment.[70] Christensen and colleagues[71] showed that *P. aeruginosa* biofilms in an *in vivo* mouse model can be disrupted by the antibiotic tobramycin and several QS molecules, including furanone and horseradish juice extract. Synergy was seen between both treatments, and the presence of quorum-sensing inhibitor molecules increased the susceptibility of the *P. aeruginosa* biofilm to tobramycin.

Such methods create a less favourable surface for biofilms to reside on, and they reduce biofilm pathogenicity using QS inhibitors, demonstrating a promising and exciting potential avenue for further exploration. However, it is clear that more work needs to be done to incorporate these ideas into an *in vivo* environment, particularly in the case of biofilm formation, as *in vitro* biofilm models may not mimic complex *in vivo* conditions.

CONCLUSION

Biofilms are of great importance in HCAI because of their increased virulence and tolerance to antimicrobial treatment. It is clear from the literature that the identification of medical biofilms must be through a combination of advanced microscopy, traditional culture techniques and more sophisticated molecular methods. Any of these alone may not identify all micro-organisms that reside in the biofilm. In fact, the knowledge that mixed-species biofilms may interact with each other to increase virulence and resistance to antimicrobial action enhances this theory. Nevertheless, preventative measures, such as antimicrobial coating and surface alterations of medical devices and inhibitors of QS molecules, provide promising opportunities in the prevention and treatment of biofilm formation in HCAI.

Key Points

- HCAI are a direct result of medical care or treatment in hospital or a community care setting.
- Micro-organisms in a biofilm are more tolerant to antimicrobial treatment and mechanical force.
- Microbial biofilms have been associated with several HCAI, including VAP, central-line septicaemia, catheter-associated UTIs and surgical site infections.
- Mixed use of culture techniques and molecular methods are key to the identification of the mixed-species biofilms.
- The potential use of modified materials for implanted medical devices may reduce the likelihood of biofilm formation.

REFERENCES

1. van Kleef E, Robotham JV, Jit M, Deeny SR, Edmunds WJ. Modelling the transmission of healthcare-associated infections: a systematic review. *BMC Infect Dis* 2013;**13**:294.
2. HPA. *English National Point Prevalence Survey on Healthcare-associated Infections and Antimicrobial Use.* Preliminary data. London: Health Protection Agency; 2012.
3. NAO. *The Management and Control of Hospital Acquired Infection in Acute NHS Trusts in England.* National Audit Office; 2000.
4. Graves N. Economics and preventing hospital-acquired infection. *Emerg Infect Dis* 2004;**10**(4):561.
5. Hocevar SN, Edwards JR, Horan TC, Morrell GC, Iwamoto M, Lessa FC. Device-associated infections among neonatal intensive care unit patients: incidence and associated pathogens reported to the National Healthcare Safety Network, 2006-2008. *Infect Control Hosp Epidemiol* 2012;**33**(12):1200–6.
6. Saint S, Greene MT, Kowalski CP, Watson SR, Hofer TP, Krein SL. Preventing catheter-associated urinary tract infection in the United States: a national comparative study. *JAMA Intern Med* 2013; **173**(10):874–9.
7. Stone J, Gruber TJ, Rozzelle CJ. Healthcare savings associated with reduced infection rates using antimicrobial suture wound closure for cerebrospinal fluid shunt procedures. *Pediatr Neurosurg* 2010; **46**(1):19–24.
8. Gamble HP, Duckworth GJ, Ridgway GL. Endoscope decontamination incidents in England 2003–2004. *J Hosp Infect* 2007;**67**(4):350–4.
9. Stewart PS, Costerton JW. Antibiotic resistance of bacteria in biofilms. *Lancet* 2001;**358**(9276):135–8.
10. Scali C, Kunimoto B. An update on chronic wounds and the role of biofilms. *J Cutan Med Surg* 2013;**17**(0):1–6.
11. Bjarnsholt T. The role of bacterial biofilms in chronic infections. *APMIS* 2013;**136**(Suppl):1–51.
12. Percival SL, Hill KE, Williams DW, Hooper SJ, Thomas DW, Costerton JW. A review of the scientific evidence for biofilms in wounds. *Wound Repair Regen* 2012;**20**(5):647–57.
13. de la Fuente-Núñez C, Reffuveille F, Fernández L, Hancock RE. Bacterial biofilm development as a multicellular adaptation: antibiotic resistance and new therapeutic strategies. *Curr Opin Microbiol* 2013; pp. 580–9.
14. Percival SL, Dowd S, Cutting KF, Wolcott R. Surgical-site infections—biofilms, dehiscence, and delayed healing. *US Dermato* 2008;**3**(1):56–9.

15. Abdul-Jabbar A, Berven SH, Hu SS, Chou D, Mummaneni PV, Takemoto S, et al. Surgical site infections in spine surgery: identification of microbiologic and surgical characteristics in 239 cases. *Spine* 2013;**38**:425–31.
16. NICE. *Surgical site infection: Prevention and treatment of surgical site infection*. London: National Institute for Health and Care Excellence; 2008.
17. Goyal N, Miller A, Tripathi M, Parvizi J. Methicillin-resistant *Staphylococcus aureus* (MRSA): Colonisation and pre-operative screening. *Bone Joint J* 2013;**95-B**(1):4–9.
18. Garner JS, Jarvis WR, et al. CDC definitions for nosocomial infections. *Am J Infect Control* 1988; **16**(3):128–40.
19. Horan TC, Gaynes RP, et al. CDC definitions of nosocomial surgical site infections, 1992: a modification of CDC definitions of surgical wound infections. *Inf Cont Hosp Epidemiol* 1992; **13**(10):606–8.
20. Hall-Stoodley L, Costerton JW, et al. Bacterial biofilms: from the natural environment to infectious diseases. *Nat Rev Microbiol* 2004;**2**(2):95–108.
21. Dower R, Turner ML. Pilot study of timing of biofilm formation on closed suction wound drains. *Plast Reconstr Surg* 2012;**130**(5):1141–6.
22. Donlan RM, Costerton JW. Biofilms: survival mechanisms of clinically relevant microorganisms. *Clin Microbiol Rev* 2002;**15**(2):167–93.
23. Donlan RM. Biofilms: microbial life on surfaces. *Emerg Infect Dis* 2002;**8**(9):881.
24. Frade JP, Arthington–Skaggs BA. Effect of serum and surface characteristics on *Candida albicans* biofilm formation. *Mycoses* 2011;**54**(4):e154–62.
25. Lindsay D, Von Holy A. Bacterial biofilms within the clinical setting: what healthcare professionals should know. *J Hosp Infect* 2006;**64**(4):313–25.
26. Murphy CN, Clegg S. *Klebsiella pneumoniae* and type 3 fimbriae: nosocomial infection, regulation and biofilm formation. *Future Microbiol* 2012;**7**(8):991–1002.
27. Costerton J, Lewandowski Z, et al. Biofilms, the customized microniche. *J Bacteriol* 1994;**176**(8):2137.
28. Kaplan J. Biofilm dispersal: mechanisms, clinical implications, and potential therapeutic uses. *J Dent Res* 2010;**89**(3):205–18.
29. Lindsay D, Von Holy A. Bacterial biofilms within the clinical setting: what healthcare professionals should know. *J Hosp Infect* 2006;**64**(4):313–25.
30. Driscoll JA, Brody SL, et al. The epidemiology, pathogenesis and treatment of *Pseudomonas aeruginosa* infections. *Drugs* 2007;**67**(3):351–68.
31. Hentzer M, Wu H, et al. Attenuation of *Pseudomonas aeruginosa* virulence by quorum-sensing inhibitors. *EMBO J* 2003;**22**(15):3803–15.
32. Leid JG, Willson CJ, et al. The exopolysaccharide alginate protects *Pseudomonas aeruginosa* biofilm bacteria from IFN-γ-mediated macrophage killing. *J Immunol* 2005;**175**(11):7512–8.
33. Walker TS, Tomlin KL, et al. Enhanced *Pseudomonas aeruginosa* biofilm development mediated by human neutrophils. *Infect Immun* 2005;**73**(6):3693–701.
34. van Gennip M, Christensen LD, et al. Interactions between polymorphonuclear leukocytes and *Pseudomonas aeruginosa* biofilms on silicone implants *in vivo*. *Infect Immun* 2012;**80**(8):2601–7.
35. Chandra J, McCormick TS, et al. Interaction of *Candida albicans* with adherent human peripheral blood mononuclear cells increases *C. albicans* biofilm formation and results in differential expression of pro- and anti-inflammatory cytokines. *Infect Immun* 2007;**75**(5):2612–20.
36. Hausner M, Wuertz S. High rates of conjugation in bacterial biofilms as determined by quantitative *in situ* analysis. *Appl Environ Microbiol* 1999;**65**(8):3710–3.
37. Francolini I, Donelli G. Prevention and control of biofilm-based medical device-related infections. *FEMS Immunol & Med Microbiol* 2010;**59**(3):227–38.
38. Kim J, Hahn JS, et al. Tolerance of dormant and active cells in *Pseudomonas aeruginosa* PA01 biofilm to antimicrobial agents. *J Antimicrob Chemother* 2009;**63**(1):129–35.
39. Vandecandelaere I, Matthijs N, et al. Assessment of microbial diversity in biofilms recovered from endotracheal tubes using culture dependent and independent approaches. *PloS One* 2012;**7**(6):e38401.
40. Harriott MM, Noverr MC. *Candida albicans* and *Staphylococcus aureus* form polymicrobial biofilms: effects on antimicrobial resistance. *Antimicrob Agents Chemother* 2009;**53**(9):3914–22.

41. Peters BM, Ovchinnikova ES, et al. *Staphylococcus aureus* adherence to *Candida albicans* hyphae is mediated by the hyphal adhesin Als3p. *Microbiol* 2012;**158**(Pt 12):2975–86.
42. Bryers JD. Medical biofilms. *Biotechnol Bioeng* 2008;**100**(1):1–18.
43. Høgdall D, Hvolris JJ, et al. Improved detection methods for infected hip joint prostheses. *Apmis* 2010;**118**(11):815–23.
44. Arciola CR, Montanaro L, et al. New trends in diagnosis and control strategies for implant infections. *Int J Artif Organs* 2011;**34**(9):727.
45. Weinstein RA, Gaynes R, et al. Overview of nosocomial infections caused by Gram-negative *bacilli*. *Clin Infect Dis* 2005;**41**(6):848–54.
46. Peleg AY, Hooper DC. Hospital-acquired infections due to gram-negative *bacterisa*. *N Engl J Med* 2010;**362**(19):1804–13.
47. Baghal Asghari F, Nikaeen M, et al. Rapid monitoring of *Pseudomonas aeruginosa* in hospital water systems: a key priority in prevention of nosocomial infection. *FEMS Microbiol Lett* 2013;**343**:77–81.
48. Murphy CN, Mortensen MS, et al. Role of *Klebsiella pneumoniae* type 1 and type 3 fimbriae in colonizing silicone tubes implanted into the bladder of mice as a model of catheter-associated urinary tract infections. *Infect Immun* 2013;**81**(8):3009–17.
49. Hancock V, Ferrieres L, et al. Biofilm formation by asymptomatic and virulent urinary tract infectious *Escherichia coli* strains. *FEMS Microbiol Lett* 2007;**267**(1):30–7.
50. Kluytmans J, Van Belkum A, et al. Nasal carriage of *Staphylococcus aureus*: epidemiology, underlying mechanisms, and associated risks. *Clin Microbiol Rev* 1997;**10**(3):505–20.
51. Kiedrowski MR, Horswill AR. New approaches for treating staphylococcal biofilm infections. *Ann NY Acad Sci* 2011;**1241**(1):104–21.
52. DeLeo FR, Otto M, et al. Community-associated meticillin-resistant *Staphylococcus aureus*. *Lancet* 2010;**375**(9725):1557–68.
53. Bouza E. Consequences of *Clostridium difficile* infection: understanding the healthcare burden. *Clin Microbiol Infect* 2012;**18**(s6):5–12.
54. Đapa T, Leuzzi R, et al. Multiple factors modulate biofilm formation by the anaerobic pathogen *Clostridium difficile*. *J Bacteriol* 2013;**195**(3):545–55.
55. Pannanusorn S, Fernandez V, et al. Prevalence of biofilm formation in clinical isolates of *Candida* species causing bloodstream infection. *Mycoses* 2012;**56**(3):264–72.
56. Wang X, Lünsdorf H, et al. Characteristics of biofilms from urinary tract catheters and presence of biofilm-related components in *Escherichia coli*. *Curr Microbiol* 2010;**60**(6):446–53.
57. Singhai M, Malik A, et al. A study on device-related infections with special reference to biofilm production and antibiotic resistance. *J Global Infect Dis* 2012;**4**(4):193.
58. Larsen MK, Thomsen TR, et al. Use of cultivation-dependent and -independent techniques to assess contamination of central venous catheters: a pilot study. *BMC Clin Pathol* 2008;**8**(1):10.
59. Stoodley P, Conti SF, et al. Characterization of a mixed MRSA/MRSE biofilm in an explanted total ankle arthroplasty. *FEMS Immunol Med Microbiol* 2011;**62**(1):66–74.
60. Thomsen TR, Hall-Stoodley L, et al. *The role of bacterial biofilms in infections of catheters and shunts in biofilm infections*. Springer; 2011. p 91-109.
61. Rohacek M, Weisser M, et al. Bacterial colonization and infection of electrophysiological cardiac devices detected with sonication and swab culture. *Circulation* 2010;**121**(15):1691–7.
62. Biel MA, Sievert C, et al. Reduction of endotracheal tube biofilms using antimicrobial photodynamic therapy. *Lasers Surg Med* 2011;**43**(7):586–90.
63. Ren W, Sheng X, et al. Evaluation of detergents and contact time on biofilm removal from flexible endoscopes. *Am J Infect Control* 2013
64. Leaper D, McBain AJ, et al. Healthcare-associated infection: novel strategies and antimicrobial implants to prevent surgical site infection. *Ann R Coll Surg Engl* 2010;**92**(6):453.
65. Raad II, Mohamed JA, et al. The prevention of biofilm colonization by multidrug-resistant pathogens that cause ventilator-associated pneumonia with antimicrobial-coated endotracheal tubes. *Biomaterials* 2011;**32**(11):2689–94.
66. Machado MC, Tarquinio KM, et al. Decreased *Staphylococcus aureus* biofilm formation on nanomodified endotracheal tubes: a dynamic airway model. *Int J Nanomedicine* 2012;**7**:3741.

67. Ma Y, Chen M, et al. Inhibition of *Staphylococcus epidermidis* biofilm by trimethylsilane plasma coating. *Antimicrob Agents Chemother* 2012;**56**(11):5923–37.
68. Hentzer M, Wu H, et al. Attenuation of *Pseudomonas aeruginosa* virulence by quorum-sensing inhibitors. *EMBO J* 2003;**22**(15):3803–15.
69. Rasmussen TB, Givskov M. Quorum-sensing inhibitors as anti-pathogenic drugs. *Int J Med Microbiol* 2006;**296**(2–3):149.
70. Bjarnsholt T, Givskov M. Quorum-sensing inhibitory drugs as next generation antimicrobials: worth the effort? *Curr Infect Dis Rep* 2008;**10**(1):22–8.
71. Christensen LD, van Gennip M, et al. Synergistic antibacterial efficacy of early combination treatment with tobramycin and quorum-sensing inhibitors against *Pseudomonas aeruginosa* in an intraperitoneal foreign-body infection mouse model. *J Antimicrob Chemother* 2012;**67**(5):1198–206.

CHAPTER ELEVEN

Biofilms' Role in Intravascular Catheter Infections

Emma Woods and Steven L. Percival

INTRODUCTION

Several of the chapters in this book describe the incidence and impact of micro-organisms on various types of indwelling medical devices. The intravascular (IV) catheter is one such device. It is known that micro-organisms commonly attach to both biotic and abiotic surfaces. In the case of IV catheters we have an example of a wound tissue environment in combination with an inert implanted surface. This chapter discusses types of IV catheters and the incidence of infection associated with them, as well as research on the role of biofilm formation in these devices.

Intravascular catheters are an important tool in modern clinical practice, in both patients with acute critical injuries and those with chronic illnesses. Table 11.1 lists the types of IV catheters, their uses and durations of use. It is estimated that approximately 90% of patients admitted to hospital will require some form of short- or long-term IV catheter therapy, which is used to deliver medications, nutritional supplements and blood products directly into a patient's bloodstream; typically, this is via the patient's venous system through either a peripheral or a central venous catheter (CVC). Arterial IV catheters are also used to continuously sample and monitor a patient's blood pressure and blood gases and are often used during surgical procedures when a patient is under anaesthesia. Central venous catheters, as well as some other long-term IV devices, have integrated cuffs which anchor them in place to reduce mechanical damage at the insertion point. More sophisticated vascular catheters also include multi-lumen devices (e.g., a Hickman line) which allow both drug administration and blood sampling.

In the United States it is estimated that hospitals and clinics purchase >300 million IV devices annually.[1] Peripheral venus catheters (PVCs) make up the majority of these devices, with CVCs accounting for about 3 million. Most epidemiological studies on the use and associated complications of IV catheters tend to focus on the intensive care environment. These units have high levels of catheter use in an accessible patient group.[1] However, it should be remembered that IV devices are critical across a broad range of clinical settings for vascular access and that their use can put patients at risk of localised and systemic complications in all environments.

Biofilms in Infection Prevention and Control
Doi: http://dx.doi.org/10.1016/B978-0-12-397043-5.00011-6

Table 11.1 Main Types and Uses of IV Catheters

Type	Duration of use before removal/replacement*	Examples of use**
Central venous catheter (CVC)	Short and long term	Pain management, chemotherapy, delivery of antibiotics Monitoring of central venous pressure Dialysis
Peripherally inserted central catheter (PICC)	Long term	When repeated peripheral intravenous delivery fails Alternative to CVC
Peripheral venous catheter (PVC)	Short term (≤3 days)	Delivery of fluids and medication to acutely ill patients
Pulmonary arterial catheter (PAC)	Short term	Monitoring of heart function, signs of sepsis; evaluation of drug regimes
Peripheral arterial catheter	Short term	Monitoring of patient's heamo-dynamic status (e.g., while under anaesthesia)
Tunnelled central venous catheter	Long term	Surgically implanted; delivery of medication and nutritional supplements to chronically ill patients
Implanted IV ports	Long term	

*Duration of time before removal/replacement is dependent on individual patient needs. Local guidelines should be referred to for recommended duration of use at various centres.
**Healthcare providers should refer to local guidelines for use recommendations at their centres.

COMPLICATIONS OF INTRAVASCULAR CATHETERS

Complications arising from the use of IV catheter devices include local damage and more serious systemic ones. *Phlebitis* is inflammation of the blood vessel into which the cannula of the IV device is inserted. Estimates of the incidence of phlebitis in the United Kingdom indicate that 20 to 80% of patients with a PVC develop the condition. Phlebitis can be caused by movement of the catheter at the site of insertion and is thus often associated with peripheral catheters.[2]

When a catheter is used to administer medication, chemical irritation can occur and induce vascular inflammation at the insertion site. Phlebitis can also occur as a result of local microbial colonisation at the insertion site and can lead to more serious systemic complications such as sepsis if undetected. Indicators of phlebitis include erythema and swelling along the blood vessel track, redness, heat and at times localised pain, particularly on administration of medication. The occurrence of phlebitis at the site of insertion should act as an early indicator for healthcare staff to ensure that the appropriate measures are taken to reduce residual effects.

Table 11.2 Catheter-Related Infections

Type	Description
Exit site infection	Observation of exudate from the catheter exit site which is positive for microbial cultures with or without associated bloodstream infection
Tunnel infection	Pain and/or erythema along the subcutaneous tract of the catheter and >2 cm from the catheter exit site with or without associated bloodstream infection
Pocket infection	Infected fluid observed in the subcutaneous pocket of an implanted IV port device often with pain and/or erythema, rupture, drainage and necrosis of the overlying skin tissue with or without associated bloodstream infection
Bloodstream infection: infusate-related	Isolation of the same organism from infusate from the catheter site and cultures taken from percutaneous blood sample with no other source of infection
Bloodstream infection: catheter-related	Isolation and positive culture of the same organism(s) from catheter segments following removal and concordantly isolated via a peripheral blood sample where no other source of infection is present (other than the IV device); systemic symptoms of infection include fever and/or hypotension

The most common complication associated with IV devices is catheter-related bloodstream infections (CRBSIs); some types of these infections are listed in Table 11.2. In the United States, more than 250,000 CRBSIs are recorded each year,[3] with approximately 80,000 associated with CVCs.[4] European data show that >60% of nosocomial infections are related to the use of IV catheters, with the majority being linked specifically to CVCs.[4] It is estimated that the infection rate when CVCs are used is in the region of 3 to 8%, with attributable mortality at 5 to 25%.[5] In a point prevalence study carried out across a total of 151 European hospitals in 2001, it was shown that 21.7% of all catheter samples processed were positive for microbial colonisation.[6] Although it is difficult to accurately estimate the overall impact on patient morbidity and mortality, it is clear that catheters add a significant burden to the duration of stay and cost for patients in intensive care units.[4]

ROUTES OF INTRAVASCULAR CATHETER INFECTION

Microbial colonisation associated with an IV catheter can occur at the site of insertion, along the subcutaneous tract of tunnelled catheters and on the body and tip of the catheter within the vascular lumen.[7] Subsequent microbial infection is evidenced by phlebitis, as described previously, as well as cellulitis along the device tract and formation of pus.[7] These visual symptoms provide an indicator for the removal of the catheter.

Two main sources of microbial contamination can lead to catheter-related infections[8]: contact with skin surface organisms, resulting in colonisation along the external catheter surface, and transfer of micro-organisms from the patient or healthcare worker to the hub of the catheter, leading to luminal colonisation. CRBSIs in the instance of short-term catheter use are most often associated with contact with the skin surface at the time of catheter insertion.[9] Following contamination, micro-organisms migrate along and colonise the external catheter surface. Contamination of the catheter hub and resulting internal surface colonisation can occur at any time following insertion and thus are linked most frequently to CRBSIs resulting from long-term IV catheter use.[10]

In both scenarios of catheter contamination, it is clear that strict adherence to infection control measures is key to avoiding complications resulting from microbial infections. These measures should include thorough handwashing by all healthcare staff at all points of contact with IV devices, cleaning the patient's skin with an appropriate antimicrobial agent prior to inserting the device and educating patients and visitors to ensure that adequate hygiene is observed at all times. In addition, the incidence of infection is far lower with devices that are implanted surgically, such as tunnelled catheters and IV ports, when compared to short-term CVCs.[11]

DIAGNOSIS OF INFECTION

A number of guidelines have been published to address the complications associated with the use of intravenous catheters. These guidelines focus on diagnosis and treatment with a specific emphasis on complications linked to bacteraemia and fungaemia. For healthcare professionals it is important to adhere to the specific guidelines and recommendations recognised by their individual hospitals and clinics. Regular visual assessment of the catheter exit site and surrounding tissue for signs of local infection (e.g., phlebitis or cellulitis), as described previously is generally required for patients with IV catheters.

The Infectious Diseases Society of America (IDSA), the American College of Critical Care Medicine and the Society for Healthcare Epidemiology of America, as well as the Centers for Disease Control and Prevention (CDC) and the Healthcare Infection Control Practices Advisory Committee (HICPAC), have published and continue to update guidelines for the management of IV-related infections[12–14] These documents provide recommendations for the diagnosis, management and prevention of infections linked to the various types of IV devices.

Clinical observation of symptoms provides an indicator for the presence of device-related infection, but because of their lack of specificity and sensitivity these symptoms are unreliable for accurate diagnosis. There are a number of specific protocols and recommendations for each type of catheter; however, diagnosis ultimately depends on the effective isolation and culture of concomitant organisms from both the intravenous device and the bloodstream of the patient. Isolation is carried out either by removing the catheter and carrying out quantitative or semi-quantitative culture methods or, in cases where removal of the device is to be avoided (e.g., in patients with limited vascular access), by blood culturing methods.

One of the most frequently used semi-quantitative device culture methods is the roll-plate method.[15] This involves rolling a section of the removed catheter across the surface of an agar plate and counting the colonies which grow following an overnight incubation. Quantitative analysis involves flushing the catheter with growth media and vortexing or sonicating catheter sections, followed by serial dilutions and growth on agar. A study comparing the sensitivity of sonication, flushing and roll-plate methodologies showed that sonication had the best sensitivity (80%) followed by roll-plating (60%) and then flush culture (40–50%).[16]

The type of culture method and isolation technique should be based on the type and duration of catheter use. For example, for long-term catheters in which intraluminal colonisation resulting from contamination at the catheter hub is the most common cause of CRBSI, a method which most effectively cultures organisms from the internal surface is appropriate. In this case roll-plate isolation would be less appropriate than a quantitative isolation by flushing or sonication.[12] Following semi-quantitative culture from a catheter segment, a yield of ≥ 15 cfu or $\geq 10^2$ cfu from a quantitative culture, together with signs of local or systemic infection, indicates a catheter-related infection.[12]

For concomitant determination of CRBSI, isolation of the same organism from the catheter and from a percutaneous blood sample obtained from a position away from the device site, with no other identifiable route of infection, provides a reliable diagnosis. In cases where the catheter cannot be removed (i.e., because of a lack of alternative catheter placement sites in patients who have inadequate vascular access), quantitative blood culture methods have been developed to compare peripheral blood cultures with blood cultures obtained through the catheter hub. Generally, a catheter sample is expected to yield a 5- to 10-fold greater cfu value than the peripheral blood sample.[17] A CVC blood sample generating a value of ≥ 100 cfu/ml can also be used to diagnose an IV catheter infection without a comparative peripheral sample.[18]

The micro-organisms most commonly associated with IV catheter colonisation and infection are coagulase-negative *Staphylococci*, *Staphylococcus aureus*, *Enterococcus*, Gram-negative bacilli, *Candida albicans* and some species of mycobacteria. In most patients without additional clinical complications, removal of the colonised catheter together with appropriate antibiotic treatment is sufficient to manage an infection.

Coagulase-negative *Staphylococci* (CNS) are responsible for the majority of catheter-related infections.[19,20] Clinically these infections are associated with fever and inflammation localised to the site of the catheter. Isolation of *S. aureus* from a catheter raises concerns about the increased risk of endocarditis,[21] and thus extended treatment regimes and follow-up blood cultures and echocardiography may need to be performed.

Gram-negative bacilli are most commonly associated with contaminated infusate and are linked to CRBSIs in immuno-suppressed patients with tunnelled devices.[3] Similar to bacillus infections, incidences of *Candida* colonisation of an IV catheter are most often seen in patients with reduced immune function. As with bacteremia, removal of the catheter together with treatment with an antifungal is warranted.[22]

Of the 50 species of rapidly growing mycobacteria, around 20 are opportunistic pathogens in humans, including *Mycobacterium abscessus*, *Mycobacterium mucogenicum* and *Mycobacterium chelonae*. They can be found throughout the natural environment in soil, dust and water. Diagnosis of CRBSIs caused by mycobacteria is challenging, as cultivation and identification of the organisms within 72 hours is difficult. Furthermore, catheter segments that produce a negative result at 48 to 72 hours are often discarded. Thus fulfilling the ISDA criteria for CRBSIs is not always achievable when the infective agent is of this type. In catheters, CRBSIs involving mycobacterial infection are not considered to be as virulent as infection by Gram-positive or Gram-negative organisms; however, they are still a risk, particularly to immuno-compromised patients.[23]

EVIDENCE AND IMPACT OF BIOFILM FORMATION

It has been demonstrated that a major factor in the pathogenesis of microbial surface colonisation is the formation of biofilms, which provide protection from host immune defences and reduce the efficacy of antimicrobial treatments. Reports suggest that bacterial colonisation can occur within the first 24 hours following catheter insertion,[24] with biofilm formation taking place within three days.[25] As described in previous chapters, the key steps involved in biofilm formation include attachment, growth and dispersal; thus by considering the characteristics and use of IV catheter devices alongside each of these stages, we can understand the role that biofilms play in the prevalence of catheter-related infections. The use of indwelling devices (e.g., IV catheters) provides both a route of infection via acute injury to a patient's protective skin defences and a device surface on which micro-organisms can adhere and colonise.

An important consideration with respect to biofilm formation in catheters is the nature and composition of the fluid administered. Often these fluids provide limited nutrients to sustain and promote microbial growth and, in the case of drug administration, antibiotic therapies may select for the survival of some organisms over others. It has been shown that intravenous fluids do not support the growth of *S. aureus* and *S. epidermidis*, whereas they do support Gram-negative organisms such as *P. aeruginosa* and *Enterobacter*.[5]

Stage 1: Attachment

In vitro studies of attachment of micro-organisms to materials which are frequently found in IV catheters have revealed that certain biomaterials support adherence and colonisation more than others. Materials, such as silicone, latex and PVC, have a greater propensity for microbial colonisation than materials like stainless steel and titanium.[26]

The physical characteristics of the biomaterial, which determine the initial reversible attachment properties, include surface topography, charge and material hydrophobicity. Topographical studies of biomaterials have shown that rough and irregular surfaces tend to support attachment more than do smoother surfaces.[26]

The majority of bacterial cells are negatively charged when in an aqueous solution; thus, materials which are negatively charged have a repulsive effect and can reduced microbial attachment. This has been demonstrated in a study using a polyurethane surface grafted with acrylic acid, and it was shown that this negatively charged surface reduces the adherence of *S. epidermidis*.[27] Microbial cells frequently have hydrophobic cell surfaces and are thus attracted to the hydrophobic surfaces of many of the biomaterials currently used in IV catheter manufacture.[28]

Biomaterial characteristics not only have a direct affect on microbial attachment but also have an indirect affect through their impact on conditioning film formation. Indwelling devices (e.g., IV catheters) inevitably come into contact with blood and other biological secretions, which form a conditioning film on the material's surface. The nature and composition of the conditioning film is linked both to the composition of the biological fluid in contact with the device and to the biomaterial surface characteristics such as charge and hydrophobicity. The protein- and glycoprotein-rich films change the original surface properties of the catheter material.

A conditioning film formed by contact with biological fluids can contain a broad range of proteins, including fibronectin, laminin, fibrin, albumin, elastin, collagen and immunoglobulins, some of which provide receptor sites for the attachment of micro-organisms.[29] Fibronectin, fibrinogen and fibrin, in particular, are the most important for microbial adherence, and studies have shown that they enhance the adherence of organisms including Gram-positive *Staphylococci* and *C. albicans*.[30,31] Following initial reversible adherence to a material surface that has suitable physical characteristics, receptor–ligand interactions between micro-organisms and the surface-adsorbed conditioning film result in irreversible adherence.

Stage 2: Growth

Once adhered to the IV catheter surface, micro-organisms proliferate and form micro-colonies. During this growth phase, bacteria and other colonising microbes begin to deposit and encase themselves in an extracellular matrix, also referred to as an extracellular polymeric substance (EPS) or slime. Typically this substance is composed of proteins, nucleic acids and polysaccharides.

Mycobacteria are unique in that they do not produce exopolysaccharides but instead secrete primarily glycopeptidolipids and fatty acids.[32] Growth of pioneer organisms can subsequently develop suitable environments for secondary species, giving rise to multi-species biofilms. As well as providing protection for encased microbes from host immune defences and antimicrobial treatments, biofilms containing Gram-negative bacteria can be a source of endotoxins which trigger host immunity and lead to a pyogenic reaction.

Stage 3: Dispersion

In nature, the detachment and subsequent dispersal of cells from a biofilm has evolved to allow continued survival and propogation of a bacterial species. When related to colonisation of IV catheters, this stage results in the progression of a localised infection at the site of a catheter to a systemic infection. As micro-organisms detach from the biofilm as either individual planktonic cells or cell clusters, they are released into a patient's bloodstream, thus leading to a CRBSI. Microbial cells are dispersed from the surface of the biofilm during active growth as a consequence of a change in nutrient levels or quorum sensing, or because of shearing as a result of hydrodynamic effects.[33]

PREVENTION AND TREATMENT OF INFECTIONS

The clinical reliance on IV devices continues to rise, and with this will come increased risk of catheter-related infections and associated mortality. After understanding the key causes of catheter-related infections, it can be seen that the implementation of robust infection control measures plays a central role in their prevention. In addition, there are clinical strategies in place to prevent CRBSIs, including the use of catheter lock solutions and devices impregnated with antimicrobial agents. The specific strategies relating to the prevention or elimination of biofilms, while predominantly experimental, nonetheless provide a basis for the development of future treatments.

Antimicrobial Lock Therapy

Antimicrobial lock therapy (ALT) has been used as a means to prevent colonisation and to control CRBSIs. ALT may be considered when a case of CRBSI represents a low to moderate risk of a poor outcome[34] and thus can be used as an approach to preserve long-term tunnelled catheters.[35] This strategy involves the infusion of a solution in the catheter following use. The solution is allowed to settle (lock) in the lumen while the catheter is not in use. The solutions used contain antibiotic drugs or other antimicrobial agents together with an anti-coagulant such as heparin.

Using antimicrobials at a concentration that is significantly higher (100–1000-fold) than the minimum inhibitory concentration (MIC) or usual target systemic concentration aims to eradicate microbial biofilms on the internal surface. Concerns have arisen surrounding the routine use of these solutions and the development of resistant micro-organisms.[36] Several studies have demonstrated the effectiveness of vancomycin as an ALT; however, evidence of the emergence of vancomycin resistance in clinically relevant bacterial strains means that this approach is not recommended.[37–39]

The addition of chelating agents, such as ethylene diaminetetraacetic acid (EDTA), to a catheter lock solution has demonstrated clinical efficacy against biofilms. As well as exhibiting some antimicrobial properties, these destabilise the biofilm by chelating metal ions such as Ca^{2+}, Mg^{2+} and Fe^{2+}. Disodium EDTA in combination with minocycline has been shown to be effective against biofilms both *in vitro* using explanted catheter tips[40] and *in vivo* in the treatment of catheter-related bacteremia in hemodialysis.[41] In an *in vitro* study conducted by Percival et al.,[42] it was demonstrated that 40 mg/ml of tetrasodium EDTA alone can significantly reduce or potentially eradicate CVC-associated biofilms of clinically relevant organisms.

An alternative chelating agent, sodium citrate, has been shown to exhibit antimicrobial and anti-biofilm activity *in vitro* against several strains of *S. aureus* and CNS.[43] A combination of 4% trisodium citrate and 30% ethanol has been demonstrated to prevent biofilm formation by several organisms associated with CRBSIs, including *S. aureus* and *S. epidermidis*.[44]

The use of ethanol alone in 20 to 80% concentrations has been shown to be an effective treatment for CRBSIs in a number of case studies. For example, reduction in bloodstream infection in immuno-suppressed patients with infected tunnelled catheters, when a 70% ethanol ALT lock is used, has been demonstrated.[45] *In vitro* studies have supported this outcome and demonstrated that a 24-hour exposure to 20% ethanol is successful in eradicating biofilms of *S. epidermidis*.[46] In a study using a meticillin-resistant strain of *S. aureus*, it was shown that the combination of 25% ethanol, minocycline (2 mg/ml) and EDTA (30 mg/ml) was successful at eliminating *in vitro* biofilms.[47]

Other anti-biofilm agents may offer the potential to be used as ALT solutions. Oxidising biocides, such as chlorine, surfactants or enzymes, can disrupt biofilms and lead to cell dispersal.[48] In an *in vitro* study, an unsaturated fatty acid produced by *P. aeruginosa*, known as *cis*-2-decenoic acid (CDA), was shown to induce dispersal of biofilms including *S. aureus* and *C. albicans*.[49] Nitric oxide has also been shown to have a dispersal effect on biofilms as well as being able to prevent colonisation when used as a pre-treatment for biomaterial surfaces.[50] The combination of dispersing agents with other antimicrobial agents to kill the released cells, preventing further colonisation and bloodstream infection, may thus be potential treatments for CRBSI.

Antimicrobial Catheters

Catheter devices that incorporate antimicrobial agents have been developed as a means to prevent the colonisation of micro-organisms and thus reduce the risk of CRBSI. Devices impregnated with chlorhexidine and silver sulfadiazine have demonstrated some positive effects when used in clinical studies. Early devices, which where coated on the external surface, had some success in reducing the incidence of CRBSI linked to catheter use for <7 days; thus, they may be indicated in clinical settings where the risk of short-term catheter complications is high.[51]

A meta-analysis of clinical studies using newer devices, which have these same agents impregnated on both the external surface and the catheter lumen, demonstrated that these catheters can halve the risk of developing CRBSIs.[52] However, in a comparative study of antimicrobial IV devices it was shown that catheters impregnated with minocycline and rifampin are less likely to be colonised by micro-organisms than are the chlorhexidine/sulfadiazine devices and are associated with a lower CRBSI incidence.[53]

To ease concerns about the development of antimicrobial resistance, IV catheters incorporating the anti-metabolite drug 5-fluorourracil (5-FU) have been developed. The concentrations used are well below those used for cancer therapy but have been shown to inhibit the growth of several clinically relevant strains of micro-organisms, including *S. epidermidis* and *C. albicans*. Furthermore, in a multi-centre controlled trial these devices were demonstrated to have a lower risk of colonisation and a reduced incidence of CRBSI when compared to catheters coated with chlorhexidine and silver sulfadiazine.[54]

Alternative Strategies

There are a number of novel strategies which use alternative non-chemical means to treat and prevent IV catheter colonisation, biofilm formation and subsequent catheter-related infections. Ultrasonic energy has a number of effects on biofilms, including abrogation of microbial adherence to surfaces; disruption of nutrient, oxygen and signalling pathways within the biofilm; and mechanical damage.[55] *In vitro* ultrasonication has been shown to significantly increase the transport of antibiotics within biofilms of *P. aeruginosa* and *E.coli*, resulting in increased cell death.[56]

It also has been hypothesised that an additional effect of ultrasonic treatment is the improvement of oxygen and nutrient transport to cells within a biofilm not conducive to the sessile state.[57] Fine-tuning of the acoustic frequencies prevents biofilm formation on various types of indwelling devices, including catheters.[55] The overall affect of acoustic energy on biofilms without the presence of antibiotics was found to depend predominantly on the intensity of the energy used rather than on the specific frequency.[58]

Electromagnetic radiation in the ultraviolet (UV) range (328–210 nm) has also been suggested as a means to eradicate biofilms.[59] Optimal bacteriocidal effects occur at 280 to 250 nm, causing accumulation of photoproducts and irreversible DNA damage leading to cell death.[60] The use of optical fibres embedded in the biomaterial of the catheter allows UV light to be transmitted along the catheter while *in situ* and can be used to both prevent colonisation and treat established biofilms. Considerations, such as the UV stability of the device material, the appropriate source of UV light and the consequences of UV light exposure on the cells and tissue of the intravascular space, present significant obstacles. However, overcoming these may provide a potential therapeutic strategy to prevent CRBSI.

CONCLUSION

Microbial contamination and subsequent colonisation and biofilm formation present a significant risk to patients who require monitoring and administration of treatments via intravascular catheters. The development of guidelines, which provide protocols for the diagnosis and management of IV catheter infections, is an important resource for the clinical community in recognising, detecting and providing the appropriate treatment as soon as possible. The impact of biofilms on these devices is clearly an area that has not yet been fully addressed. Continued research into methods to prevent biofilm formation, together with effective ways to remove and treat biofilms on catheter surfaces, should ultimately aim to reduce the incidence and morbidity associated with catheter-related infections.

REFERENCES

1. Hockenhull JC, Dwan K, Boland A, Smith G, Bagust A, Dundar Y, et al. The clinical effectiveness and cost-effectiveness of central venous catheters treated with anti-infective agents in preventing bloodstream infections: a systematic review and economic evaluation. *Health Technol Assess* 2008;**12**(iii-iv, xi-xii):1–154 (Winchester, England).
2. Higginson R, Parry A. Phlebitis: treatment, care and prevention. *Nurs Times* 2011;**107**:18–21.
3. Maki DG, Mermel LA. Infections due to infusion therapy. In: Bennett JV, editor. *Hospital infections.* Lipponcott/Raven; 1998. p. 689–724.
4. Leonidou L, Gogos CA. Catheter-related bloodstream infections: catheter management according to pathogen. *Int J Antimicrob Agents* 2010;**36**:S26–32.
5. Donlan RM. Biofilm formation: a clinically relevant microbiological process. *Clinical Infectious Diseases—an official publication of the Infectious Diseases Society of America* 2001;**33**:1387–92.
6. Bouza E, San Juan R, Munoz P, Pascau J, Voss A, Desco M.—Cooperative Group of the European Study Group on Nosocomial I. *A European perspective on intravascular catheter-related infections: report on the microbiology workload, aetiology and antimicrobial susceptibility (ESGNI-005 Study). Clinical Microbiology and Infection*—the official publication of the European Society of Clinical Microbiology and Infectious Diseases 2004;**10**:838–42.
7. Edgeworth J. Intravascular catheter infections. *J Hosp Infect* 2009;**73**:323–30.
8. Raad I, Bodey G. Infectious complications of indwelling vascular catheters. *Clin Infect Dis* 1992;**15**:197–210.
9. Maki DG. Infections caused by intravascular devices used for infusion therapy: pathogenesis, prevention, and management. In: Bisno AL, Waldvogel FA, editors. *Infections associated with indwelling medical devices*, 2nd ed. ASM Press; 1994. p. 155–212.
10. Linares J, Sitgesserra A, Garau J, Perez J, Martin R. Pathogenesis of catheter sepsis-a prospective-study with quantitative and semiquantitative cultures of catheter hub and segments. *J Clin Microbiol* 1985;**21**:357–60.
11. Maki DG, Kluger DM, Crnich CJ. The risk of bloodstream infection in adults with different intravascular devices: a systematic review of 200 published prospective studies. *Mayo Clin Proceed* 2006;**81**:1159–71.
12. Mermel LA, Farr BM, Sherertz RJ, Raad II, O'Grady N, Harris JS, et al.—Infectious Diseases Society of America, American College of Critical Care Medicine, Society for Healthcare Epidemiology of America. *Guidelines for the management of intravascular catheter-related infections. Clinical Infectious Diseases*—an official publication of the Infectious Diseases Society of America 2001;**32**:1249–72.
13. Mermel LA, Allon M, Bouza E, Craven DE, Flynn P, O'Grady NP, et al. Clinical practice guidelines for the diagnosis and management of intravascular catheter-related infection: update by the

Infectious Diseases Society of America. *Clinical Infectious Diseases*—an official publication of the Infectious Diseases Society of America 2009;**49**:1–45.

14. O'Grady NPAM, Burns LA, Dellinger EP, Garland J, Heard SO, Lipsett PA, et al.—Healthcare Infection Control Practices Advisory Committee (HICPAC). *Guidelines for the Prevention of Intravascular Catheter-Related Infections*. Centers for Disease Control and Prevention (CDC), Atlanta; 2011.
15. Maki D, Weise C, Sarafin H. Semiquantitative culture method for identifying intravenous-catheter-related infection. *N Engl J Med* 1977;**296**:1305–9.
16. Sherertz R, Heard S, Raad I. Diagnosis of triple-lumen catheter infection: comparison of roll plate, sonication, and flushing methodologies. *J Clin Microbiol* 1997;**35**:641–6.
17. Fan S, Teohchan C, Lau K. Evaluation of central venous catheter sepsis by differential quantitative blood culture. *Eur J Clin Microbiol Infect Dis* 1989;**8**:142–4.
18. Capdevila J, Planes A, Palomar M, Gasser I, Almirante B, Pahissa A, et al. Value of differential quantitative blood cultures in the diagnosis of catheter-related sepsis. *Eur J Clin Microbiol Infect Dis* 1992;**11**:403–7.
19. Sattler F, Foderaro J, Aber R. Staphylococcus-epidermidis bacteremia associated with vascular catheters—An important cause of febrile morbidity in hospitalized-patients. *Infect Cont Hosp Epidemiol* 1984;**5**:279–83.
20. Huebner J, Goldmann DA. Coagulase-negative staphylococci: role as pathogens. *Annu Rev Med* 1999; **50**:223–36.
21. Rosen A, Fowler V, Corey G, Downs S, Biddle A, Li J, et al. Cost-effectiveness of transesophageal echocardiography to determine the duration of therapy for intravascular catheter-associated *Staphylococcus aureus* bacteremia. *Ann Intern Med* 1999;**130**:810.
22. Rex J, Bennett J, Sugar A, Pappas P, Serody J, Edwards J, et al. Intravascular catheter exchange and duration of candidemia. *Clin Infect Dis* 1995;**21**:994–6.
23. El Helou G, Viola GM, Hachem R, Han XY, Raad II. Rapidly growing mycobacterial bloodstream infections. *Lancet Infect Dis* 2013;**13**:166–74.
24. Osma S, Kahveci SF, Kaya FN, Akalin H, Ozakin C, Yilmaz E, et al. Efficacy of antiseptic-impregnated catheters on catheter colonization and catheter-related bloodstream infections in patients in an intensive care unit. *J Hosp Infect* 2006;**62**:156–62.
25. Anaissie E, Samonis G, Kontoyiannis D, Costerton J, Sabharwal U, Bodey G, et al. Role of catheter colonization and infrequent hematogenous seeding in catheter-related infections. *Eur J Clin Microbiol Infect Dis* 1995;**14**:134–7.
26. Darouiche RO. Device-associated infections: a macroproblem that starts with microadherence. *Clinical Infectious Diseases*—an official publication of the Infectious Diseases Society of America 2001; **33**:1567–72.
27. Kohnen W, Jansen B. Polymer materials for the prevention of catheter-related infections. Zentralblatt fur Bakteriologie. *Internat J Med Microbiol* 1995;**283**:175–86.
28. Schierholz JM, Beuth J. Implant infections: A haven for opportunistic bacteria. *J Hosp Infect* 2001; **49**:87–93.
29. Dickinson GM, Bisno AL. Infections associated with indwelling devices: concepts of pathogenesis; infections associated with intravascular devices. *Antimicrob Agents Chemother* 1989;**33**:597–601.
30. Pascual A. Pathogenesis of catheter-related infections: lessons for new designs. *Clinical Microbiology and Infection*—the official publication of the European Society of Clinical Microbiology and Infectious Diseases. 2002:**8**:256–264.
31. Murga R, Miller JM, Donlan RM. Biofilm formation by gram-negative bacteria on central venous catheter connectors: effect of conditioning films in a laboratory model. *J Clin Microbiol* 2001; **39**:2294–7.
32. Zambrano MM, Kolter R. Mycobacterial biofilms: a greasy way to hold it together. *Cell* 2005; **123**:762–4.
33. Donlan R. Biofilms: Microbial life on surfaces. *Emerg Infect Dis* 2002;**8**:881–90.
34. Raad I, Hanna H. Intravascular catheter-related infections—New horizons and recent advances. *Arch Intern Med* 2002;**162**:871–8.

35. Krishnasami Z, Carlton D, Bimbo L, Taylor ME, Balkovetz DF, Barker J, et al. Management of hemodialysis catheter-related bacteremia with an adjunctive antibiotic lock solution. *Kidney Int* 2002; **61**:1136–42.
36. Mermel L. Prevention of intravascular catheter-related infections. *Ann Intern Med* 2000;**132**:391–402.
37. Schwalbe R, Stapleton J, Gilligan P. Emergence of vancomycin resistance in coagulase-negative staphylococci. *N Engl J Med* 1987;**316**:927–31.
38. Kaplan A, Gilligan P, Facklam R. Recovery of resistant enterococci during vancomycin prophylaxis. *J Clin Microbiol* 1988;**26**:1216–8.
39. Sieradzki K, Leski T, Dick J, Borio L, Tomasz A. Evolution of a vancomycin-intermediate Staphylococcus aureus strain *in vivo*: multiple changes in the antibiotic resistance phenotypes of a single lineage of methicillin-resistant *S. aureus* under the impact of antibiotics administered for chemotherapy. *J Clin Microbiol* 2003;**41**:1687–93.
40. Raad I, Buzaid A, Rhyne J, Hachem R, Darouiche R, Safar H, et al. Minocycline and ethylenediaminetetraacetate for the prevention of recurrent vascular catheter infections. *Clinical Infectious Diseases*—an official publication of the Infectious Diseases Society of America 1997;**25**:149–51.
41. Campos RP, do Nascimento MM, Chula DC, Riella MC. Minocycline-EDTA lock solution prevents catheter-related bacteremia in hemodialysis. *J Am Soc Nephrol* 2011;**22**:1939–45.
42. Percival S, Kite P, Eastwood K, Murga R, Carr J, Arduino M, et al. Tetrasodium EDTA as a novel central venous catheter lock solution against biofilm. *Infect Cont Hosp Epidemiol* 2005;**26**:515–9.
43. Shanks RMQ, Sargent JL, Martinez RM, Graber ML, O'Toole GA. Catheter lock solutions influence staphylococcal biofilm formation on abiotic surfaces. *Nephrol Dialy Transplant* 2006;**21**:2247–55.
44. Takla TA, Zelenitsky SA, Vercaigne LM. Effectiveness of a 30% ethanol/4% trisodium citrate locking solution in preventing biofilm formation by organisms causing haemodialysis catheter-related infections. *J Antimicrob Chemother* 2008;**62**:1024–6.
45. Sanders J, Pithie A, Ganly P, Surgenor L, Wilson R, Merriman E, et al. A prospective double-blind randomized trial comparing intraluminal ethanol with heparinized saline for the prevention of catheter-associated bloodstream infection in immunosuppressed haematology patients. *J Antimicrob Chemother* 2008;**62**:809–15.
46. Qu Y, Istivan TS, Daley AJ, Rouch DA, Deighton MA. Comparison of various antimicrobial agents as catheter lock solutions: preference for ethanol in eradication of coagulase-negative staphylococcal biofilms. *J Med Microbiol* 2009;**58**:442–50.
47. Raad I, Hanna H, Dvorak T, Chaiban G, Hachem R. Optimal antimicrobial catheter lock solution, using different combinations of minocycline, EDTA, and 25-percent ethanol, rapidly eradicates organisms embedded in biofilm. *Antimicrob Agents Chemother* 2007;**51**:78–83.
48. Chen X, Stewart P. Biofilm removal caused by chemical treatments. *Water Res* 2000;**34**:4229–33.
49. Davies DG, Marques CNH. A fatty acid messenger is responsible for inducing dispersion in microbial biofilms. *J Bacteriol* 2009;**191**:1393–403.
50. Nablo B, Prichard H, Butler R, Klitzman B, Schoenfisch M. Inhibition of implant-associated infections via nitric oxide. *Biomaterials* 2005;**26**:6984–90.
51. Walder B, Pittet D, Tramer MR. Prevention of bloodstream infections with central venous catheters treated with anti-infective agents depends on catheter type and insertion time: evidence from a meta-analysis. *Infection Control and Hospital Epidemiology*—the official Journal of the Society of Hospital Epidemiologists of America 2002;**23**:748–56.
52. Hockenhull JC, Dwan KM, Smith GW, Gamble CL, Boland A, Walley TJ, et al. The clinical effectiveness of central venous catheters treated with anti-infective agents in preventing catheter-related bloodstream infections: a systematic review. *Crit Care Med* 2009;**37**:702–12.
53. Darouiche RO, Raad II, Heard SO, Thornby JI, Wenker OC, Gabrielli A, et al. A comparison of two antimicrobial-impregnated central venous catheters. Catheter Study Group. *N Engl J Med* 1999;**340**:1–8.
54. Walz JM, Avelar RL, Longtine KJ, Carter KL, Mermel LA, Heard SO, Group FUCS Anti-infective external coating of central venous catheters: a randomized, noninferiority trial comparing 5-fluorouracil with chlorhexidine/silver sulfadiazine in preventing catheter colonization. *Crit Care Med* 2010;**38**:2095–102.

55. Dror N, Mandel M, Hazan Z, Lavie G. Advances in microbial biofilm prevention on indwelling medical devices with emphasis on usage of acoustic energy. *Sensors (Basel)* 2009;**9**:2538–54.
56. Carmen JC, Nelson JL, Beckstead BL, Runyan CM, Robison RA, Schaalje GB, Pitt WG. Ultrasonic-enhanced gentamicin transport through colony biofilms of *Pseudomonas aeruginosa* and *Escherichia coli*. *Journal of Infection and Chemotherapy*—the official journal of the Japan Society of Chemotherapy 2004;**10**:193–9.
57. Pitt W, Ross S. Ultrasound increases the rate of bacterial cell growth. *Biotechnol Prog* 2003; **19**:1038–44.
58. Pitt WG. Removal of oral biofilm by sonic phenomena. *Am J Dent* 2005;**18**:345–52.
59. Dale BA. Intravascular-catheter-related infections. *Lancet* 1998;**351**:1739.
60. Bridges BA. Survival of bacteria following exposure to UV and ionising radiations. In: Gray TGR, Postgate PJ, editors. *The survival of vegetative microbes*. Cambridge University Press; 1976. p. 183–208.

CHAPTER TWELVE

Ventilator-Associated Pneumonia, Endotracheal Tubes and Biofilms

Steven L. Percival and David W. Williams

INTRODUCTION

Ventilator-associated pneumonia (VAP) is defined as a pneumonia that occurs in patients 48 to 72 hours after endotracheal (ET) intubation and mechanical ventilation (MV). Importantly, it is second only to urinary tract sepsis as a hospital-acquired infection in patients receiving intensive care.[1–5] A key factor in the acquisition of VAP is the presence of an endotracheal tube, an essential medical device for the maintenance of an airway and the promotion of gaseous exchange during MV. However, an ET tube also increases a patient's risk of VAP 6- to more than 20-fold.[4–7]

Because VAP prevalence rates are often reported across different studies, a VAP diagnosis is extremely problematic and depends on many factors which will be discussed throughout this chapter. Incidence rates of 4 to 50 cases per 100 patients[4–6] and mortality rates as high as 76% have been reported with VAP.[4,5,8] Longer hospital stays associated with VAP are also significant and can lead to increased hospital costs of approximately $40,000 to $51,157.[8–10] Unsurprisingly, VAP prevention is very important, not only for the intubated patient, but also for the healthcare provider.

The ET tube provides a conduit from the oral cavity to the lower airway which circumvents the normally protective host defences such as the cough reflex and mucociliary action. These mechanisms ordinarily reduce the entry of micro-organisms into the lungs, thereby reducing the risk of infection.[11] The ET tube is retained in the trachea by means of an inflated cuff, above which microbial-containing subglottic fluids can accumulate. Often, because of the type of biomaterial used in the cuff's development, micro-channels form, allowing the sublgottic fluids below the cuff to leak into the lower airway. This fluid is rich in potentially infective micro-organisms that not only may directly enter the lungs, but also may be drawn into the ET tube's inner lumen.

Inside the ET tube, both endogenously and exogenously derived micro-organisms colonise and grow as a biofilm,[3,12–14] in which they are protected from host defence mechanisms. As the biofilm develops, these micro-organisms may detach and translocate

Biofilms in Infection Prevention and Control
Doi: http://dx.doi.org/10.1016/B978-0-12-397043-5.00012-8

to the lungs to induce infection. The biofilm provides a continuous reservoir for infectious agents in MV patients.[15,16] The ET tube has been reported to become rapidly colonised (within hours) by micro-organisms once a patient is intubated.[4,17]

This aim of this chapter is to provide an overview of VAP, the micro-organisms involved, the role of biofilms play and the methods employed to both prevent and manage it.

DIAGNOSIS AND COMPLICATIONS

At present, a VAP diagnosis is limited to the Centers for Disease Control and Prevention (CDC) definition and available diagnostics employed by specific hospitals. However, its aetiology is often polymicrobial, and because there is no way to differentiate between colonisation and infection, both diagnosis and risk determination are significantly hindered. For this reason, there is no 'gold standard' of diagnosis; this consequently has led to unnecessary use and administration of antibiotics for both prevention and management.

Presently, to diagnose a VAP episode, an autopsy is performed or a direct lung tissue culture is taken; however, in many situations, particularly those involving paediatric patients, this is not possible. Therefore, for neonates a culture has to be obtained from the ET tube which can also help to predict infant late-onset sepsis.[18] Slagle and colleagues[18] found that in the first week of life, ET tube samples are culture-negative, indicating a 'sterile' tube. However, over time the ET tube becomes colonised with Gram-positive and then, additionally, Gram-negative bacteria. Further research in this same study compared the patterns of culture between infants who developed late-onset sepsis and those who did not. It was found in the study that the incidence of late-onset sepsis was 23% and the mean time to infection was 41 days. It was evident, however, that while 54% of the micro-organisms the study identified by blood culture were not present in ET tube samples, in 19% of cases there was a match between the bloodstream isolate and isolates from the ET tube.

Perkins and colleagues[19] used quantitative PCR (qPCR) and gene surveys targeting the 16S rRNA genes of bacteria to quantify and identify the biofilm community present in extubated ET tubes. Eight of the ET tubes, obtained from patients who had been incubated between 12 hours and 23 days in surgical and medical intensive care, were sampled for evidence of biofilms and the microbial community. Quantitative PCR data showed that the ET tubes were colonised in 24 hours. A large variation among patients was evident in respect to bacterial load, but positive correlations between bacterial load and intubation period were not possible. The researchers found more than 70% of sequences analysed to be associated with the genera of the normal oral microflora and 6% to be associated with the gastrointestinal flora. The most common genera identified in this study were *Streptococcus* followed by *Prevotella* and then *Neisseria*.

The Perkins study also showed that as intubation periods increase, so does the opportunity for pathogenic micro-organisms to proliferate in the ET tube biofilm. For ET tubes in place for 23 days, 95% of the sequences were found to be from *Pseudomonas aeruginosa*. More recently, Cairns and colleagues[20] analysed 24 ET tube biofilms from 20 patients, using a combination of traditional micro-biological culture, species-specific PCR and denaturing gradient gel electrophoresis (DGGE) profiling. The results highlighted the high number of micro-organisms present (>10^8 cfu per cm^2 section of tube), the polymicrobial nature of the biofilms (up to 22 distinct DGGE bands detected on occasion, with a mean of 6 bands per tube) and the presence of micro-organisms normally associated with the oral cavity (e.g., *Streptococcus mutans*, *Porphyromonas gingivalis, Candida albicans*).

Epidemiological surveillance is vital in monitoring complications in MV patients. A recent study by Skrupky and colleagues[21] found complications in 1.2 to 8.5 cases per 1000 ventilator-days (0.6–4%). Unsurprisingly, high on the list of complications was VAP, which, as previously mentioned, is highly prevalent in MV patients and is associated with increased mortality and longer intensive care stays. Its associated mortality rates are 27 to 76%.[4,5] Causative agents of VAP are wide ranging, with numerous microbial species implicated, including multi-drug-resistant (MDR) opportunistic pathogens such as meticillin-resistant *Staphylococcus aureus* (MRSA), *Acinetobacter* species, *Escherichia coli* and *Pseudomonas aeruginosa*.[22,23]

BIOFILMS AND ENDOTRACHEAL TUBES

One potential source/reservoir of micro-organisms associated with VAP is the ET tube itself. When micro-organisms adhere to the inner lumen of the ET tube, they form a biofilm (Figure 12.1) that then becomes a source of infectious agents in the intubated patient.[24] A polymicrobial biofilm develops very quickly, with studies showing antibiotic-resistant bacteria in ET tube biofilms occurring in less than 24 hours.[13,25] Since dissemination of micro-organsims from the ET tube to the lower airway is of major concern, preventing microbial attachment and utilising effective infection control procedures are vital measures.

Biofilms on ET tubes are composed of complex communities of micro-organisms that are attached to the tube's surface and encased in an extracellular polymer material composed of polysaccharides, proteins and nucleic acids. They are the source of micro-organisms that disseminate into the lungs and lead to pneumonia.[17,3] As mentioned earlier, studies have demonstrated the presence of oral micro-organisms among others in the tube's polymcrobial biofilm.[20] The importance of this finding is that, while oral microbes may act directly as respiratory pathogens, they may also facilitate the introduction of traditional respiratory pathogens into the tube. Indeed, pioneer colonisers of teeth, such as certain *Streptococcus* species (e.g., *S. mutans*), are adept at producing extracellular glucans which act as a 'microbial glue' for recruitment of other micro-organisms into the plaque biofilm. Theoretically, this same situation may also occur in the ET tube.

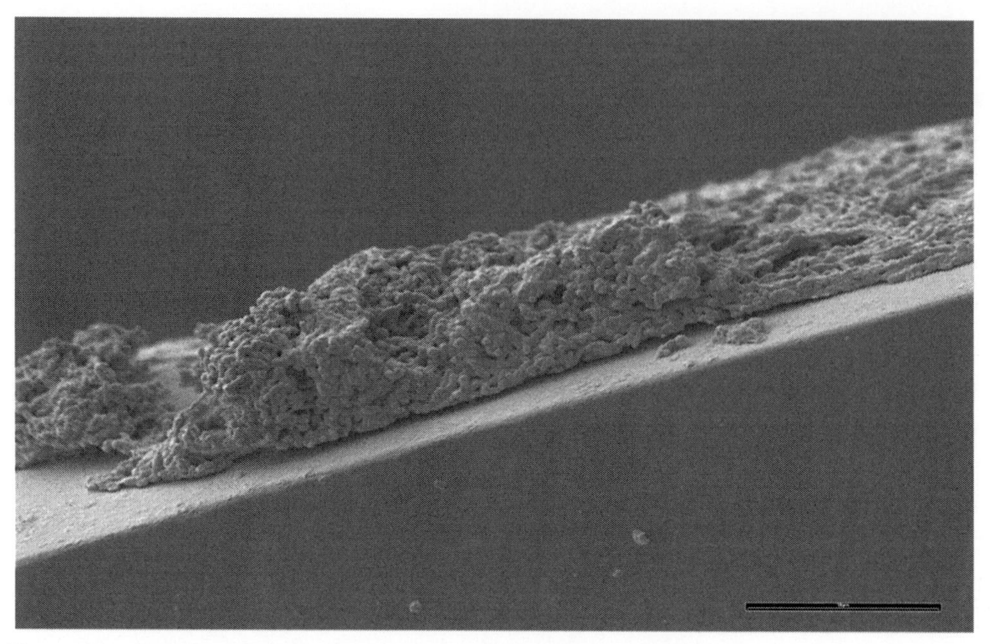

Figure 12.1 ***Scanning electron micrograph showing biofilm formation on the inner lumen of an ET tube.*** *Source: Image courtesy of Kirsty Sands, School of Dentistry, Cardiff, Wales.*

Adair and colleagues[3] have documented that, of 20 patients sampled, 70% become infected following intubation because of the presence of biofilms. Lee and colleagues[26] investigated whether patients intubated for prolonged periods of time are more likely to have bacterial biofilm on their ET tubes. They collected tubes from patients at extubation and searched for evidence of biofilms by scanning with electron microscopy. Of 32 ET tubes sampled from patients who had been intubated for 6 days or longer, a significantly higher percentage of bacterial biofilms was evident compared with individuals intubated for less than 6 days (88.9% versus 57.1%).

Wilson et al.[27] hypothesised a relationship between the stage of biofilm development on the ET tube and the development of pneumonia. In this study, 32 ET tubes were analysed for biofilms and staged in the development of pneumonia, duration of intubation and comorbidities; microbiological information was recorded. The presence of pneumonia was indicated by fever, white blood cell count (WBC) >12K or <4K, infiltrate on chest x-rays and purulent sputum with positive lower airway culture (bronchoalveolar lavage or brush). Also, the mean intensive care stay was 13 days and the mean length of intubation was 7.4 days. Half of the patients studied developed pneumonia while intubated; thus, it was concluded that a relationship exists between incidence of pneumonia and increasing biofilms. However, no relationship was found linking duration of intubation, patient age or hospital stay to biofilm stage.

Interestingly, it has been found that with 70% of patients the VAP micro-organisms identified on the ET tube are identical to those found in the lung.[3]

As discussed, in general the ET tube plays a role in introducing micro-organisms and therefore infection of the lungs. Because of this, much research is being undertaken in the prevention of this route of infection in intubated patients.

VAP AND BIOFILM CONTROL

A number of strategies have been employed to prevent VAP; these include care bundles encompassing universally accepted components such as elevation of beds, non-invasive ventilation when possible, selective decontamination for patients who are MV for longer than 48 hours, sedation vacation, daily weaning assessment and adequate and regular oral care.[28]

As discussed previously, the ET tube is a major concern and risk factor for VAP because its surface provides both a site for biofilm development and a source of infectious micro-organisms.[29,3,30] Correlations between the micro-organisms found in the lower respiratory tract and the presence of an ET tube are well documented.[3] Sottile et al.[17] found that micro-organisms in the tube can colonise in a few hours, with the formation of a biofilm occurring quickly afterward. The risk of a biofilm forming on an ET tube increases significantly for patients with other underlying conditions (e.g., chronic pulmonary disease, blood transfusion, acute respiratory distress and neurologic disorders in particular) or those who are severely immuno-compromised or on chemotherapy (e.g., antibiotics).

Efforts have been made to inhibit microbial contamination of ET tubes through biomaterial modification. Changes in ET tube surfaces at the nano-scale level have also been attempted.[31–34] Machado et al.[33] reported that a 1.5 log reduction in the total number of *S. aureus* could be achieved on a nano-modified ET tube compared with conventional ET tubes after 24 hours of airflow. As a consequence, microbial colonisation in this situation could be reduced without antimicrobial treatment. Overall, this study showed that chemical etching can create nano-rough surface features on PVC that helps to suppress *S. aureus* growth.

Other strategies to reduce biofilm formation in the ET's inner lumen include the Venner™ PneuX P.Y™-VAP Prevention System (previously known as the Young LoTrach™ System). This is not an antimicrobial approach per se but a multi-factorial strategy to prevent pulmonary aspiration via a specialised cuff made of an ultrathin material less prone to material folding, which can also be maintained at a constant pressure. Both properties reduce micro-channel leakage of subglottic secretions below the cuff. Subglottic drainage above the cuff also occurs through specialised ports in the system.

Other methods to prevent and control biofilms have been documented. For example, mechanical removal of already populated biofilms has been shown to have a significant impact on biofilm removal and reduce the incidence of infection.[35–37]

In addition to modification of the surface of ET tubes and mechanical removal of subglottic secretions, an area of intensive research is impregnation of the tube with antimicrobials. Numerous researchers have looked at ways to prevent or control biofilm formation using antibiotics and antiseptics incorporated into the biomaterial of the ET tube to decrease subsequent lung colonisation.[38–41] Silver has been investigated as an antimicrobial agent, and in 2008 C. R. Bard marketed the first antimicrobial ET tube under the name Agento™.

A recent randomised control study involving more than 2000 intubated patients reported that those intubated for 24 hours or longer with a silver-coated ET tube indeed had significantly lower colonisation rates than control groups.[42] However, the study found no differences in length of stay for any patient in either the intensive care unit or the hospital. Results did, however, confirm that reduction in microbial bacteria colonisation and biofilm formation reduced the incidence of VAP.[42]

Other studies have shown that silver-impregnated ET tubes can have a positive effect in reducing the colonisation and biofilm development of many bacteria, including, as an example, *P. aeruginosa*.[41,43] However, Kollef et al.[42] demonstrated that breakthrough VAP can still occur despite the use of a silver coating. It is well known that silver is an antimicrobial with a broad spectrum of activity, which has been shown to be efficacious on biofilms, applicable to medical device-related infections. In animal studies, silver-impregnated ET tubes have been shown to delay microbial colonisation[40]—a phenomenon that has been replicated in human studies.[41,39]

Despite promising results with silver-impregnated ET tubes, such tubes are not universally used; perhaps not specifically because of clinical efficacy, but because they are more expensive than uncoated ET tubes. Consequently, the quest continues for antimicrobials or agents/methods other than silver for controlling ET tube colonisation and biofilm formation.

Other ET antimicrobials have been investigated: specifically chlorhexidine (CHX) in combination with sulfadiazine.[38] CHX has also been combined with gentian violet.[44] Reitzel et al.[45] investigated the synergistic effects of combining brilliant green and CHX. This combination was found to be effective against well-known biofilm-forming microorganisms such as *C. albicans* and *Enterococcus faecium*. Gentian violet and CHX have been used to coat central venous and urinary catheters as well as ET tubes. This combination has been proven to provide antimicrobial efficacy against MRSA, *E. coli* and *P. aeruginosa*.[44]

Antimicrobial photodynamic therapy has been reported to have significant results after a single treatment.[46] The procedure employed by Biel et al.[46] involved spraying a photosentiser into the lumen of the ET tube and then irradiating the ET tube at 664 nm via an optical fiber.

Raad and colleagues[47] employed an *in vitro* biofilm colonisation model to compare the activity and durability of gardine- and gendine-coated ET tubes with those of silver-coated ET tubes in preventing adherence of drug-resistant bacteria and yeast commonly associated with VAP. They found that the gardine/gendine-coated tubes completely inhibited adherence of MRSA, MDR Gram-negative bacteria and *Candida*. They also found that this combination was significantly more efficacious than silver, and associated it with antimicrobial durability against MRSA, noting that it was more effective than silver against this infection. Shorr and colleagues[48] found a coated ET tube to be more cost effective in the prevention of VAP: they reported that a hospital could save approximately $12,840 per case of VAP prevented.

Interestingly, oral health can be correlated with the aetiology of VAP, with mounting evidence indicating that good oral hygiene may help to reduce its incidence.[49,50] Studies have shown that the oral hygiene of patients admitted to intensive care is often low compared to that of healthy individuals and declines during the stay.

As previously mentioned, some normal oral microflora enhance biofilm formation by potential respiratory pathogens in the ET tube. There is also evidence that dental plaque in MV patients becomes colonised with potential respiratory pathogens during hospital stay. The reason for this 'microbial shift' in the composition of dental plaque remains unclear, but it might relate to changes in the oral cavity caused by altered salivary flow, changes in saliva biochemistry or a patient's immunity associated with administered therapies or underlying illness. Given that the oral cavity is comparatively more accessible to healthcare providers than the ET tube, interventions that improve oral hygiene of patients susceptible to VAP are obviously attractive.

Existing healthcare bundles do indeed advocate CHX in oral healthcare regimes for MV patients. This approach follows research findings that have shown its benefits (often in cardiac patients) through either single or repeated application in the mouths of MV patients.[51,52] Needleman et al.[53] further found that the use of a powered toothbrush together with 0.2% CHX mouthwash significantly lowered plaque levels in 46 MV patients compared with controls that employed a sponge toothette. However, the effect of this approach on VAP reduction was not reported.[53]

CONCLUSION

The colonisation of micro-organisms and biofilm formation in ET tubes are proposed as a mechanism for the pathogenesis of VAP. Numerous papers have reported evidence that reducing microbial colonisation and biofilm formation on an ET tube can lower occurrence or even prevent VAP. This has very important implications for morbidity and associated healthcare cost increases.

The technologies available to achieve VAP reductions include changing the ET tube cuff material and shape and impregnating the tube with antimicrobials. Effective impregnated antimicrobial agents include silver and chlorhexidine as well as photodynamic agents. Results suggest that silver-coated ET tubes are effective in VAP prevention, although this remains an area of continuing research.

Overall, it is clearly evident that lowering the incidence of VAP requires the reduction of microbial colonisation and biofilm formation via multiple interventions. Research needs to focus on strategies to reduce biofilm formation by modifying the ET tube surface or by incorporating antimicrobial agents that have clinical significance and efficacy and are, at the same time, cost effective.

REFERENCES

1. ATS/IDSA. Guidelines for the management of adults with hospital-acquired,ventilator-associated, and healthcare-associated pneumonia. *Am J Respir Crit Care Med* 2005;**171**:388e–416e.
2. Koerner RJ. Contribution of endotracheal tubes to the pathogenesis of ventilator-associated pneumonia. *J Hosp Infect* 1997;**35**:83–9.
3. Adair CG, et al. Implications of endotracheal tube biofilm for ventilator-associated pneumonia. *Intensive Care Med* 1999;**25**:1072–6.
4. Augustyn B. Ventilator-associated pneumonia: risk factors and prevention. *Crit Care Nurse* 2007;**27**:32.
5. Heo SM, Haase EM, Lesse AJ, Gill SR, Scannapieco FA. Genetic relationships between respiratory pathogens isolated from dental plaque and bronchoalveolar lavage fluid from patients in the intensive care unit undergoing mechanical ventilation. *Clin Infect Dis* 2008;**47**:1562–70.
6. Depuydt P, Myny D, Blot S. Nosocomial pneumonia: aetiology, diagnosis and treatment. *Curr Opin Pulm Med* 2006;**12**:192–7.
7. Amin A. Clinical and economic consequences of ventilator-associated pneumonia. *Clin Infect Dis* 2009; **49**:S36–43.
8. Palmer LB. Ventilator-associated infection. *Curr Opin Pulm Med* 2009;**15**:230–5.
9. Diaz E, Rodriguez AH, Rello J. Ventilator-associated pneumonia: issues related to the artificial airway. *Respir Care* 2005;**50**:900–6.
10. Jackson WL, Shorr AF. Update in ventilator-associated pneumonia. *Curr Opin Anaesthesiol* 2006; **19**:117–21.
11. Pneumatikos IA, Dragoumanis CK, Bouros DE. Ventilator-associated pneumonia or endotracheal tube-associated pneumonia?. *Anesthesiol* 2009;**110**:673–80.
12. Inglis TJ, Millar MR, Jones JG, Robinson DA. Tracheal tube biofilm as a source of bacterial colonization of the lung. *J Clin Microbiol* 1989;**27**(9):2014–8.
13. Bauer TT, Torres A, Ferrer R, Heyer CM, Schultze-Werninghaus G, Rasche K. Biofilm formation in endotracheal tubes. Association between pneumonia and the persistence of pathogens. *Monaldi Arch Chest Dis* 2002;**57**:84–7.
14. Vandecandelaere I, Matthijs N, Van Nieuwerburgh F, Deforce D, Vosters P, De Bus L, et al. Assessment of microbial diversity in biofilms recovered from endotracheal tubes using culture-dependent and independent approaches. *PLoS One* 2012;**7**(6):e38401.
15. Edwards C. Problems posed by natural environments for monitoring micro-organisms. *Mol Biotechnol* 2000;**15**:211–23.
16. Luna CM, Sibila O, Agusti C, Torres A. Animal models of ventilator-associated pneumonia. *Eur Respir J* 2009;**33**:182–8.
17. Sottile FD, Marrie TJ, Prough DS, Hobgood CD, Gower GJ, et al. Nosocomial pulmonary infection: possible etiologic significance of bacterial adhesion to endotracheal tubes. *Crit Care Med* 1986; **14**:265–70.
18. Slagle TA, Bifano EM, Wolf JW, Gross SJ. *Arch Dis Child* 1989;**64**(1 Spec No):34–8.
19. Perkins SD, Woeltje KF, Angenent LT. Endotracheal tube biofilm inoculation of oral flora and subsequent colonization of opportunistic pathogens. *Int J Med Microbiol* 2010;**300**(7):503–11.
20. Cairns S, Thomas JG, Hooper SJ, Wise MP, Frost PJ, Wilson MJ, et al. Molecular analysis of microbial communities in endotracheal tube biofilms. *PLoS One* 2011;**14**(6(3)):e14759.
21. Skrupky LP, McConnell K, Dallas J, Kollef MH. A comparison of ventilator-associated pneumonia rates as identified according to the National Healthcare Safety Network and American College of Chest Physicians criteria. *Crit Care Med* 2012;**40**(1):281–4.

22. Richards MJ, Edwards JR, Culver DH, Gaynes RP. The National Nosocomial Infections Surveillance System Nosocomial infections in pediatric intensive care units in the United States. *Pediatrics* 1999; **103**(4):e39.
23. Gaynes R, Edwards JR. Overview of nosocomial infections caused by gram-negative bacilli. *Clin Infect Dis* 2005;**41**:848–54.
24. Tablan OC, Anderson LJ, Besser R, Bridges C, Hajjeh R. CDC Healthcare Infection Control Practices Advisory Committee Guidelines for preventing health-care-associated pneumonia, 2003: recommendations. *MMWR Recomm Rep* 2004;**53**(RR-3RR-3):1–36.
25. Dissemi Ramirez P, Ferrer M, Torres A. Prevention measures for ventilator-associated pneumonia: a new focus on the endotracheal tube. *Curr Opin Infect Dis* 2007;**20**:190–197D.
26. Lee JM, Hashmi N, Bloom JD, Tamashiro E, Doghramji L, Sarani B, et al. Biofilm accumulation on endotracheal tubes following prolonged intubation. *J Laryngol Otol* 2012;**126**(3):267–70.
27. Wilson A, Gray D, Karakiozis J, Thomas J. Advanced endotracheal tube biofilm stage, not duration of intubation, is related to pneumonia. *J Trauma Acute Care Surg* 2012;**72**(4):916–23.
28. Rello J, Lode H, Cornaglia G, Masterton R., and the VAP Care Bundle Contributors. A European care bundle for prevention of ventilator-associated pneumonia. *Intensive Care Med* 2010;**36**:773–80.
29. Inglis TJ, Millar MR, Jones JG, Robinson DA. Tracheal tube biofilm as a source of bacterial colonization of the lung. *J Clin Microbiol* 1989;**27**:2014–8.
30. Adair CG, Gorman SP, O'Neill FB, McClurg B, Goldsmith EC, Webb CH. Selective decontamination of the digestive tract (SDD) does not prevent the formation of microbial biofilms on endotracheal tubes. *J Antimicrob Chemother* 1993;**31**(5):689–97.
31. Murthy SK. Nanoparticles in modern medicine: state of the art and future challenges. *Int J Nanomedicine* 2007;**2**:129–41.
32. Machado MC, Cheng D, Tarquinio KM, Webster TJ. Nanotechnology: pediatric applications. *Pediatr Res* 2010;**67**(5):500–4.
33. Machado MC, Tarquinio KM, Webster TJ. Decreased *Staphylococcus aureus* biofilm formation on nanomodified endotracheal tubes: a dynamic airway model. *Int J Nanomedicine* 2012;**7**:3741–50.
34. Durmus NG, Taylor EN, Inci F, Kummer KM, Tarquinio KM, Webster TJ. Fructose-enhanced reduction of bacterial growth on nanorough surfaces. *Int J Nanomedicine* 2012;**7**:537–45.
35. Kolobow T, Li Bassi G, Curto F, Zanella A. The Mucus Slurper: a novel tracheal tube that requires no tracheal tube suctioning. A preliminary report. *Intensive Care Med* 2006;**32**(9):1414–8.
36. Berra L, Curto F, Li Bassi G, Laquerriere P, Baccarelli A, Kolobow T. Antibacterial-coated tracheal tubes cleaned with the Mucus Shaver: a novel method to retain long-term bactericidal activity of coated tracheal tubes. *Intensive Care Med* 2006;**32**(6):888–93.
37. Berra L, Coppadoro A, Bittner EA, et al. A clinical assessment of the Mucus Shaver: a device to keep the endotracheal tube free from secretions. *Crit Care Med* 2012;**40**(1):119–24.
38. Berra L, De Marchi L, Yu ZX, Laquerriere P, Baccarelli A, Kolobow T. Endotracheal tubes coated with antiseptics decrease bacterial colonization of the ventilator circuits, lungs, and endotracheal tube. *Anesthesiology* 2004;**100**(6):1446–56.
39. Berra L, Kolobow T, Laquerriere P, et al. Internally coated endotracheal tubes with silver sulfadiazine in polyurethane to prevent bacterial colonization: a clinical trial. *Intensive Care Med* 2008;**34**(6):1030–7.
40. Olson ME, Harmon BG, Kollef MH. Silver-coated endotracheal tubes associated with reduced bacterial burden in the lungs of mechanically ventilated dogs. *Chest* 2002;**121**(3):863–70.
41. Rello J, Kollef M, Diaz E, et al. Reduced burden of bacterial airway colonization with a novel silver-coated endotracheal tube in a randomized multiple-center feasibility study. *Crit Care Med* 2006; **34**(11):2766–72.
42. Kollef MH, Afessa B, Anzueto A, Veremakis C, Kerr KM, Margolis BD, et al. Silver-coated endotracheal tubes and incidence of ventilator-associated pneumonia: the NASCENT randomized trial. *JAMA* 2008;**300**:805e–13e.
43. Hartmann M, Guttmann J, Muller B, Hallmann T, Geiger K. Reduction of the bacterial load by the silver-coated endotracheal tube (SCET), a laboratory investigation. *Technol Health Care* 1999;**7**:359e–70e.
44. Chaiban G, Hanna H, Dvorak T, Raad I. A rapid method of impregnating endotracheal tubes and urinary catheters with gendine: a novel antiseptic agent. *J Antimicrob Chemother* 2005;**55**:51e6.

45. Reitzel RA, Dvorak TL, Hachem RY, Fang X, Jiang Y, Raad I. Efficacy of novelantimicrobial gloves impregnated with antiseptic dyes in preventing the adherence of multidrug-resistant nosocomial pathogens. *Am J Infect Control* 2009;**37**:294e–300e.
46. Biel MA, Sievert C, Usacheva M, et al. Reduction of endotracheal tube biofilms using antimicrobial photodynamic therapy. *Lasers Surg Med* 2011;**43**(7):586–90.
47. Raad II, Mohamed JA, Reitzel RA, Jiang Y, et al. The prevention of biofilm colonization by multi-drug-resistant pathogens that cause ventilator-associated pneumonia with antimicrobial-coated endo-tracheal tube. *Biomaterials* 2011;**32**:2689–94.
48. Shorr AF, Zilberberg MD, Kollef M. Cost-effectiveness analysis of a silver-coated endotracheal tube to reduce the incidence of ventilator-associated pneumonia. *Infect Control Hosp Epidemiol* 2009; **30**(8):759–63.
49. Tantipong H, Morkchareonpong C, Jaiyindee S, Thamlikitkul V. Randomized controlled trial and meta-analysis of oral decontamination with 2% chlorhexidine solution for the prevention of ventilator-associated pneumonia. *Infect Control Hosp Epidemiol* 2008;**29**:131–6.
50. Chan EY. Oral decontamination for ventilator-associated pneumonia prevention. *Aus Crit Care* 2009; **22**:3–4.
51. Grap MJ, Munro CL, Hamilton VA, Elswick Jr RK, Sessler CN, Ward KR. Early, single chlorhexi-dine application reduces ventilator-associated pneumonia in trauma patients. *Heart Lung* 2011; **40**(5):e115–22.
52. Munro CL, Grap MJ, Jones DJ, McClish DK, Sessler CN. Chlorhexidine, tooth brushing, and pre-venting ventilator-associated pneumonia in critically ill adults. *Am J Crit Care* 2009;**18**(5):428–37.
53. Needleman IG, Hirsch NP, Leemans M, Moles DR, et al. Randomized controlled trial of tooth-brushing to reduce ventilator-associated pneumonia pathogens and dental plaque in a critical care unit. *J Clin Periodontol* 2011;**38**(3):246–52.

CHAPTER THIRTEEN

Antimicrobial Chemotherapy

Significance to Healthcare

Tomoari Kuriyama, Tadahiro Karasawa and David W. Williams

INTRODUCTION

Antimicrobial chemotherapy is used to cure an infectious disease by combating causative pathogens using drugs with selective toxicity against them. The selective toxicity of these drugs does significant damage to pathogens with minimal harmful effects to the host. The feasibility of such selective toxicity depends on the existence of exploitable biochemical differences between the pathogens and host cells. Chemotherapeutic agents can target pathogen functions which are unique to the pathogens and not present in the host, or which are shared by the host, but vary in importance between the pathogen and the host, or which are similar but not identical to those in the host. For example, bacteria possess a cell wall, whereas mammalian cells do not. Accordingly, drugs that interfere with the production of the bacterial cell wall are highly toxic to bacteria but harmless to the host.[1] Furthermore, the bacterial ribosome consists of two subunits—the larger 50 S subunit and the smaller 30 S subunit—while the eukaryotic ribosome consists of the larger 60 S subunit and the smaller 40 S subunit. This difference imparts selectivity in the targeting of certain chemotherapeutic agents.

Antimicrobial agents are drugs that kill or inhibit the growth of micro-organisms. It is important to recognise that antimicrobial agents work selectively against micro-organisms and that they are different from disinfectants and antiseptics, which can have high toxicity to both the host cell and the micro-organism.

Antimicrobial agents have a wide range of antibacterial, antifungal and antiviral uses. Strictly speaking, antibacterial drugs are classified as antibiotics, synthetic agents and semi-synthetic agents. Originally, the term 'antibiotic' described a natural substance that was released by bacteria or fungi into the environment as a means of inhibiting other organisms at low concentrations. Penicillin, which is an antibiotic derived from the *Penicillium* mold, may be the best-known example. However, common usage often extends the term 'antibiotic' to include synthetic antibacterial agents that are not the products of microbes. Although some antibiotics are effective against fungi and viruses,

Biofilms in Infection Prevention and Control
Doi: http://dx.doi.org/10.1016/B978-0-12-397043-5.00013-X

the large majority of these are used to combat bacterial infection. Therefore, in practice the term 'antibiotic' is often used synonymously with 'antibacterial drug'.

This chapter primarily discusses general considerations of antibiotic therapy for bacterial infections, with additional discussion of clinically important antifungal agents.

DEVELOPMENT OF ANTIMICROBIAL THERAPY

Infections used to account for a very large proportion of diseases as a whole and were the most likely cause of death. In the latter half of the nineteenth century, micro-organisms were found to be responsible for a number of infectious diseases.

Modern chemotherapy has been dated to the work of Paul Ehrlich (1854–1915) in Germany, whose initial interest was in the field of immunology. Ehrlich perceived an analogy between antigen–antibody interaction and the selective toxicity of chemical agents against infecting organisms.[2,3] He hypothesised that substances that selectively bind only to pathogens and not the host cell would be available as drugs that are effective against the pathogens without producing harmful effects on the host. The metaphorical term 'magic bullet' was given to such an ideal chemotherapeutic agent. This was indeed the first concept of chemotherapy.

In 1909, Ehrlich's laboratory discovered arsphenamine (Salvarsan), an arsenical compound and the first effective treatment for syphilis.[2,3] At that time, syphilis was widespread in Europe and mercury salts were the only known treatment. Arsphenamine was more effective than mercury salts, with a much reduced toxicity, although its safety and efficacy were extremely low compared with syphilis medicines that are available today. It is not surprising that Ehrlich is currently recognised as the 'Father of Chemotherapy'.

In 1928, penicillin was discovered by the Scottish biologist Alexander Fleming at St. Mary's Hospital in London.[4,5] This discovery came from a fortuitous accident. When Fleming cultured *Staphylococcus aureus* in a petri dish in the laboratory, the culture plate was mistakenly contaminated with a blue mold. Fleming noticed that the growth of staphylococci was inhibited in a zone surrounding the contaminating blue mold, and he concluded that the contaminating micro-organism was producing a substance that could inhibit the growth of the inoculated *S. aureus*.[4,5] The antibiotic derived from this mold (*Penicillium notatum*) was named 'penicillin'. Fleming did not get to the stage of purifying and testing the effects of this agent against human bacterial infections, and penicillin did not come into immediate clinical use after its discovery. It was not until the early 1940s that the true potential of penicillin was realised and was put to practical use.

In 1935, the first commercially available antibacterial drug, Prontosil, which was a brilliant orange-red compound primarily used as an industrial dye, was developed by a research team led by Gerhard Domagk at Bayer Laboratories—part of the German IG Farben conglomerate.[5] Unlike arsphenamine, Prontosil was effective against several types of bacteria. In later years, it was found that its antibacterial activity resulted from sulfonamide, which was generated from Prontosil in the human body.[6] However, Bayer's Prontosil was a synthetic compound and had limitations in terms of safety and efficacy.

Around the same time, the German-born British biochemist Ernst Chain and the Australian pharmacologist Howard Florey investigated Fleming's original discovery while at the University of Oxford.[5,6] In 1940, their research group succeeded in purifying the first penicillin (penicillin G), and demonstrated its therapeutic effect through animal experiments and clinical trials. Development of penicillin is historically significant as it was the first antibiotic and medicine to be effective against many 'life-threating' infections with minimal toxicity. By this time, the Second World War had broken out and penicillin was being mass-produced in the United States and saved many wounded Allied soldiers.[6] After the end of war, penicillin became widely available outside the military.

The development of penicillin led to renewed interest in the search for antibiotic compounds with similar efficacy and safety. In 1943, the American microbiologists Selman Waksman and Albert Schatz discovered streptomycin (the first aminoglycoside) from the soil organism *Streptomyces*.[6] The term 'antibiotic' was actually coined by Waksman. Subsequently, chloramphenicol, tetracycline, macrolides, glycopeptides and polyenes (antimycotics) were developed from these soil micro-organisms. The first-generation cephalosporin agents were developed during the 1960s.[7] Since that time, many new antibiotics with improvements in safety and pharmacodynamics, in addition to enhancement of antimicrobial spectrum and activity, have become available for clinical use.

With the development and use of antibiotics has come recognition of the capacity of micro-organisms to acquire resistance to antimicrobial agents, and this has been significant. Because bacteria have the ability to rapidly change and adapt to different environments, they often become resistant to new antibiotics soon after their production. For example, *S. aureus* was initially susceptible to sulfonamides and penicillin. However, it rapidly acquired resistance to sulfonamides when the agent was available for clinical use, and the prevalence of strains resistant to penicillin by production of penicillinase increased during the 1950s.[7]

Although meticillin, a penicillin-class agent that was stable to penicillinase, was developed in 1960, as early as the following year, meticillin-resistant *S. aureus* (MRSA) was isolated in the United Kingdom.[7] Bacteria can acquire resistance even against the latest antibiotic that overcomes pre-existing antibiotic resistances. The vicious circle of antibiotic-resistant bacteria appearing shortly after release of a new antibiotic is a matter of serious concern for human healthcare.

PRINCIPLES OF ANTIBIOTIC THERAPY

The concepts of antibiotic pharmacology and antibiotic therapy, including the mechanisms of action, spectra of activity and important pharmacologic properties are discussed in the following subsections.

Mode of Antibiotic Action

Although animals, plants and fungi are eukaryotic organisms, bacteria are prokaryotic and thus there are fundamental differences in their cellular structure and biochemistry. In addition to lacking a defined nucleus and other membrane-bound organelles, the bacterial cell also differs from the human cell with respect to its chromosomal and ribosomal structure. Bacteria also differ from human cells through the possession of a distinct peptidoglycan cell wall outside the cell membrane which protects the cell from adverse environmental conditions. Many micro-organisms, including bacteria, biosynthesise folate, which is required for nucleic acids synthesis, while humans lack this function and acquire folate through their diet. Several of these distinctions from animal cells provide the basis for selective antimicrobial action. Most antibiotics act by selectively interfering with the synthesis of one of the large-molecule constituents of the cell-cell wall, proteins or nucleic acids (Figure 13.1).

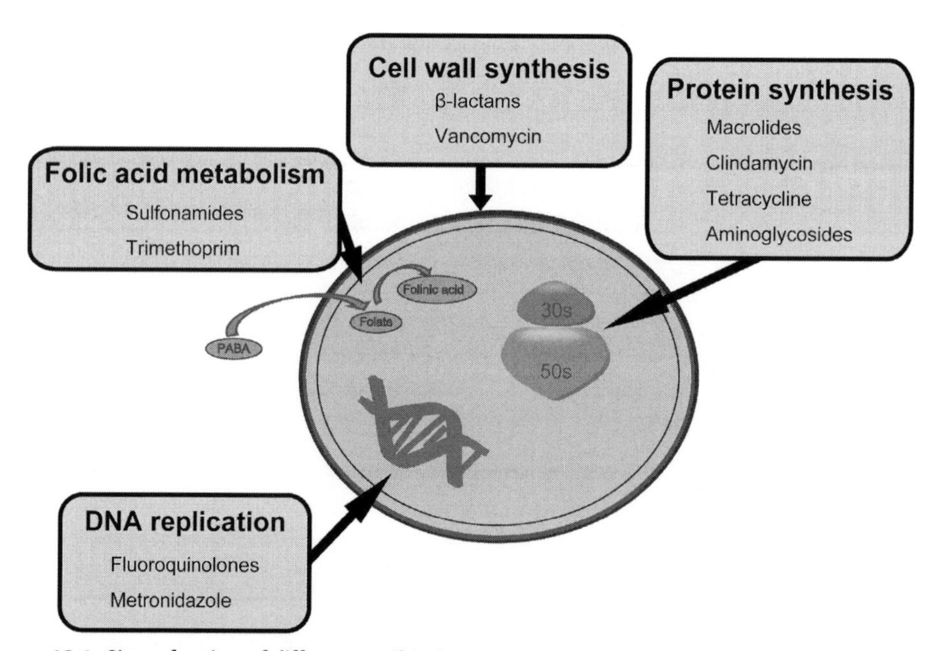

Figure 13.1 ***Sites of action of different antibiotic types.*** Antimicrobial agents act by interfering with cell wall synthesis, nucleic acid synthesis (DNA replication or folate synthesis) or ribosomal function (protein synthesis).

Table 13.1 Antibacterial Properties of Antibiotics

Drug	Property
Penicillins	Bactericidal
Cephalosporins	Bactericidal
Carbapenems	Bactericidal
Monobactams	Bactericidal
Macrolides	Bacteriostatic
Clindamycin	Bacteriostatic
Tetracyclines	Bacteriostatic
Aminoglycosides	Bactericidal
Fluoroquinolones	Bactericidal
Glycopeptides	Bactericidal
Metronidazole	Bactericidal
Sulfonamides	Bacteriostatic

Bactericidal and Bacteriostatic Properties

Antimicrobial properties are divided into bactericidal and bacteriostatic activities. Bactericidal activity directly kills bacteria, while bacteriostatic activity interferes with bacterial reproduction and it is the host immune system that subsequently eliminates the bacteria.[8] In general, the activity of antibiotics that inhibits bacterial cell wall synthesis is bactericidal. Although the majority of antibiotics responsible for inhibiting synthesis of bacterial proteins have bacteriostatic activity, some of them exhibit bactericidal activity (Table 13.1).

Antimicrobial Activity/Antimicrobial Susceptibility

In vitro antimicrobial activity and *in vitro* antimicrobial susceptibility are synonymous terms. These terms are commonly expressed as the 'minimum inhibitory concentration' (MIC). The MIC is the lowest concentration of an antimicrobial that inhibits the growth of a particular bacterial species *in vitro*. An antibiotic that exhibits a low MIC value is generally regarded effective against the target organism, as the growth of the bacteria should be inhibited by low antibiotic concentrations. A micro-organism with growth that is inhibited by a low antibiotic MIC is generally regarded to be highly susceptible to the antibiotic.

The susceptibility of individual strains of the same bacterial species to an antibiotic can be highly variable. A large difference in susceptibility results in a wide, but not normal, MIC distribution. Therefore, when addressing the susceptibility of bacterial strains, the terms *MIC50* and *MIC90*, which indicate the antibiotic concentrations that inhibit 50% and 90% of the tested strains, respectively, are often employed as indices of evaluation.

Antimicrobial Spectrum

The antimicrobial spectrum describes the range of microbes that are susceptible to a particular antimicrobial agent. The antibiotic may be classified as narrow-spectrum or broad-spectrum based on the range of bacterial types that it is effective against. Narrow-spectrum antibiotics are effective against a relatively restricted range of bacterial types, whereas broad-spectrum antibiotics are active against a wider range.

Bacteria are commonly classified by their morphology (i.e., bacilli, cocci and spirochaetes), reaction to Gram stain (Gram-positive or Gram-negative) and oxygen needs for respiration (i.e., aerobic, facultative and anaerobic) (Table 13.2). Such classifications,

Table 13.2 Classification of Clinically Significant Bacteria

Gram stain	Morphology	Organisms
Positive	Cocci	*Staphylococcus*
		Streptococcus pyogenes
		Streptococcus pneumoniae
		Viridans streptococci (α-*Streptococcus*)
		Enterococcus
	Bacilli	*Corynebacterium*
		Bacillus
		Mycobacterium
		*Clostridium**
Negative	Cocci	*Neisseria*
		Moraxella
	Bacilli	*Acinetobacter*
		Escherichia†
		Citrobacter†
		Salmonella†
		Shigella†
		Klebsiella†
		Enterobacter†
		Serratia†
		Proteus†
		Vibrio
		Pseudomonas
		Haemophilus
		Campylobacter
		Legionella
		Helicobacter
		*Prevotella**
		*Bacteroides**

*Strict anaerobes (other organisms are aerobes or facultative organisms).
†Members of Enterobacteriaceae.

especially whether Gram-positive or Gram-negative, are often linked to antimicrobial susceptibility. Gram-negative bacteria possess a unique structure—an outer membrane—which surrounds the peptidoglycan layer of the cell wall. This outer membrane is often a permeability barrier to an antibiotic molecule. As a consequence, Gram-negative bacteria generally have reduced susceptibility to antibiotics compared with Gram-positive bacteria.

Pharmacokinetic and Pharmacodynamic Aspects

Pharmacokinetics is the effect of a body on a drug, whereas pharmacodynamics is the effect of a drug on a body. Such pharmacological properties of antibiotics significantly influence their *in vivo* efficacy as well as their *in vitro* antimicrobial activity.

Pharmacokinetics

Once an antibiotic is administered, it moves from the site of administration into the blood's circulation, and then disperses and disseminates within body fluids and tissues. In general, intravenous administration provides a well-distributed antibiotic with high levels of the agent in the plasma. In the case of oral administration, absorption in the gut is necessary and is affected by various factors, including stomach and intestine acidity/alkalinity, food in the stomach and the chemical properties of the antibiotic. An antibiotic entering the blood may bind plasma protein and, since only a non-protein-bound antibiotic is pharmacologically active, plasma protein binding (PPB) reduces the free fraction of antibiotic available for bacterial killing.[9] Therefore, *in vivo* activity of the antibiotic with a higher degree of PPB may be reduced compared with that anticipated based on *in vitro* activity.

Systemically administrated antibiotics are distributed unevenly throughout the body. However, the drug concentration in each organ, tissue and fluid is generally in proportion to the antibiotic's blood plasma concentration.[10] Consequently, dosage and route of administration are often determined based on the antibiotic's likely concentration in the blood plasma. It is important to recognise that many antibiotics are restricted in their penetration into specific organs and tissues, including the central nervous system (CNS), bone, prostate and urinary tract.[10] In the treatment of infections in these organs, a specific antibiotic with good tissue penetration should be used.

After distribution in the tissues and body fluids, metabolising enzymes in the body may degrade the antibiotic. The liver is the principal organ of metabolism, although any biological tissue can metabolise drugs. By metabolic processes, the drug is inactivated and converted into a more readily excreted substance. However, not all antibiotics are readily converted into metabolites and, on occasion, the metabolites may retain activity.

Both unchanged antibiotics and antibiotic metabolites are removed from the body by excretion pathways. Excretion from the kidneys (as urine) is a major route of drug elimination; excretion into the bile (as feces) from the liver is another. With regards to the time course of drug concentration in the body following administration, once the concentration has reached maximum (peak concentration), the drug is gradually eliminated from the bloodstream by the process of metabolism and excretion.

The half-life of a drug is the time it takes for its plasma concentration to reach half of its maximum concentration (Figure 13.2). As described previously, the antibiotic concentration at an infection site generally correlates with its blood plasma concentration. Therefore, the half-life of a drug may be available to estimate how long it takes for half of it to be eliminated from the infection site. Drugs with a short half-life need frequent dosage, whereas long half-life drugs may be suitable for twice- or once-daily dosing. The decline speed of the drug concentration from the blood primarily depends on the metabolism and excretion function. Therefore, impairment of metabolism or excretion can lengthen the half-life of a drug. It should be noted that drugs with a long half-life lengthen the duration of not only antimicrobial effects but also any adverse effects.

Pharmacodynamics

Antimicrobial pharmacodynamics is the relationship between the concentration of an antibiotic and its ability to inhibit the growth of micro-organisms. The principle pharmacodynamic parameter is the MIC of the target organisms for a drug.

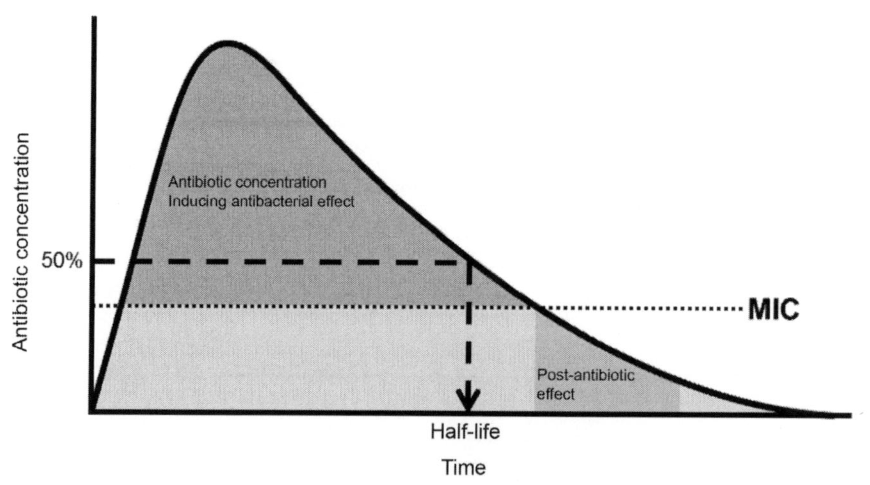

Figure 13.2 ***Time course of antibiotic concentration in the bloodstream following administration of a dose of antibiotic.*** The half-life of a drug is the time it takes for its plasma concentration to reach half of its maximum concentration. Although antibiotics inhibit the growth of a causative organism at the concentration above the MIC for the organism, some antibiotics have prolonged persistent effects (post-antibiotic effect, or PAE) in which the antibacterial action continues for a period of time after the antibiotic level falls below its MIC.

The antibiotic concentration at the site of the infection must exceed the MIC for the target bacteria in order to obtain an antimicrobial effect. Interestingly, some antibiotics have prolonged persistent effects (post-antibiotic effect) in which the antibacterial action continues for a period of time after the antibiotic level falls below its MIC.[11] The ability of an antibiotic to induce a PAE is an attractive property, as antibiotic concentrations can fall below the MIC for the bacterium yet retain effectiveness in suppressing microbial growth.

Pharmacokinetic/Pharmacodynamic Modelling

During antimicrobial therapy, the concentration of the antibiotic, the time that it remains at the infection site (pharmacokinetic factors) and the MIC of the antibiotic against the target organism (pharmacodynamic factors) primarily determine *in vivo* efficacy. Although dosage regimens previously were made empirically, it has become apparent that those determined based on pharmacokinetic (PK) and pharmacodynamic (PD) properties of an antibiotic can provide optimal therapeutic effect while minimising the likelihood of drug toxicity and the risk of drug resistance developing during treatment.[11]

The parameters created by combining the PK and PD properties (PK/PD parameters) are currently used to predict the *in vivo* effects of antimicrobial chemotherapy. Figure 13.3 shows these parameters, which include the ratio of the peak antibiotic

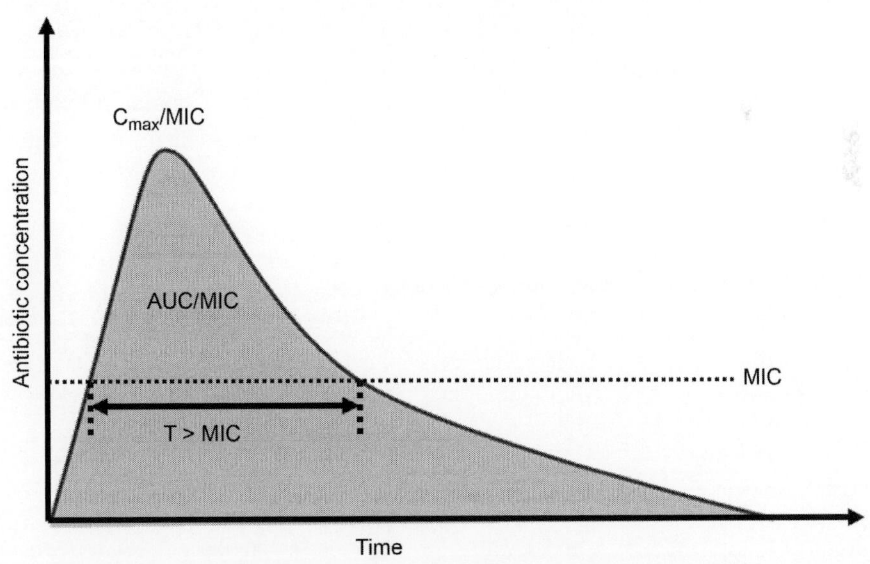

Figure 13.3 ***Pharmacokinetic/pharmacodynamic predictors of efficacy.*** C_{max}/MIC, the ratio of peak antibiotic concentration and MIC; AUC/MIC, the area under the concentration-time curve and MIC ratio; T > MIC, the time that free antimicrobial concentrations remain above the MIC for the organism.

concentration and MIC (C_{max}/MIC),[2] the area under the concentration-time curve and the MIC ratio (AUC/MIC), and[3] the time that free antimicrobial concentrations remain above the MIC for the micro-organism (T > MIC).[12]

The action of an antibiotic is either time- or concentration-dependent. In the case of agents that exhibit time-dependent bactericidal activity, antimicrobial efficacy correlates with the percentage of time within a dosing interval that the concentration of the antimicrobial remains above the MIC for the organism. This type of agent has little relationship to the magnitude of concentration, as long as the concentration is above a minimally effective level. In the case of agents that exhibit concentration-dependent bactericidal activity, a high concentration relative to the MIC is important for antimicrobial efficacy and thus dose size is critical. Significantly, concentration-dependent antibiotics usually have relatively large PAE, while time-dependent antibiotics have various PAE sizes and lengths.[11] Depending on the antimicrobial action of an antibiotic and the PAE length, specific PK/PD parameters may correlate with *in vivo* efficacy[13] (Table 13.3).

Safety

Although antibiotics are designed to exert selective toxicity against only the target pathogen, they may provide a range of toxicity or cause unwanted effects in the host even when administrated properly. The harmful and undesired effects of antibiotics are referred to as being 'adverse' and range from mild to fatal. All antibiotics carry the possibility of adverse effects. Sometimes they interact with other administered drugs, and this interaction can result in harmful and undesired outcomes. Antibiotics that have a minimum risk of adverse effects and drug interactions are largely regarded as safe agents.

Almost all antimicrobial agents are capable of crossing the placenta and exhibiting a range of teratogenic effects. Further discussion of adverse effects, drug interaction and teratogenesis is provided in a later section.

Table 13.3 Pharmacodynamic Properties of Antibiotics

Pattern of activity	Antibiotic	Goal of therapy	PK/PD parameter
Concentration-dependent killing and prolonged persistent effects	Aminoglycosides Fluoroquinolones	Maximise concentrations	AUC/MIC C_{max}/MIC
Time-dependent killing and minimal persistent effects	β-lactams Erythromycin	Maximise duration of exposure	T > MIC
Time-dependent killing and moderate to prolonged persistent effects	Clindamycin Azithromycin Tetracyclines Vancomycin	Maximise amount of drug	AUC/MIC

Source: Adapted from Akita.[13]

CLASSES OF ANTIBIOTICS

Antibiotics are usually classified based on chemical structure. The following subsections describe antibiotics that are considered to have significance in clinical practice.

β-Lactams

β-lactam antibiotics have been widely used in the prevention and treatment of a variety of human bacterial infections. Although they can be classified into penicillin, cephalosporin, carbapenem and monobactam subclasses, all have a chemical structure called a β-lactam ring and carry out bactericidal activity through binding to penicillin-binding proteins and inhibiting synthesis of the bacterial peptidoglycan cell wall.[14,15]

Some types of bacteria can produce β-lactamases, which are enzymes capable of destroying and inactivating β-lactam antibiotics. Production of β-lactamases is one of the prime mechanisms for bacterial resistance to β-lactam antibiotics. β-lactamases have different properties and preferred substrates (antibiotics). For example, some are specific for penicillins (i.e., penicillinases) and some preferably destroy cephalosporins (i.e., cephalosporinases). To date, more than 200 different β-lactamases have been described. The molecular classification based on their amino sequences is presented in Table 13.4.[16,17]

Table 13.4 Molecular Classification of β-Lactamases

Class	Notable characteristics
A	Serine-based hydrolytic mechanism (serine-β-lactamase) Inactivate primarily penicillins Capable of hydrolyzing early-generation cephalosporins Generally incapable of inactivating the third or later generation of cephalosporins, carbapenems, and monobactams Extended spectrum β-lactamases (ESBLs) primarily belong to this class as mutant-types; unlike standard types of class A enzyme, inactivate most cephalosporins, including third- and fourth-generation cephalosporins and monobactams Susceptible to β-lactamase inhibitors
B	Require zinc ions for catalytic activity; therefore, usually called 'metallo-β-lactamases' Able to hydrolyze most β-lactam agents, including carbapenems Resistant to β-lactamase inhibitors
C	Serine-β-lactamases Inactivate primarily cephalosporins Capable of hydrolyzing penicillins Generally unable to inactivate fourth-generation cephalosporins and carbapenems Resistant to β-lactamase inhibitors
D	Serine-β-lactamases Similar to class A enzymes Inactivate not only penicillins, but also oxacillin—a penicillinase-resistant penicillin agent Can become ESBL by mutation

Source: Adapted from Ambler.[17]

β-lactam antibiotics are generally regarded as safe agents because they target the bacteria's cell wall, which does not exist in human cells. However, hypersensitivity to antibiotics should be considered. The most likely form of hypersensitivity is dermatological reaction. Anaphylactic reaction is rare but can, under certain conditions, be serious and even fatal.

Penicillin

The penicillin subclass of β-lactam antibiotics has a long history and remains one of the most important groups of antibiotics. Penicillin agents are derived directly or indirectly from strains of fungi of the genus *Penicillium* and other soil-inhabiting fungi. Penicillin antibiotics are generally divided into two categories: natural (biosynthetic) and semi-synthetic. Natural penicillins include penicillin G and penicillin V.[1] These are not expensive and are still widely used in clinical practice.

The natural penicillins are not stable to penicillinase and have a narrow spectrum of activity. They are active mainly against Gram-positive bacteria including penicillin-susceptible staphylococci, *Streptococcus pneumoniae*, *Streptococcus pyogenes* and oral streptococci.[1,10,14,15,18,19] Among the Gram-positive organisms, however, enterococci are resistant, and the increased prevalence of penicillin-resistant isolates of *S. pneumoniae* is also a matter of concern. Many aerobic and facultative Gram-negative bacilli are also resistant to penicillin. The natural penicillins are the drugs of choice for syphilis. Penicillin G is incompletely absorbed, so it is used mainly as an intravenous drug. Penicillin V is tolerant to gastric acid and is the preferred oral form.

Semi-synthetic penicillins have greater resistance to penicillinases or an extended spectrum of activity. Penicillinase-resistant penicillins include meticillin, nafcillin and oxacillin.[1,10,14,15,18,19] These are primarily used in the treatment of infection caused by penicillinase-producing staphylococci. Ampicillin was the first broad-spectrum penicillin and has a broader antibacterial range of action than that of penicillin G. Ampicillin is effective against many Gram-negative bacilli including *Escherichia coli*, *Haemophilus*, *Shigella* and *Proteus*. However, it is not effective against *Pseudomonas*, *Klebsiella* and *Serratia*. Similar to natural penicillins, ampicillin is not resistant to penicillinase. Although amoxicillin has a similar spectra of activity to those of ampicillin, it is better absorbed and provides a higher plasma level, with a longer duration of action, following oral administration, compared with other oral penicillins.

Some semi-synthetic penicillins have anti-pseudomonal activity; carbenicillin and piperacillin are included in this type.[1,10,14,15,18,19] These agents generally possess the same spectrum of activity as ampicillin with additional activity against aerobic Gram-negative bacteria, including *Klebsiella*, *Enterobacter* and *Pseudomonas*, although they are not stable to penicillinase.

Penicillin antibiotics are generally well distributed throughout the body with a sufficiently high concentration for therapeutic purposes in many organs and tissues; however, they do not cross the blood–brain barrier unless the meninges are inflamed with a resulting decline in barrier function.[10,19] Penicillin antibiotics are excreted into the urine from the kidneys in a non-metabolite form.[10]

Members of the penicillin group have minimal direct toxicity. Hypersensitivity reactions are the most common adverse effect. In general, penicillin agents are regarded as the safest antibiotics to receive during pregnancy.

Cephalosporin

The cephalosporin subclass is another important group of β-lactam antibiotics; these antibiotics have the same mechanism of action as penicillins. However, they have a wider antibacterial spectrum with increased stability to many types of β-lactamase and have improved pharmacokinetic properties.

The cephalosporin antibiotics include various types of agents, and at present they are often classified into four generation classes by their antimicrobial spectrum properties[15] (Table 13.5). In general, later generations are more resistant to β-lactamases and are characterised by extended spectra.

Table 13.5 Classification of Cephalosporins by Generation

Generation	Characteristics	Antibiotics
1	Good activity against Gram-positive bacteria (e.g., streptococci, staphylococci); relatively modest activity against Gram-negative bacteria	Cefazolin Cephalothin Cephapirin Cephalexin Cefadroxil Cephradine
2	Somewhat increased activity against Gram-negative bacteria, but weaker activity against Gram-positive bacteria compared with first-generation agents	Cefamandole Cefuroxime Cefoxitin Cefotetan Cefmetazole Cefaclor Cefprozil Cefurozime
3	Generally even less activity against Gram-positive bacteria than second-generation agents, but much more active against the Gram-negative bacteria including some of the β-lactamase-producing organisms (e.g., *Pseudomonas aeruginosa*, *Enterobacter*)	Cefotaxime Ceftriaxone Ceftizoxime Ceftazidime Cefoperazone Cefpodoxime Cefixime Cefdinir
4	Similar spectrum to third-generation agents, but enhanced activity against Gram-positive bacteria and more stable to β-lactamases, particularly class C β-lactamases	Cefepime Cefpirome

Although pharmacological properties vary between agents, cephalosporin antibiotics are generally well distributed to many body sites; however, rarely penetrate the blood–brain barrier.[10] Most cephalosporin agents are excreted primarily in the urine, although cefoperazone and ceftriaxone have significant biliary excretion.[10]

The toxicity level of cephalosporin antibiotics is low and, as with penicillin, hypersensitivity is the most common adverse effect. The majority of allergic reactions to cephalosporins are rashes, although anaphylaxis can occur. Because of the similarity in structure of the penicillin and cephalosporin groups, patients who are allergic to one class of them may manifest cross-reactivity when a member of the other class is administered. It has been reported that 5 to 10% of penicillin-allergic patients also have allergic reactions to cephalosporins.[20] The use of cephalosporin antibiotics in patients with possible penicillin allergy requires careful consideration.

Carbapenem

The carbapenems have an extremely broad spectrum of antimicrobial activity and are highly resistant to a variety of β-lactamases. Antimicrobial activity of carbapenem antibiotics is extremely high.[1,10,14,15,18,19] Many multi-drug-resistant hospital-acquired bacteria are often sensitive to carbapenems. However, expanded use of carbapenems has resulted in some carbapenem resistance in Gram-negative organisms such as certain Enterobacteriaceae and *Pseudomonas*.[18] The carbapenems are distributed widely in the body and are mostly excreted by the kidneys.[10]

The adverse effects of carbapenems are similar to those of other β-lactam antibiotics. As carbapenem antibiotics are structurally related to penicillin, caution should be used when administering them to a penicillin-allergic patient.

Monobactam

Monobactam agents have a narrow and characteristic spectrum of activity. They work only against Gram-negative bacilli, including *Pseudomonas aeruginosa*, and are not active against Gram-positive bacteria and anaerobes.[1,10,14,15,18,19]

Monobactam agents are relatively stable to β-lactamases.[19] However, the emergence of resistant organisms is an increasing problem with their frequent use. Monobactam is widely distributed in body tissues and fluids. It is regarded as a safe agent and has a similar toxicity profile to that of other β-lactam antibiotics.[15] Because there are no cross-hypersensitivity reactions with penicillin, these agents can be administered to patients who have a penicillin allergy.

β-Lactamase Inhibitors

The β-lactamase inhibitors bind to β-lactamases and inactivate them. Commercially available inhibitors include clavulanic acid, sulbactam and tazobactam. The β-lactamase inhibitors themselves have little direct antimicrobial activity; however, when combined with an antibiotic they extend the antibiotic's spectrum of activity and increase stability against β-lactamases.[14] Augmentin® is a product of amoxicillin combined with clavulanate, while Unasyn® comprises ampicillin and sulbactam. Tazocin® and Zosyn® are combination antibiotics containing piperacillin and tazobactam. Unfortunately, the available β-lactamase inhibitors do not inhibit all types of β-lactamases.[16,21]

Macrolides

The macrolide antibiotics have the common structure of a macrocyclic lactone ring to which are attached one or more deoxy sugars. The macrolide antibiotics are bacteriostatic agents that inhibit bacterial protein synthesis by binding reversibly to 50S ribosomal subunits of sensitive micro-organisms. The prototypic macrolide is erythromycin; other clinically important macrolides include clarithromycin and azithromycin.

The antimicrobial spectra of the macrolide agents are similar to those of penicillin.[10,19] These agents are active against Gram-positive cocci, including streptococci and staphylococci and spirochaetes, but are not active against enterococci, penicillin-resistant staphylococci and most Gram-negative bacteria with the notable exception of *Neisseria gonorrhoeae*. Erythromycin has weak activity against *Haemophilus influenzae*, but clarithromycin and azithromycin exhibit considerably better activity against this micro-organism. Consequently, macrolides are often used in the treatment of infections caused by Gram-positive bacteria as alternatives for patients who are allergic to penicillin.

The macrolide antibiotics are generally active against strict anaerobes. They are also effective against chlamydia, *Legionella pneumophila* and mycoplasma, against which many types of antibiotics, including β-lactams, are ineffective.[18,19] Macrolides are commonly administrated orally, although erythromycin can be given parenterally. Erythromycin is somewhat unstable in the presence of gastric acid, while clarithromycin and azithromycin are more acid-stable.[10,19] Macrolides diffuse readily into most tissues, but do not cross the blood–brain barrier.

The plasma half-life of clarithromycin and azithromycin is 3 times and 8 to 16 times longer than that of erythromycin's 90 minutes, respectively.[19] The macrolides enter and are concentrated within phagocytes.[19] Especially in the case of azithromycin, phagocytes may act as important vehicles for delivering the antibiotic to the infection site and sustaining its high concentration in the tissue. Because of this unique property and the extremely long elimination half-life, the clinical effects of this agent, with once-daily dosing for only 1 to 3 days, can be maintained for 7 days or more.[22]

The macrolides are inactivated in the liver, and the major route of elimination is in the bile. They have low toxicity, and serious untoward effects are rarely encountered. Possible adverse effects include hypersensitivity reactions, hepatitis, elevation of liver enzymes and gastrointestinal disturbance (e.g., diarrhea, nausea, vomiting).[1,10,19]

Clindamycin

Clindamycin binds exclusively to the 50S subunit of bacterial ribosomes and suppresses protein synthesis. The spectrum of activity for clindamycin is generally similar to that of macrolides. Importantly, clindamycin is highly active against oral streptococci and strictly anaerobic bacteria, although resistance has emerged among these organisms in some regions.[15,23,24] Therefore, clindamycin is used primarily to treat anaerobic infections, including acute dental abscess, acute sinusitis and aspiration pneumonitis. Clindamycin is not active against aerobic and facultatively anaerobic Gram-negative bacilli.

Clindamycin can be administrated orally or parenterally and is widely distributed in body fluids, organs and tissues, including bone, but it does not cross the blood–brain barrier.[19,24] This agent is also concentrated within phagocytes.[19] Clindamycin is metabolised in the liver and is excreted in both bile and urine.[19]

The common adverse effect of clindamycin is gastrointestinal disturbance and, in particular, diarrhoea. Pseudomembranous colitis is the most striking adverse effect. Significantly, however, it has been reported that there is not a notable difference in the risk of antibiotic-related colitis between clindamycin and other antibiotics such as the β-lactams.[25] One other possible adverse effect is skin rash.

Tetracyclines

The tetracyclines have a moderate broad-spectrum antimicrobial activity that is generally bacteriostatic by inhibition of bacterial protein synthesis through reversible binding to the 30S ribosomal subunit. The members of tetracyclines include tetracycline, doxycycline and minocycline.

Tetracyclines are active against many Gram-positive and Gram-negative bacteria, mycoplasma, rickettsia, chlamydia and spirochaetes.[1,10,19] However, their general usefulness has declined because of widespread resistance, although they remain the first-line agent for treatment of some specific infections such as rickettsial infection.[19] Tetracycline agents are often used in patients allergic to β-lactams and macrolides.

Frequently, tetracyclines are given orally, but parenteral administration is available. Dairy products; antacids containing calcium, aluminum, zinc, magnesium or silicate; and vitamins with iron can all interfere with the absorption of tetracyclines in the gastrointestinal tract when taken simultaneously with the drug.[19]

Tetracyclines are widely distributed in tissues except the cerebrospinal fluid.[19] They cross the placenta and are incorporated into fetal bone and teeth. Tetracyclines are metabolised by the liver and excreted primarily in the urine, but doxycycline is excreted via the biliary tract.[19]

Possible adverse effects include gastrointestinal irritation and photosensitivity.[10,19] The effects on teeth may be the most recognisable as a result of systemic administration of tetracyclines to children (≤8 years). In these situations a permanent brown discoloration of the teeth can occur.[1,10,19] In the case of pregnant women receiving tetracyclines, tooth discoloration may subsequently occur in their babies. As a consequence, members of the tetracycline group of antibiotics should not be prescribed to pregnant women or children.

Aminoglycosides

Structurally, aminoglycosides contain two or more amino sugars linked by glycosidic bonds to an aminocyclitol ring nucleus.[1] They are bactericidal inhibitors of protein synthesis through binding to the 30S subunit of the bacterial ribosome; however, the majority of bacterial protein synthesis inhibitors are bacteriostatic. These antibiotics include streptomycin, neomycin, kanamycin, amikacin, gentamicin and tobramycin.

The aminoglycosides carry out high antimicrobial activity against a wide range of aerobic and facultative Gram-negative bacteria, staphylococci and mycobacteria.[1,10,15,18,19,26] They are used primarily to treat infections caused by aeróbic and facultative Gram-negative bacteria (e.g., *P. aeruginosa*, *Acinetobacter* and *Enterobacter*) which tend to be resistant to multiple antibiotics.

Moreover, although not intrinsically active against enterococci, an aminoglycoside added to penicillin or vancomycin induces antibacterial synergy and results in strong bactericidal activity to these organisms.[1] Due to the synergistic effects, aminoglycosides are often administrated in combination with penicillin in the treatment of severe infections caused by Gram-negative bacteria. In contrast, since the aminoglycosides enter bacterial cells via an oxygen-dependent transport system, they have minimal action against anaerobic bacteria.[26]

Since aminoglycosides are not absorbed from the gut, they are usually administered intravenously and intramuscularly.[1,19,26] Some are used in topical preparations for wounds. After intravenous administration, aminoglycosides are freely distributed in the extracellular space, but poorly penetrate the cerebrospinal fluid, the vitreous fluid of the eye, the biliary tract, prostate and tracheobronchial secretions.[19] The pharmacodynamic properties of aminoglycosides include concentration-dependent activity and significant post-antibiotic effect. As a result of these properties, once-daily dosing is advocated instead of divided doses.[27] Aminoglycosides are excreted primarily by the kidney.[19]

Aminoglycosides are relatively toxic compared with the other classes of antibiotics, having two major toxicities: nephrotoxicity (renal toxicity) and ototoxicity (damage to the inner ear).[1] The risk of these is both dose- and duration-dependent. Nephrotoxicity is more common, but it can be reversed if use of the drug is stopped.[1] The likelihood of nephrotoxicity is enhanced in patients with pre-existing renal impairment. Ototoxicity can affect not only the cochlea (hearing) but also the vestibular (balance) system. It is often permanent and may occur even after discontinuation of the agent; it is cumulative with repeated courses of the agent.[15]

Fluoroquinolone

The first generation of quinolones is of relatively minor interest now because of their limited antimicrobial spectrum and therapeutic use, and the rapid development of bacterial resistance. However, the introduction of fluorinate 4-quinolones extended the antimicrobial activity and resulted in a particularly important therapeutic advance. The term *quinolone* usually refers to fluoroquinolone.

Fluoroquinolones exhibit concentration-dependent bactericidal activity through inhibiting bacterial DNA replication and transcription. As was mentioned earlier, the quinolone antibiotics have a broad antibacterial spectrum and are effective against both Gram-positive and Gram-negative organisms, which include Enterobacteriaceae, *Haemophilus*, *Moraxella catarrhalis* and, in the case of ciprofloxacin, *P. aeruginosa*.[1,15,18,19]

Fluoroquinolones have become an increasingly popular class of antibiotics for use for a variety of infections—in particular, infections due to aerobic and facultative anaerobic Gram-negative bacilli that are not susceptible to other agents. However, MRSA is usually resistant to these antibiotics.[15,18,19] The majority of fluoroquinolone agents are not highly active against streptococci and anaerobes. As their use has increased, resistance has developed among Enterobacteriaceae, *Pseudomonas*, *S. pneumoniae* and *Neisseria*.[15,18] However, newer fluoroquinolones have improved activity against streptococci, including *S. pneumoniae,* with reduced penicillin sensitivity and anaerobes.[1] These agents are often called 'respiratory fluoroquinolones' and include moxifloxacin.

The fluoroquinolones are well absorbed after oral administration, although parenteral administration is available for some agents. They are widely distributed in most extracellular and intracellular fluids, particularly in the kidneys, the prostate and them.[19] Aluminum and magnesium antacids interfere with the absorption of them.[19] Newer fluoroquinolones tend to have a longer half-life. The majority of fluoroquinolones are metabolised in the liver. They are primarily eliminated via the kidneys, although some are excreted in bile.

Unwanted effects occur infrequently; the most common include gastrointestinal, skin and central nervous system reactions (e.g., headache, dizziness, confusion, insomnia, mood alteration and agitation).[19] Despite being controversial, use of non-steroidal anti-inflammatory drugs during therapy with fluoroquinolones may enhance the agent's CNS stimulatory effects.[14,19] It has also been suggested that arthralgias and joint damage can develop in children receiving fluoroquinolones.[19] Therefore, clinical use of them for pre-pubertal children and pregnant women is not advocated.

Glycopeptide

A glycopeptide antibiotic is composed of glycosylated cyclic or polycyclic non-ribosomal peptides. Vancomycin is among the most important glycopeptides, although teicoplanin is also available. Glycopeptide antibiotics act primarily by inhibiting cell wall synthesis of bacteria. Vancomycin and teicoplanin have antimicrobial activity against almost all types of Gram-positive organisms including MRSA although their spectra of activity tend to be limited to Gram-positive organisms.[1,10,15,18,19]

Glycopeptide antibiotics—in particular, vancomycin—were historically regarded as the last effective line of defense. However, there are increasing reports of resistant enterococci strains.[28] Moreover, glycopeptide-resistant staphylococcal strains have increasingly been isolated from clinical specimens.[29,30] Nevertheless, vancomycin plays a significant role in the treatment of serious, life-threatening infections by Gram-positive bacteria which are unresponsive to other, less toxic antibiotics. Vancomycin is also indicated for treating pseudomembranous colitis caused by *Clostridium difficile*, against which a limited number of antibiotics are effective.[15]

The glycopeptide antibiotics are primarily eliminated via the kidneys.[15] Although vancomycin has traditionally been considered a highly nephrotoxic agent, it has recently been suggested that renal impairment caused by use of this agent is less likely than previously postulated.[31] However, the risk of nephrotoxicity is enhanced when vancomycin is administered to patients with impaired renal function/critical illnesses, when administered with concomitant nephrotoxic agents, or when patients are undergoing prolonged therapy.[32]

Vancomycin and teicoplanin can cause two types of hypersensitivity reactions: 'red man syndrome' and anaphylaxis.[33] Red man syndrome may be associated with rapid infusion of the first dose of the drug.[33] Patients typically develop symptoms 5 to 10 minutes after administration. The syndrome is characterised by itching and flushing of the face, neck and torso. It is not a true allergy but a non-specific mast cell degranulation, despite symptoms being similar to an allergic reaction. Slow intravenous administration should minimise its risk.

Metronidazole

Although metronidazole is used as an anti-protozoan agent, it is also available for the treatment of infections involving anaerobic bacteria.[15,18,19] Metronidazole exhibits high antimicrobial activity against almost all anaerobic bacteria, although it is not active against aerobes and facultative bacteria.[15,18,19] Therefore, metronidazole is effective for the treatment of anaerobic infections. This agent is often used as an adjunct to antibiotics with an aerobic spectrum of activity in treatment of a mixed aerobic (or facultative) and anaerobic infection. Metronidazole is active against *C. difficile*; therefore it is often used as the first-line agent in treatment of pseudomembranous colitis.[27]

Occurrences of serious adverse effects are rare. The most common complications are gastrointestinal symptoms (e.g., nausea and diarrhoea) and an unpleasant metallic taste (with oral therapy).[15] Metronidazole has a disulfiram-like affect, resulting in some patients experiencing abdominal distress, vomiting, flushing or headache if they drink alcohol during this drug therapy.

Trimethoprim/Sulfamethoxazole

Sulfonamides were the first modern anti-infective drugs. These agents competitively inhibit conversion of *p*-aminobenzoic acid to dihydropteroate in the process of microbial folate synthesis, and their effect is bacteriostatic.[1,19,34] Sulfonamides are relatively toxic.[1] Moreover, since many bacteria readily acquire resistance, sulfonamides have been supplanted by more effective and less toxic antibiotics for many bacterial infections. However, although sulfonamides alone are no longer recommended, they are still used in combination with dihydrofolate reductase inhibitors (especially trimethoprim) in treating several infections, as this combination has a synergistic antimicrobial effect and prevents bacterial resistance to either component alone.[1,34]

The antibiotic preparation trimethoprim/sulfamethoxazole (that is, co-trimoxazole, TMP-SMZ) combines trimethoprim and sulfamethoxazole (a sulfonamide-class agent) at a ratio of 1:5. TMP-SMZ blocks two different steps of the folate biosynthetic pathway; sulfamethoxazole acts as a folate synthesis inhibitor, and trimethoprim blocks bacterial nucleotide synthesis that is achieved in the presence of folate. This preparation is active against many Gram-positive and Gram-negative organisms, although it has poor activity against *P. aeruginosa* and strict anaerobes.[1] Moreover, this agent effectively inhibits the growth of some types of protozoa and fungi. TMP-SMZ is very active against *Pneumocystis jirovecii*, so it is significant in the prevention and treatment of *Pneumocystis* pneumonia.[1,34] It is also the first-line agent in the treatment of *Stenotrophomonas maltophilia* and *Burkholderia cepacia* infections.

TMP-SMZ is available in both oral and intravenous formulations. It is well absorbed from the gastrointestinal tract and widely distributed in tissues and fluids. In particular, it concentrates in the urinary tract and the prostate.[19] For these reasons, TMP-SMZ is usually the drug of choice for urinary tract infections and prostatitis.[1,19,34]

Adverse effects of TMP-SMZ are not common but are more likely to occur in patients with acquired immune deficiency syndrome.[34] The common adverse effects include gastrointestinal disturbances (e.g., nausea, vomiting, anorexia) and allergic skin reactions.[35] Other possible adverse effects include Stevens-Johnson syndrome, toxic epidermal necrolysis, haematologic toxicities and disorders (e.g., agranulocytosis, aplastic anemia, thrombocytopenia, leukopenia, neutropenia), hepatic and renal impairments and significant electrolyte disturbance (e.g., hyponatremia and hyperkalemia).[34,35] Clinical use of TMP-SMZ in pregnant women, neonates and infants should be avoided.[27]

MICROBIOLOGICAL EXAMINATION

Microbiological examination plays an important role in the diagnosis and control of infectious disease. As such examination identifies micro-organisms likely to be involved in the disease and their susceptibility to chemotherapeutic agents, case-specific antimicrobial regimens can be made based on the results. However, results obtained with improper collection, poor technique, inappropriate transport and inadequate management of the specimen may contribute to misdiagnosis and inappropriate antimicrobial therapy.

Collection of Specimen

Clinical material for microbiological testing should be collected from a site representative of the active disease process. Sites of inflammation that are free of contaminating micro-flora are optimal. Since large numbers of micro-organisms reside in the human body, specimens often may be contaminated with indigenous micro-organisms that are not involved in the infection. To reduce this risk, any contact with surrounding tissues and fluids during collection must be avoided or minimised. In addition, avoid contamination of specimens with any micro-organisms colonising the hospital environment and associated healthcare workers. Collection of specimens must be in accordance with the principles presented in Table 13.6.

When collecting specimens, the site of sampling, visual features (e.g., color, turbidity and viscosity if the material is a liquid), odor, status of the collection site (e.g., in the case of an abscess, whether or not it has drained), method of collection, patient condition, suspected diagnosis and details of the present antimicrobial chemotherapy must be recorded. Such information can be helpful for preliminary diagnosis. For example, a green purulent exudate often suggests involvement of *P. aeruginosa*. A 'rotten' or gangrene-like odor indicates likely involvement of anaerobes.

Transport of Specimen

Specimens should be transported to the laboratory as soon as possible. Excessive delay or exposure to extreme temperatures can compromise results and must be avoided.

Some micro-organisms are sensitive to *in vitro* environments. For example, oxygen kills strictly anaerobic micro-organisms, and drying kills most bacteria when left exposed to the air. Conversely, some types of aerobic organisms can multiply when stored for several hours at room temperature. For these reasons, inadequate management

Table 13.6 Principles of Specimen Collection for Microbiological Examination

Select the proper anatomic site for collection
Avoid contamination with indigenous flora
Disinfect surface; then aspirate and biopsy tissue
Collect sufficient volume of material to enable all test requests to be performed satisfactorily
Insufficient volume may yield false-negative results
Use a specimen container designed to promote survival of pathogenic bacteria, eliminate leakage and allow safe handling during transport and processing

Source: U.S. Food and Drug Administration.[40]

of specimens after collection or delayed transport to the laboratory may result in deterioration and/or overgrowth of micro-organisms within specimens and may produce incorrect results. An insufficient sample volume may also provide inaccurate results.[36]

Detection and Identification of Bacteria

Once a specimen is collected and sent to the laboratory, it is incubated on an agar medium to culture any bacteria or fungi present for subsequent isolation of microbial colonies. Aerobic and facultative bacteria and fungi may grow relatively rapidly on agar media, while sufficient growth of strictly anaerobic organisms requires at least a few days. Therefore, microbiological examination, particularly involving the culture of anaerobic organisms, usually takes several days.

Colonies of bacteria or fungi grown on agar plates are identified at the genus or species level. Identification is usually based on growth characteristics under various conditions, the results of biochemical tests and/or the results of commercial identification kits, together with microscopic determination of colony and cellular morphology.

Automated instruments for bacterial identification and antimicrobial susceptibility are currently available. These instruments culture the micro-organism and identify it once a sufficient quantity is present to generate a detectable biological signal. Despite limited flexibility due to significant costs for installation, automated instruments can provide a dramatic reduction in labor and time for examination compared with traditional manual methods.

Antimicrobial Susceptibility Test

The prime objective of susceptibility testing is to predict the likely outcome of treatment with tested antibiotics. The results of this test also provide a guide to evaluate the effectiveness of the patient's present antimicrobial chemotherapy.

Susceptibility tests are either quantitative or qualitative. The quantitative tests provide MIC values for antibiotics against isolates. In general, the lower the MIC, the greater the *in vitro* activity of an antibiotic. However, an MIC cannot be interpreted based on absolute values alone. Interpretive criteria are required to determine whether the antibiotic is likely to work for a given patient's specific infection. An antibiotic breakpoint is a maximum MIC threshold for predicting successful antibiotic therapy; these have been established by regulatory authorities such as the US Clinical and Laboratory Standards Institute (CLSI).

Breakpoints for each antibiotic are established using many parameters, including microbiological, pharmacokinetic, pharmacodynamic and clinical data. Thus, the breakpoint differs for drug, target organism type and species. Based on the breakpoint, the MIC for a tested micro-organism is interpreted as susceptible, intermediate or resistant. The implication of a 'susceptible' result is that the infection caused by the

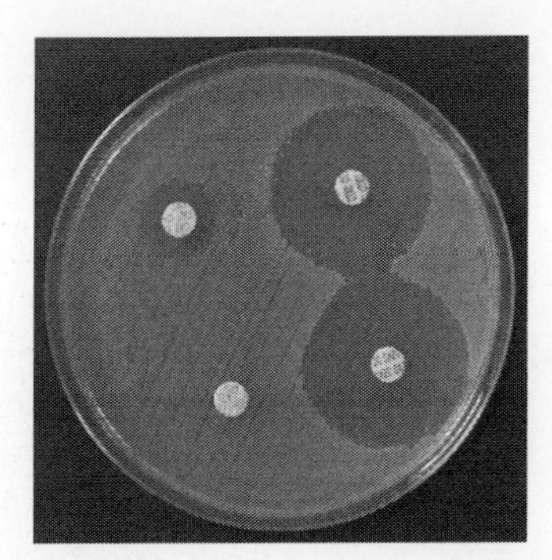

Figure 13.4 ***The disk diffusion method.*** The paper antibiotic disks produce a range of zones of bacterial growth inhibition. A large zone of inhibition indicates high susceptibility of bacteria.

micro-organism is likely to respond to treatment with antibiotic therapy at the recommended dosage.[37] Conversely, the 'resistant' result implies that an infection will not be successfully treated with that antimicrobial agent. An 'intermediate' result suggests that the infection is possibly treatable with a high concentration of the antibiotic but not at the recommended dosage.

It should be noted that selected breakpoints for a given drug against a given micro-organism differ among countries because of differences in drug dosing, laboratory methods for determining susceptibility and philosophies surrounding breakpoints.[38]

In practice, the disk diffusion method, which is not quantitative but qualitative, is often employed to determine antibiotic susceptibility because of its relatively good flexibility and cost effectiveness. In the disk diffusion test, commercially prepared paper antibiotic disks of fixed antibiotic concentrations are placed on an agar surface inoculated with the test isolate.

After incubation, the diameters of growth inhibition zones around each of the antibiotic disks are measured to the nearest millimeter (Figure 13.4). The diameter of the zone is related to the susceptibility of the isolate. The lower the MIC, the larger the zone of inhibition. The zone diameters of each drug are interpreted as susceptible, intermediate or resistant, using the criteria, which were originally determined by regulatory authorities, provided by disk manufacturers.

Interpretation of Results

Specimens often may be contaminated with indigenous micro-flora or micro-organisms colonising the local environment, while opportunistic infections usually involve commensal or environmental organisms. Therefore, it is sometimes difficult to identify the exact causative organism. In general, in cultures of specimens from sites likely to be contaminated (e.g., sputum, urine and those of superficial wounds), potential pathogens should outnumber contaminated organisms.

In circumstances where culture results are far from the clinical picture of the patient, or where clinicians feel that the results may be unreliable, the sample site should be retested at a subsequent time point. Good communication between the clinician and the microbiologist is essential to ensure appropriate sample collection and interpretation of results.

In biofilm infections, interpretation of susceptibility results needs to be made with caution. Biofilm formation provides micro-organisms with enhanced tolerance to antibiotics. It has been reported that the MICs for bacteria living in biofilms can exceed 1000-fold those of their free-living counterparts.[39] It should be emphasised that *in vitro* susceptibility results are only one aspect of a complex picture and should be considered with all other factors when planning antimicrobial chemotherapy regimens.

INDICATION AND SELECTION OF ANTIBIOTICS

Although antibiotic therapy plays a significant role in the prevention and management of bacterial infections, it can also have negative aspects. As mentioned previously, all antibiotics may exhibit a variety of adverse effects or may interact with other systemically administered medicines, and some of these effects are potentially serious and fatal. As will be discussed later, inappropriate antibiotic therapy can also increase the chances of bacterial resistance. The cost of the antibiotic also might be an important matter for consideration. Therefore, antibiotics should be used when it is actually necessary, and they must be selected and administered appropriately.

Necessity

Antibiotics must be used only to treat an infection that likely involves bacteria, or to prevent bacterial infection in patients with a high risk of developing a serious infection (e.g., prevention of surgical site infection in an immuno-compromised host). Certain bacterial infections (e.g., abscesses and infections with foreign bodies) require surgical intervention and may not respond to antibiotic therapy alone.

Patients could have an underlying debilitation of their normal immune defenses, such as is evident in uncontrolled diabetes. Moreover, patients are sometimes debilitated with varying degrees of dehydration and/or nutritional and electrolyte imbalances. Since such factors may interfere with the effect of antimicrobial chemotherapy and delay recovery, they should initially be assessed and corrected.

Antibiotic therapy works only against bacterial infections. It is therefore important to restrict the use of antibiotics to situations where the involvement of bacterial infection has been confirmed or is highly probable. Regrettably, however, antibiotics are often prescribed in cases of viral infection such as common colds and influenza and undifferentiated fever and sickness. These are obviously inappropriate uses for antibiotic therapy and should be avoided.

Selection

Once a bacterial infection has been confirmed or is highly suspected, antibiotic selection and its regimen should be tailored to individual cases. The following subsections describe the factors to be considered.

Spectrum

The choice of antibiotic therapy should be based on the result of microbiological examination whenever available. However, in practice, such results are often unavailable for several days. As a consequence, an initial choice is made on an empirical basis; that is, the antibiotic needs to be chosen according to the most likely pathogens before laboratory data are available. The likely pathogens are often predictable through careful physical examination of the patient and observation of visual features and the odor of pus or purulent exudate. Furthermore, statistical probabilities based on the site of the infection, details of the host (e.g., age, underlying disease, pre-existing therapy) and susceptibility trends in the local hospital or community setting might further indicate the likely pathogens and their antimicrobial susceptibility.

Since an antibiotic is distributed throughout the body when administered systemically, it can affect micro-organisms of the normal micro-flora. Therefore, ideal antibiotics target only the causative organisms without affecting other commensal ones. Obviously, use of broad-spectrum antibiotics is more likely to disrupt such microbial homeostasis and is more likely to trigger the development of antibiotic resistance compared with narrow-spectrum antibiotics. Thus, whether or not chosen according to the results of microbiological examination, those antibiotics with the narrowest spectrum of activity that can control the infection should be used. Antibiotics that exhibit a broader antimicrobial spectrum should be reserved for situations where narrow-spectrum antibiotics are unsuitable.

Broad-spectrum antibiotics are advocated when the causative bacteria have not been identified and there is a wide range of possible causative agents, or when the bacteria are resistant to narrow-spectrum antibiotics. Nevertheless, the validity of the broad-spectrum agent should be reassessed and the possibility of de-escalation to a narrow-spectrum antibiotic should be considered during ongoing therapy.

Use of two or more specific antibiotics in combination can provide synergistic activity against micro-organisms or can enhance the antibacterial spectrum. Therefore, a multi-drug regimen of antibiotics is advocated for management of patients with severe infection or a high likelihood of a serious complication. It should be remembered that not all combinations of antibiotics act synergistically. Some are antagonistic or have little or no advantageous effects.

Pharmacokinetics and Pharmacodynamics

In vivo effectiveness is influenced by antibiotic concentration achieved at the infected site and the time course of antibacterial effects exerted by drug levels in the blood and at the site of infection. Most antibiotics are distributed throughout the body, but their transfer to specific organs and tissues may be restricted depending on type. For example, only a few antibiotics penetrate the CNS at sufficient concentrations for therapy of meningitis or brain abscess. It is essential that the administered antibiotic reach the site of infection at a concentration above the MIC of the pathogen. Pharmacokinetics and pharmacodynamics of the antibiotic should be considered in selection of the agent.

Safety

All antibiotics have various degrees of toxicity and adverse effects. In general, β-lactam antibiotics, especially penicillins, are considered safe. Macrolides, fluoroquinolone and metronidazole exhibit minor toxicity with a low incidence of adverse effects. In contrast, aminoglycosides exhibit high renal toxicity.

Antibiotics sometimes interact with other drugs and cause harmful and undesirable effects. Antibiotics may provide an unwanted enhancement or reduction of the effect of systemically administered medicines.

Patient Condition

Almost all antimicrobial agents are capable of crossing the placenta and have a range of teratogenic effects. The systemic use of antibiotics in pregnant women involves an evaluation of risk versus benefit. Any unnecessary use of medications during pregnancy, in particular during the first trimester, must be avoided.

The US Food and Drug Administration (FDA) has established a fetal risk summary which divides drugs into categories based on accumulated safety data from animal and human studies,[40] as shown in Table 13.7. Use of agents in category A or B for pregnant

Table 13.7 Pharmaceutical Agent Risk Categories during Pregnancy and Usage Criteria

Category	Usage criteria
A	Adequate and well-controlled studies in pregnant women show no increased risk of fetal abnormalities in any trimester
B	Animal studies reveal no evidence of foetal harm; no adequate, well-controlled studies in pregnant women
	Animal studies show adverse effect, but adequate, well-controlled studies in pregnant women fail to demonstrate foetal risk
C	Animal reproduction studies show adverse effect on fetus; no adequate, well-controlled studies in humans
	No animal studies conducted; no adequate, well-controlled studies in pregnant women
D	Positive evidence of human foetal risk based on adverse reaction data from investigational or marketing experience or studies in humans; potential benefits may warrant use of drug in pregnant women despite potential risks
X	Adequate, well-controlled or observational studies in animals or pregnant women show positive evidence of foetal abnormalities; use is contra-indicated in women who are, or may become, pregnant

Source: US Food and Drug Administration.[40]

women and infants is recommended. Category C agents should be used only when there is no safer alternative, or when the benefit outweighs the risk. Use of category D drugs may be advocated for life-threatening or serious infections for which safer drugs cannot be used or prove ineffective. Drugs listed as category X are contra-indicated for use during pregnancy.

It should be noted that most antibiotics administered to the lactating mother are also detectable within her breast milk; this must be considered when administering antibiotics to breastfeeding women. Penicillin may be the most suitable candidate during pregnancy and for infants and children in the tooth- and skeletal-development stages (Table 13.8). In contrast, clinical use of tetracyclines (intrinsic dental staining), aminoglycosides (nephrotoxicity and ototoxicity) and fluoroquinolones (chondrotoxicity in growing cartilage) for these patients is not advocated.

The kidneys' compensatory capacity is relatively restricted, and drugs cleared by them in a form retaining strong pharmacological activity may present a significant burden. Consequently, when these drugs are administered in patients with renal failure, impairment of drug clearance by the kidney could result in abnormal accumulation of antibiotic in the body, enhancing the likelihood of adverse effects. Therefore, in patients with renal impairment or who are undergoing dialysis, a dose adjustment may be required depending on the severity of renal failure and type of antibiotic.[27]

Antibiotic doses and scheduling may also need to be adjusted for infants and the elderly, as the drug metabolism and excretion in these patients may be insufficient (in infants) or significantly deteriorated (in the elderly).

Table 13.8 Risk Categories of Antibiotics during Pregnancy

Antibiotic and/or drug	Categories
Penicillins	B
Cephalosporins	B
Carbapenems	
Imipenem	C
Meropenem	B
Monobactam	B
Macrolides	
Erythromycin	B
Azithromycin	B
Clarithromycin	C
Clindamycin	B
Aminoglycosides	D
Tetracyclines	D
Vancomycin	C
Fluoroquinolones	C
Metronidazole	B
Trimethoprim/sulfamethoxazole	C

Source: US Food and Drug Administration.[40]

Route of Administration

Antibiotics are commonly administered orally because it is easy, painless and cost effective. However, the level of drug absorption in the gastrointestinal tract differs greatly between individuals. Intravenous administration may ensure sufficient and consistent levels of antibiotic in the plasma and at the infection site. In the case of severe infection or an infection with a high risk of serious complications, antibiotics should be administered an intravenously. Intravenous administration is preferred if oral antibiotics cannot be tolerated, when oral antibiotics cannot be absorbed because of impairment in intestinal absorbance or when no oral formulation is available.

Continuous intravenous administration of drugs may be difficult with patients who arbitrarily remove intravenous access devices (e.g., needle, catheter) that have been inserted for antimicrobial therapy. This situation is often encountered in treatment of patients with certain types of dementia, mental disorder or severe mental retardation. In such cases, intramuscular injection may be the choice for administration, although not all injection drugs can be given intramuscularly.

Cost

The cost of therapy may be an important factor in the selection of an antibiotic, and less expensive antibiotics are, of course, preferable. Newer drugs (e.g., broad-spectrum cephalosporins and carbapenems) tend to be more expensive, while older antibiotics, including penicillin and agents that are available as generics, may be more cost effective. It should also be remembered that the cost of antimicrobial chemotherapy is not merely the expense of the drug itself but also the cost of administration, monitoring (if required) and retreatment in the case of treatment failure. Intravenous administration is generally more expensive than is the use of oral antibiotics.

DRUG INTERACTIONS AND ADVERSE EFFECTS

Many antimicrobial agents can interact with co-administered drugs. Despite the fact that some drug interactions are beneficial and available for therapeutic purposes, many often result in detrimental effects. In general, 'drug interaction' usually refers to the unwanted and harmful actions of one drug arising from concomitant administration of another drug or chemical substance. A home remedy, birth-control pill, alcohol and food type can be the cause. Drug interactions commonly result in an undesired change in drug concentration in the body due to the affect of a co-administered agent on absorption, distribution, metabolism or elimination of the drug; unwanted enhancement or reduction of drug activity resulting from synergistic or antagonistic effect with a co-administered agent; or physical or chemical incompatibility (chemical inactivation or precipitation) when two agents are mixed in the same intravenous fluid.[41]

All drugs that provide a proven therapeutic benefit can cause adverse effects, which range from mild to fatal. Notable adverse effects associated with antimicrobial chemotherapy are described next.

Diarrhoea and Colitis

Antibiotics administrated systemically can affect not only the target bacteria at the infection site but also the body's commensal bacterial micro-flora, which are actually beneficial to health. The disruption of the species composition in the commens micro-flora can result in undesirable effects. The most common clinical effect is antibiotic-associated diarrhoea, which results from an imbalance in the colonic microbiota caused by antimicrobial chemotherapy. Any type of antibiotic can be associated with such diarrhoea. The majority of cases are not severe and usually appear as loose stools or mild diarrhoea. This commonly ends spontaneously once administration of the antibiotic has ceased. However, additional treatment may be required depending on the severity of diarrhoea and the patient's underlying condition.

Diarrhoea on rare occasions, is associated with pseudomembranous colitis. This type of colitis involves an overgrowth of *C. difficile* (which is ordinarily a minor member of the intestinal flora) caused by a disruption of the normal bacterial content of the large intestine resulting in a loss of the normal healthy bacteria because of antimicrobial chemotherapy.[42] Prognosis of this type of colitis is generally good, but in severe cases might be fatal. Use of clindamycin, penicillin and cephalosporins for a long duration are the most common cause of this condition.[42] Although rare, diarrhoea can result from an allergic inflammatory reaction to the drug in the intestinal membrane.

Drug Rashes

Drug rashes are an adverse effect of an agent that manifests as a skin reaction. These usually result from an allergic reaction. Drug rashes vary in severity from mild redness with 'pimples' over a small region to complete skin peeling. They may appear suddenly within minutes after administration of a drug, or they may be delayed for hours, days or weeks.

Although drug rashes may have various clinical appearances, the maculopapular (morbilliform) type, characterised by a flat, red rash, which may include pimples similar to measles, is the most common.[43,44] This rash typically appears one to two weeks after starting the medication and begins as discrete red or pink spots on the trunk of the body. These spots gradually cluster together and merge to form sheets of flat, blotchy rashes that spread to the neck and limbs in a symmetrical pattern.

Hives (urticaria) are another important form of skin rash. They are characterised by red, raised spots with a pale center and are commonly 'itchy' and swollen and of variable size. Hives typically appear within 24 hours of administration of an antibiotic and can affect mucous membranes. They sometimes develop as a component of anaphylaxis.[43,44] Skin reaction to an antibiotic can also form a fixed eruption—a dark red or purple rash that reacts at the same site on the skin.

Any type of antibiotic can trigger a drug rash, although penicillins and cephalosporins are the most common causes. Rashes usually resolve in days to weeks after discontinuing the medication; however, they can be part of a more serious, potentially life-threatening reaction such as Stevens-Johnson syndrome, toxic epidermal necrolysis and anaphylaxis.[43]

Some antibiotics can cause photosensitivity. Patients with photosensitivity sunburn particularly easily or develop exaggerated skin reactions to sunlight. As a result, they may have an unexpected sunburn or a dry, bumpy or blistering rash on sun-exposed skin, commonly on the face, neck, arms and backs of the hands. Tetracyclines and fluoroquinolones are common causes of this type of reaction.[44]

Anaphylaxis

Despite a very low incidence, anaphylaxis (i.e., anaphylactic shock) is the most serious and life-threatening adverse effect of antibiotics. Although β-lactam antibiotics and, in particular, members of the penicillin group primarily cause this reaction, any type of antibiotic can be the trigger. Clinical signs include diffuse urticaria, stridor and angio-oedema. Rapid pulse, reduced blood pressure, weakness, heart arrhythmias, mental confusion and various abdominal symptoms (e.g., nausea, vomiting, diarrhoea and abdominal pain or cramping) may also occur.[45] Airway obstruction due to glottic (laryngeal) oedema and bronchospasm is extremely important to recognise because of potential fatal consequences.

Anaphylactic reactions usually begin 1 to 15 minutes after administration of the trigger agent and require immediate professional treatment. The priority is to transfer the patient to hospital as an emergency. The patient's vital signs must be checked and emergency management starts with the ABC (airway, breathing, circulation) of resuscitation. Adrenaline (1:1000, 0.3–0.5 ml) should be injected intramuscularly without delay to open the airways, to raise blood pressure by constricting blood vessels, and to stop the allergic reaction until professional treatment starts.[45] If available, corticosteroids (e.g., prednisone) should be given to further reduce symptoms after primary life-saving treatment is administered.

Nephrotoxicity

Antibiotics are excreted either into the bile through the liver or into the urine from the kidneys. In the process of drug excretion through the kidneys, they are exposed to a potentially toxic agent and nephrotoxic injury can develop by direct damage from this agent or because of allergic reaction.

Symptoms of nephrotoxic injury are generally similar to those of renal failure, and include proteinuria, haematuria, a range of systemic oedemas and high blood pressure (hypertension). However, in the early stages only the fluctuation of serum enzyme concentrations, related to renal function, to unacceptable levels is detected without any visual sign and symptoms.

Although many drugs that are excreted from the kidneys can cause nephrotoxic injury, aminoglycosides, glycopeptide (vancomycin), meticillin, some cephalosporins and amphotericin B are the most likely to be associated with this effect. Elderly people are more likely to have nephrotoxic injury, as their renal functions often have declined. Other risk factors include underlying kidney disease or renal insufficiency, diabetes, sepsis, severe dehydration and use of multiple nephrotoxic drugs such as NSAIDs.[46,47]

The outcome of nephrotoxic injury is determined by the severity of the damage. In mild cases, kidney function may recover once use of the causative drug is discontinued, whereas significant and permanent damage can result in chronic renal failure or even death.

ANTIBIOTIC RESISTANCE

Details of microbial resistance against antimicrobial agents are described in other chapters of this book. This section briefly discusses the basic concepts of bacterial resistance to antibiotics.

Antibiotic resistance is the ability of bacteria to resist the effects of an antibiotic agent; it is divided into natural and acquired resistance. Natural resistance means that the bacteria are 'intrinsically' resistant. Not all micro-organisms are susceptible to all antibiotics, and bacteria may be inherently resistant to an antibiotic. For example, *Streptomyces* has some genes that are responsible for resistance to the antibiotic produced by it.

Acquired resistance is said to occur when a particular micro-organism resists the activity of a particular antimicrobial agent to which it was previously susceptible. In general, 'antibiotic resistance' is acquired. It can result from the mutation of genes involved in normal physiological processes and cellular structures, or from the acquisition of foreign resistance genes, or from a combination of these two mechanisms.[48] The mechanisms of antibiotic resistance are presented in Figure 13.5 and Table 13.9.

Antibiotic-resistant bacteria interfere with an antibiotic's therapeutic effects. As a result, management of infection becomes more difficult and more expensive. A bacterium resistant to one antibiotic is not resistant to all of them. However, bacteria that acquire resistance to more than one type of antibiotic are referred to as multi-resistant. In such cases, treatment may be extremely difficult because very few antibiotics may be available that are effective in combating such bacteria.

Healthcare professionals must remember that use of an antibiotic can promote bacterial resistance: Sensitive bacteria are killed and resistant ones survive and grow. The more antibiotics are used, the greater is this 'selective pressure', favouring resistant strains. The process is a classic example of Darwinian evolution driven by the principles of natural selection.

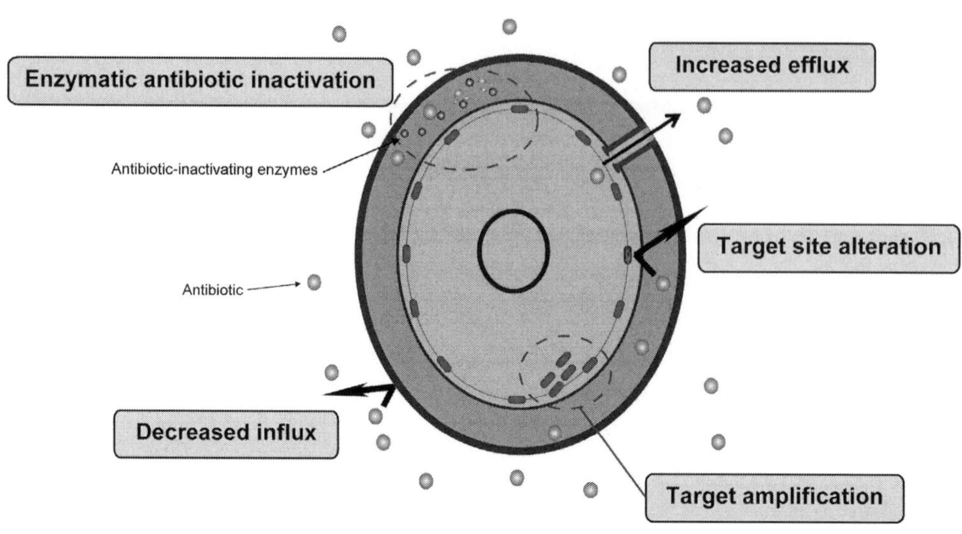

Figure 13.5 ***The prime mechanisms of bacterial resistance to antibiotics.*** These include enzymatic degradation of antibacterial drugs; alteration of bacterial proteins that are antimicrobial targets; decreased membrane permeability to antibiotic or creation of efflux systems that prevent the drug from reaching its intracellular target; overproduction of target (antibiotics entering the bacterial cell cannot bind to all targets, so targets survive). *Source: Adapted from Schmieder and Edwards.*[49]

Table 13.9 Mechanisms of Bacterial Resistance to Antibiotics

Mechanism	Antibiotics
Antibiotic inactivation	β-lactams
	Aminoglycosides
	Macrolides
	Clindamycin
	Glycopeptides
Influx decrease	β-lactams
	Aminoglycosides
	Metronidazole
Efflux increase	β-lactams
	Macrolides
	Aminoglycosides
	Tetracyclines
	Floroquinolones
Target site alteration	β-lactams
	Macrolides
	Clindamycin
	Aminoglycosides
	Tetracyclines
	Floroquinolones
	Glycopeptides
Target amplification	Sulfonamides/trimethoprim

Source: Data obtained from five references.[48–52]

Antibiotic-resistant bacteria can spread through direct contact with a person carrying the resistant micro-organism, and the hospital setting is often involved in such dissemination. The common ways in which bacteria can be spread in hospitals include direct contact with the contaminated hands of hospital staff; contact with contaminated surfaces such as door handles, overbed tables and call bells; and contact with contaminated equipment such as stethoscopes, blood pressure cuffs and the keyboard and mouse of a computer. Nursing care, such as diaper exchange and toilet and bathing care, can also be involved. The widespread use of antibiotics in animal husbandry is also creating drug-resistant bacteria, which may be transmitted to humans through the food chain.

Antibiotic resistance is ubiquitous, and associated infections may also spread in the community at large. In addition, increased globalisation can result in the spread of drug-resistant micro-organisms. For example, New Delhi metallo-β-lactamase-1 has spread from India to Europe and North America probably through increased tourism.[53]

ANTIFUNGALS

Most fungi are not affected by antibacterial agents. Antifungal agents are drugs that inhibit fungal growth. In contrast with bacteria, fungi are eukaryotic cells and therefore cytologically have higher similarity with host cells. As a consequence, there are a limited number of drugs that can affect fungal cells without significant toxicity to human cells. In general, more adverse effects are likely to occur with antifungal agents compared with antibiotic therapies.

Polyene

Amphotericin B is the best known of the polyene antifungals. Its mechanism of action is by fungal membrane disruption induced by the agent binding to a sterol moiety, primarily ergosterol, in the cell membrane of the fungi. Amphotericin B has high antifungal activity against a variety of fungal species. Its absorption from the gastrointestinal tract is very low. Therefore, intravenous administration is employed in systemic therapy. Amphotericin B is excreted in the urine over several days. The major acute reaction to intravenous amphotericin B is fever and chills,[54] with nephrotoxicity another significant side effect.[54] Topical amphotericin B is a useful candidate for treatment of superficial candidosis and is considered safe.

Nystatin is another polyene antifungal. It, too, is not absorbed from the gut. Nystatin is too toxic for parenteral administration, and its clinical use is limited to topical applications to the skin and mucous membranes.

Azoles

Azole antifungal agents are frequently used in the prevention and treatment of fungal infection because of their relatively broad spectrum of activity and low toxicity compared with polyene antifungals. Azoles impair the synthesis of the fungal cell membrane by inhibition of sterol synthesis. They include two broad classes: imidazoles and triazoles. Miconazole is an imidazole agent that is widely used clinically; fluconazole, itraconazole and voriconazole are triazoles. Although both classes share the same antifungal spectrum and mechanism of action, the systemic triazoles are more slowly metabolised and have lower toxicity to human cells compared with the imidazoles.[54]

Fluconazole and itraconazole are administered systemically in the treatment of various fungal infections. They are readily absorbed from the gastrointestinal tract. Voriconazole is a broad-spectrum triazole and is very effective in the treatment of *Aspergillus* infections.[55] It is also active against many types of fungi, including those that are not susceptible to other azole agents. Oral and intravenous forms are available.

Fluconazole and itraconazole are generally well tolerated. Gastrointestinal distress, nausea, vomiting, skin rash and headache are major side effects. Voriconazole is more toxic compared with fluconazole and itraconazole. Common adverse effects include hepatotoxicity, visual disturbances and dermatologic reactions.[56] Miconazole is usually used topically, and therapy is considered safe.

Echinocandin

Echinocandins inhibit the synthesis of glucan (an important constituent of the fungal cell wall) and have a wide antifungal spectrum of activity. They can be used to treat various forms of candidosis, aspergillosis and other mycoses. Serious side effects occur rarely, and adverse interactions with other drugs are less frequent compared with other types of antifungals.[57]

CONCLUSION

Antimicrobial chemotherapy plays a significant role in the management of infections. Many types of antimicrobial agents with different mechanisms of action, pharmacological properties and spectra of activity are available. Use of antimicrobial chemotherapy should be tailored to individual cases, with a good understanding of each drug's characteristics. It should be recognised that antimicrobial chemotherapy has potential negative effects including adverse ones. Moreover, use of antimicrobials provides intense selection pressure on the microbial populations to evolve resistance. Antibiotic resistance is currently one of the world's most pressing public health problems. It is of great concern that many clinicians still use antimicrobials inappropriately, with a poor understanding of the consequences. Proper use of these agents therefore remains an ethical duty for all healthcare professionals.

Acknowledgments We would like to thank Ms. Hiromi Matsumoto, Dr. Iyo Kimura, Professor Shuichi Kawashiri (Kanazawa University), Ms. Toshimi Hagihara, Ms. Taeko Nishimura, and Ms. Rika Ushiyama (Fujimi-Kogen Medical Center) for their kind help in collecting material for this chapter.

REFERENCES

1. Page CP, Curtis M, Sutter MC, Walker M, Hoffman B. Drug and bacteria. In: *Integrated pharmacology*, 2nd ed. Mosby; 2002. p. 111–143.
2. Mansour TE. *Chemotherapeutic targets in parasites: Contemporary strategies*. Cambridge University Press; 2002. p.1–18.
3. Winau F, Westphal O, Winau R. Paul Ehrlich—in search of the magic bullet. *Microbes Infect* 2004; **6**:786–9.
4. Garrod LP. Alexander Fleming. A dedication on the 50th anniversary of the discovery of penicillin. *Br J Exp Pathol* 1979;**60**:1–2.
5. Bentley R. Different roads to discovery: Prontosil (hence sulfa drugs) and penicillin (hence β-lactams). *J Ind Microbiol Biotechnol* 2009;**36**:775–86.
6. Rubin RP. A brief history of great discoveries in pharmacology: in celebration of the centennial anniversary of the founding of the American Society of Pharmacology and Experimental Therapeutics. *Pharmacol Rev* 2007;**59**:289–359.
7. Saga T, Yamaguchi K. History of antimicrobial agents and resistant bacteria. *Japan Med Assoc J* 2009;**52**:103–8.
8. Levison ME. Pharmacodynamics of antimicrobial drugs. *Infect Dis Clin North Am* 2004;**18**:451–65.
9. Zeitlinger MA, Derendorf H, Mouton JW, Cars O, Craig WA, Andes D, et al. Protein binding: do we ever learn? *Antimicrob Agents Chemother* 2011;**55**:3067–74.
10. Endo M, Kuriyama K, Okuma S, Tanaka T, Higuchi M. *Medical pharmacology*, 4th ed. Nanzando (in Japanese); 2005. p. 529–72.
11. Slavik RS, Jewesson PJ. Selecting antibacterials for outpatient parenteral antimicrobial therapy: pharmacokinetic-pharmacodynamic considerations. *Clin Pharmacokinet* 2003;**42**:793–817.
12. Frimodt-Møller N. How predictive is PK/PD for antibacterial agents? *Int J Antimicrob Agents* 2002; **19**:333–9.
13. Akita K. The PK-PD concept in clinical practice. *Yakkyoku* 2009;**60**:19–24 (in Japanese).
14. Petri Jr. WA. Penicillin, cephalosporins, and other β-lactam antibiotics. In: Brunton LL, Lazo JS, Parker KL, editors. *Goodman and Gilman's the pharmacological basis of therapeutics,* 11th ed. McGraw-Hill; 2006. p. 1127–54.
15. Yao JDC, Moellering Jr. RC. Antibacterial agents. In: Murray PR, Baron EJ, Jorgensen JH, Landry ML, Pfaller MA, editors. *Manual of clinical microbiology,* 9th ed. ASM Press; 2007. p. 1077–113.
16. Livermore DM. β-lactamases in laboratory and clinical resistance. *Clin Microbiol Rev* 1995;**8**:557–84.
17. Ambler RP. The structure of β-lactamases. *Philos Trans R Soc Lond B Biol Sci* 1980;**289**:321–31.
18. Murray PR, Rosenthal KS, Pfaller MA. *Medical microbiology*, 6th ed. Mosby; 2009. p. 199–208.
19. Rang HP, Dale MM, Ritter JM, Moore PK. *Pharmacology*, 5th ed. Churchill Livingstone; 2003. p. 635–47.
20. Yates AB. Management of patients with a history of allergy to β-lactam antibiotics. *Am J Med* 2008; **121**: 572–526.
21. Jacoby GA. AmpC β-lactamase. *Clin Microbiol Rev* 2009;**22**:161–82.
22. Lode H, Borner K, Koeppe P, Schaberg T. Azithromycin—review of key chemical, pharmacokinetic and microbiological features. *J Antimicrob Chemother* 1996;**37**(Suppl. C):1–8.
23. Snydman DR, Jacobus NV, McDermott LA, Golan Y, Hecht DW, Goldstein EJ, et al. Lessons learned from the anaerobe survey: historical perspective and review of the most recent data (2005–2007). *Clin Infect Dis* 2010;**50**(Suppl. 1):S26–33.
24. Chambers HF. Protein synthesis inhibitors and miscellaneous antibacteria agents. In: Brunton LL, Lazo JS, Parker KL, editors. *Goodman and Gilman's the pharmacological basis of therapeutics,* 11th ed. McGraw-Hill; 2006. p. 1173–202.
25. Bartlett JG. Historical perspectives on studies of *Clostridium difficile* and *C. difficile* infection. *Clin Infect Dis* 2008;**46**(Suppl. 1):S4–11.
26. Chambers HF. Aminoglycosides. In: Brunton LL, Lazo JS, Parker KL, editors. *Goodman and Gilman's the pharmacological basis of therapeutics,* 11th ed. McGraw-Hill; 2006. p. 1155–72.
27. Gilbert DN, Moellering Jr. RC, Eliopoulos GM, Chambers HF, Saag MS. *The Sanford guide to antimicrobial therapy,* 42th ed. Antimicrobial Therapy; 2012.

28. Hayden MK. Insights into the epidemiology and control of infection with vancomycin-resistant enterococci. *Clin Infect Dis* 2000;**31**:1058–65.
29. Biavasco F, Vignaroli C, Varaldo PE. Glycopeptide resistance in coagulase-negative staphylococci. *Eur J Clin Microbiol Infect Dis* 2000;**19**:403–17.
30. Weigel LM, Clewell DB, Gill SR, Clark NC, McDougal LK, Flannagan SE, et al. Genetic analysis of a high-level vancomycin-resistant isolate of *Staphylococcus aureus*. *Science* 2003;**302**:1569–71.
31. Levine DP. Vancomycin: a history. *Clin Infect Dis* 2006;**42**(Suppl. 1):S5–12.
32. Gupta A, Biyani M, Khaira A. Vancomycin nephrotoxicity: myths and facts. *Neth J Med* 2011; **69**:379–83.
33. Wazny LD, Daghigh B. Desensitization protocols for vancomycin hypersensitivity. *Ann Pharmacother* 2001; **35**:1458-64.
34. Masters PA, O'Bryan TA, Zurlo J, Miller DQ, Joshi N. Trimethoprim-sulfamethoxazole revisited. *Arch Intern Med* 2003;**163**:402–10.
35. Jung AC, Paauw DS. Management of adverse reactions to trimethoprim-sulfamethoxazole in human immunodeficiency virus-infected patients. *Arch Intern Med* 1994;**154**:2402–6.
36. Thomson Jr. RB. Specimen collection, transport, and processing: bacteriology. In: Murray PR, Baron EJ, Jorgensen JH, Landry ML, Pfaller MA, editors. *Manual of clinical microbiology,* 9th ed. ASM Press; 2007. p. 291–333.
37. Jorgensen JH, Ferraro MJ. Antimicrobial susceptibility testing: a review of general principles and contemporary practices. *Clin Infect Dis* 2009;**49**:1749–55.
38. Ferraro MJ. Should we reevaluate antibiotic breakpoints?. *Clin Infect Dis* 2001;**33**(Suppl. 3):S227–9.
39. Ceri H, Olson ME, Stremick C, Read RR, Morck D, Buret A. The Calgary Biofilm Device: new technology for rapid determination of antibiotic susceptibilities of bacterial biofilms. *J Clin Microbiol* 1999;**37**:1771–6.
40. Food and Drug Administration. *Federal Register* 1980; **44**:37434–67.
41. Novotný J, Novotný M. Adverse drug reactions to antibiotics and major antibiotic drug interactions. *Gen Physiol Biophys* 1999;**18**:126–39.
42. Bartlett JG. Clinical practice. Antibiotic-associated diarrhea. *N Engl J Med* 2002;**346**:334–9.
43. Diaz L, Ciurea AM. Cutaneous and systemic adverse reactions to antibiotics. *Dermatol Ther* 2012; **25**:12–22.
44. Stern RS. Clinical practice. Exanthematous drug eruptions. *Nin Engl J Med* 2012;**366**:2492–501.
45. Tintinalli JE, Kelen GD, Stapczynski JS. *Emergency medicine: a comprehensive study guide*, 6th ed. McGraw-Hill; 2004.
46. Naughton CA. Drug-induced nephrotoxicity. *Am Fam Physician* 2008;**78**:743–50.
47. Schetz M, Dasta J, Goldstein S, Golper T. Drug-induced acute kidney injury. *Curr Opin Crit Care* 2005;**11**:555–65.
48. Tenover FC. Mechanisms of antimicrobial resistance in bacteria. *Am J Infect Control* 2006;**34**(5 Suppl. 1):S3–10.
49. Rasmussen BA, Bush K, Tally FP. Antimicrobial resistance in anaerobes. *Clin Infect Dis* 1997;**24**(Suppl. 1):S110–20.
50. Schmieder R, Edwards R. Insights into antibiotic resistance through metagenomic approaches. *Future Microbiol* 2012;**7**:73–89.
51. Hayes JD, Wolf CR. Molecular mechanisms of drug resistance. *Biochem J* 1990;**272**:281–95.
52. Davies J. Inactivation of antibiotics and the dissemination of resistance genes. *Science* 1994; **264**:375–82.
53. Kumarasamy KK, Toleman MA, Walsh TR, Bagaria J, Butt F, Balakrishnan R, et al. Emergence of a new antibiotic resistance mechanism in India, Pakistan, and the UK: a molecular, biological, and epidemiological study. *Lancet Infect Dis* 2010;**10**:597–602.
54. Bennett JE. Antifungal agents. In: Brunton LL, Lazo JS, Parker KL, editors. *Goodman and Gilman's the pharmacological basis of therapeutics,* 11th ed. McGraw-Hill; 2006. p. 1225–42.
55. Scott LJ, Simpson D. Voriconazole: a review of its use in the management of invasive fungal infections. *Drugs* 2007;**67**:269–98.
56. Cronin S, Chandrasekar PH. Safety of triazole antifungal drugs in patients with cancer. *J Antimicrob Chemother* 2010;**65**:410–6.
57. Cappelletty D, Eiselstein-McKitrick K. The echinocandins. *Pharmacother* 2007;**27**:369–88.

CHAPTER FOURTEEN

An Introduction to the Biology of Biofilm Recalcitrance

Gavin J. Humphreys and Andrew J. McBain

INTRODUCTION

The interpretation of bacteria as dispersed, independently growing (planktonic) cells has dominated the discipline of microbiology since the seminal work of Robert Koch and Louis Pasteur in the late nineteenth century. Although valuable, this view fails to take into account the complexity with which micro-organisms commonly grow. It is now widely claimed that, in almost all natural environments, bacteria grow as *biofilms*—defined as coordinated communities of cells organised at an interface and bound within a matrix of extracellular polymeric substance.[1,2]

The identification of biofilms within prokaryotic fossils,[3] deep-sea hydrothermal vents and hot springs[4] supports their fundamental role as an ancient microbial survival strategy. In the healthy human body, micro-organisms grow on many exposed tissue surfaces in what has been termed the human microbiome. Such sessile micro-organisms, which may be present in a biofilm-like state, serve a number of physiological and ecological roles, including modulation of the immune system[5] and colonisation-resistance against potentially pathogenic micro-organisms.[6]

On the other hand, biofilms are often associated with problems, and in this respect a notable biofilm characteristic is the ability to survive exposure to antimicrobial agents at concentrations lethal to their planktonic counterparts.[7,8] This has obvious clinical implications through environmental persistence, where, for example, disinfection regimens may fail, and in infections, particularly those involving medical implants.[3,9,10]

EVIDENCE FOR CLINICAL RECALCITRANCE

A commonly cited statistic suggests that the biofilm phenotype affords bacteria a 10- to 1000-fold decrease in antimicrobial susceptibility.[11] In the clinical setting, biofilm growth has been demonstrated on urinary catheters,[12] ventricular drains,[13] mammary implants,[14] cochlear implants[15] and in surgical site infections.[16] While this

Biofilms in Infection Prevention and Control
Doi: http://dx.doi.org/10.1016/B978-0-12-397043-5.00014-1

list is not exhaustive, case reports have repeatedly emphasised the role of biofilms in the colonisation of indwelling medical devices, accounting for up to 50% of the healthcare-associated infections (HCAI) reported in the United States.[17]

Infections involving biofilm growth have been estimated to be as high as 40% for ventricular-assisted devices, 4% for cardiac implants (e.g., heart valves) and 2% for joint prosthetics.[17] With respect to urinary catheters, virtually all patients catheterised long term (>28 days) will experience infection as as result of microbial colonisation.[18] It is not the focus of this chapter to discuss the clinical implications of biofilm-associated infections; however, their recalcitrance makes them a daunting clinical problem with symptoms often persisting despite prolonged antimicrobial interventions. For a review, see Darouiche.[17]

Biofilm-associated infections are frequently recalcitrant to a broad range of antibiotics and biocides. For this reason, significant research attention has been directed toward understanding the mechanisms responsible and the development of improved strategies for their management. Biofilm recalcitrance has been attributed to several interconnected mechanisms, including

- Reaction diffusion limitation, mediated by the biofilm matrix.
- Development of physiological gradients leading to distinct micro-environments and heterogeneity of cell phenotypes.
- Biofilm-specific phenotypes, such as bacterial persistence, whereby a subset of cells can tolerate prolonged antimicrobial exposure probably through metabolic dormancy.

This chapter reviews these concepts and in so doing addresses current understanding of biofilm recalcitrance.

CONTRIBUTIONS OF THE BIOFILM MATRIX TO RECALCITRANCE

The biofilm matrix is a heterogeneous layer of enzymes and extracellular polymers accounting for up to 90% of the total organic carbon of the biofilm.[19] Figure 14.1 illustrates a developing *Pseudomonas aeruginosa* biofilm imaged using an environmental scanning electron microscope (SEM). Although it is clear that this structure is attached to a surface (in this case a plastic peg), individual cells are difficult to discern because they are obscured by the self-produced matrix within which they reside. Using this technique, samples can be imaged under high relative humidity, minimising the artifacts associated with the desiccation of the highly hydrated constituents of the matrix.

Investigators initially regarded the biofilm matrix as the key contributor to biofilm recalcitrance simply because of its function as a barrier to antimicrobials,[20] although current knowledge gleaned through a variety of means, including mathematical modelling,[21] suggests that this notion is simplistic. Despite this, the biofilm matrix is indeed capable of interacting directly with certain antimicrobials, retarding their rate

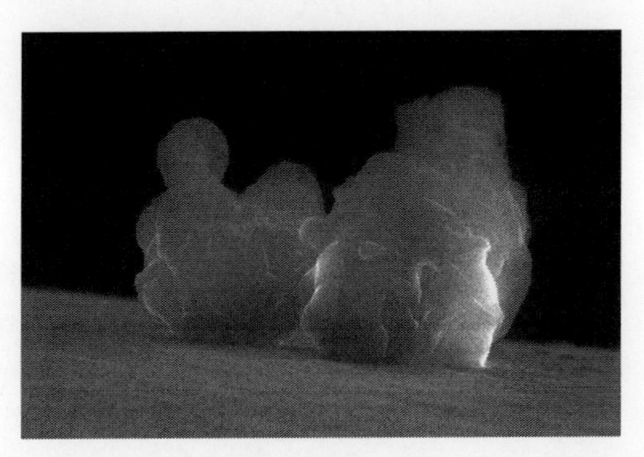

Figure 14.1 Pseudomonas aeruginosa *biofilm developing on the tip of a transposable plastic peg* (*Calgary device*). The biofilm was allowed to develop over 24 hours in a nutrient-rich broth before being imaged using an environmental scanning electron microscope. Individual cells are not visible in this image because they are enshrouded by a layer of polymeric substance (matrix). By maintaining the imaging chamber under high relative humidity, desiccation of the highly hydrated matrix is minimised.

of ingress in a process termed *reaction diffusion limitation*. Put simply, the biofilm matrix is able to interact with cationic antimicrobials (e.g., biguanides and quaternary ammonium compounds) and cationic antibiotics (e.g., aminoglycosides), leading to agent quenching.[22-25]

Functionally the biofilm matrix comprises hydrophobic and hydrophilic domains[26] as well as anchor sites onto which extracellular enzymes can attach and thus concentrate. In this respect, the biofilm matrix can be regarded not only as a structural scaffold within which microbial populations can develop but also as a means of protection against desiccation and antimicrobials.[1] Antimicrobial inactivating enzymes (e.g., β-lactamases) can concentrate within the biofilm matrix and confer protection against penicillin.[27] Similarly, formaldehyde dehydrogenase and formaldehyde lyase may contribute to biofilm recalcitrance against formaldehyde, as demonstrated in *pseudomonas* biofilms.[28]

The mechanisms by which biofilms achieve recalcitrance to specific agents through the production of antimicrobial-inactivating enzymes has been referred to as *enzyme-mediated diffusion limitation*. This can explain resistance to short-term antimicrobial exposure; however, it cannot account for long-term biofilm recalcitrance since, during prolonged exposure, it is likely that the antimicrobial occupies all available reaction sites. This renders the enzymes inactive through simple saturation kinetics.

Gilbert and McBain[7] suggested that, since the bulk phase can be in the region of 40L in the human body, there can be a large reservoir of available antimicrobial which is unlikely to be readily depleted. A particularly good example of this was demonstrated in an experiment by Huang et al.[29] in which Gram-negative biofilms were generated in a continuous-flow laboratory reactor and exposed to the biocide monochloramine (produced by adding chlorine to a solution of ammonia). Following exposure periods of two hours, a killing effect was observed only at the biofilm periphery, with respiratory activity observed in the deeper underlying regions of the community, as evidenced using epifluorescent microscopy in conjunction with a fluorescent respiratory marker (CTC formazan).[29] Such observations have led researchers to propose that physiological gradients may play a role in recalcitrance. These 'phenotypic mosaics', as they have been termed, may serve as moderators of biofilm recalcitrance and are discussed in further detail in the next section.[30]

Physiological Gradients in Biofilm Communities

Biofilms exhibit regions of oxygen limitation and nutrient deficiency that, in contrast to planktonic cultures, are non-uniformly distributed.[31] Bacteria located within starvation zones may be metabolically dormant and thus it has been suggested that growth-dependent changes in cellular processes may contribute to reduced sensitivity to many classes of antimicrobial agents.[2,7,32,33] For example, Anderl et al.[34] demonstrated reduced susceptibility of *Klebsiella pneumoniae* colony biofilms to ampicillin and fluoroquinolone exposure through nutrient limitation. Importantly, bacteria dispersed from these colony biofilms exhibited increased sensitivities to these antibiotics when introduced back into a nutrient-complete media.

Transmission electron microscopy (TEM) of colony biofilms following exposure to ampicillin showed enlargement and destruction of bacterial cells at the extremities of the biofilm but not at the mid-section, despite antimicrobial penetration. This differential bacteriostatic effect was attributed to limitations in glucose and oxygen diffusion at the centre of the colony biofilm.[34] Similar observations have been made when growing *Escherichia coli* using chemostats. Interestingly, specific growth rates of biofilm communities were able to explain much of the decreased susceptibility of *E. coli* to both cetrimide and tobramycin.[35,36]

It is now clear that both nutrient and oxygen limitation can contribute to biofilm resistance via changes in cell physiology; however, while studies have associated nutrient limitation with reduced biofilm antimicrobial susceptibility,[36,37] physiological gradients alone cannot account for biofilm recalcitrance. When outlying cells are inactivated during exposure of a biofilm to an antimicrobial, the deeper-lying cells are likely exposed to increased nutrients (including the products of lysed cells), oxygen and antimicrobial concentrations. In turn, these nutrients support increased bacterial growth rates. It is therefore likely that the eventual destruction of the biofilm community is achieved systematically from the outer layers inward during prolonged dosing with an antimicrobial.[38] If the antimicrobial were to be depleted prior to this, the biofilm could simply repopulate the substratum, which, if associated with an implanted device, would mean a resurgence of the infection.

Alternatively, during prolonged dosing physiological gradients can only account for a delayed killing effect of biofilm populations, as evidenced by mathematical modelling. For example, a 300 μm-thick biofilm was shown to have been killed following 52 hours under conditions of nutrient starvation in comparison to an estimated 2.7 hours when modelled as a planktonic culture.[39] The authors concluded that long-term biofilm recalcitrance must be achieved through a combination of protective mechanisms.

The Role of Drug-Resistant Phenotypes

Biofilms are recalcitrant to many antimicrobials through the development of physiological gradients and the matrix-mediated diffusion limitation of microbicides. This can be supported by the observation that mature biofilms are more recalcitrant to antimicrobial dosing than are younger biofilms, simply because the physiological gradients are more pronounced.[40] Although these mechanisms may offer a delayed killing effect, long-term biofilm recalcitrance likely involves a small subset of bacteria expressing antimicrobial non-susceptible phenotypes. These include inducible mechanisms, such as the hyper-expression of efflux transporters, as well as temporary biofilm-specific phenotypes such as the dormant persister cell. The remaining sections address these factors and how they relate to recalcitrance.

Efflux Pumps

Efflux pumps are transporter proteins found in bacterial membranes and are notorious in the clinical setting for their role in antibiotic resistance. Analogous systems contribute to resistance to cytotoxic drugs in cancer cells.[41] There are currently five families of efflux pump identified in bacteria:

- The multi-drug and toxic extrusion family[42]
- The ATP-binding cassette (ABC) family
- The staphylococcal multi-resistance (SMR) family
- The resistance nodulation division (RND)
- The major facilitator superfamily (MFS)[43]

Although multi-drug efflux pumps are present in a range of Gram-positive bacteria including staphylococci and enterococci,[44] they are perhaps most frequently associated with Gram-negative bacteria. For example, *E. coli* has been associated with more than 37 efflux transporters and their potential role in biofilm recalcitrance has been actively investigated.[45-47]

The role of active drug efflux in biofilm recalcitrance is unclear. Traditionally, the efflux mechanisms are associated with resistance in planktonic bacteria, whereby the expression of these transporters in cells results in marked decreases in drug sensitivity. That is, upon entry into the bacterium, the antimicrobial is rapidly and efficiently pumped back out before it can exert any cellular effect. Previously, researchers argued that such mechanisms confer no advantage to the biofilm phenotype, with no apparent protective effect associated with the major efflux pumps of *P. aeruginosa*.[48]

These data, however, are in contrast to the observations of Gillis et al.,[49] who identified several efflux mechanisms in the extrusion of macrolides (e.g., azithromycin) from biofilms. These mechanisms are more specifically referred to as the MexCD-OprJ and MexAB-OprM systems.[49] It is interesting to note that in this investigation efflux mechanisms were expressed in biofilm communities regardless of macrolide exposure, in contrast to the constitutive expression observed in planktonic cultures. Similarly, induction of mexC expression was observed throughout biofilm development but was apparently absent from planktonic cultures.

To support these observations, whole operon deletions in *P. aeruginosa* elucidated a role for a novel efflux system (called PA1874-1877) in mediating reduced sensitivities to tobramycin, gentamicin and ciprofloxacin.[50] The apparent discrepancies between these studies could be a result of varying oxygen concentrations encountered in mature and nascent biofilms—in particular, the apparent association between hypoxia and the up-regulation of efflux systems in *Pseudomonas*.[51] In the DeKievit study, biofilms were allowed to develop over a period of six hours prior to testing.[48] While more complex biofilm communities often take days to fully mature,[52] a *Pseudomonas* monospecies biofilm grown in micro-titre plates can reasonably develop within this time period. Nevertheless, the resultant oxygen gradients will likely be less pronounced than those encountered in the Gillis study[49] in which biofilms were cultured over a 24-hour period. In theory, such variations in oxygen concentration potentially account for the marked differences in efflux expression profiles between these studies.

It is arguable that, while efflux pumps are highly expressed in many biofilms, they may have a function beyond resistance modulation—that is, waste management.[53] Similarly, in *Salmonella enterica* serovar Typhimurium, multi-drug efflux expression is necessary for bacterial surface attachment and biofilm maturation through curli fibres associated with surface attachment formation.[54] Current research suggests that efflux systems represent an additional mechanism by which biofilms resist antimicrobial killing, but these effects do not fully explain long-term recalcitrance, as evidenced by the two- to three-fold reductions in sensitivity described by Zhang and Mah.[50] In addition, they fail to explain how sensitive bacteria with no known genetic basis for resistance can benefit from reduced susceptibility to antimicrobials through attachment and growth at an interface. When these cells are dispersed from the biofilm, they lose this reduced susceptibility, suggesting that recalcitrance is conferred not through a specific mutation but through a biofilm-specific phenotype.[55]

BACTERIAL PERSISTENCE AS A BIOFILM-SPECIFIC PHENOTYPE

The concept of bacterial persistence was first described by Bigger,[56] who demonstrated the limited bactericidal activity of penicillin toward a sub-population of *Streptococcus pyogenes* cells. 'As a matter of convenience', Bigger referred to these abnormal cocci as persisters to 'denote their power of surviving in the presence of sufficient penicillin to be lethal for the normal forms'.[56] Although the molecular mechanisms of bacterial persistence continue to be the subject of scrutiny and debate, it is likely that a small sub-population of cells ($\approx$1%) are persisters and utilise a multi-faceted cascade to promote limited metabolic activity and to up-regulate stress response mechanisms.[57–60] Based on these observations, it has been proposed that after removal of the challenging agent, persister cells can repopulate the biofilm, leading to relapse of a chronic infection.[60]

Determining the clinical significance of bacterial persistence has proven difficult. It is now clear that the persister phenotype is reversible, making persisters' direct isolation challenging.[59] Transcriptome analysis has led to some understanding of the molecular mechanisms underpinning persisters, in particular those involving the down-regulation of genes associated with biosynthesis and increased expression of toxin/antitoxin (TA) modules. Under optimal growth conditions, the toxin protein (e.g., HipA) is counteracted by the presence of an anti-toxin (e.g., HipB) and the bacterial cell divides normally.[61] Under certain conditions, however, such as antimicrobial stress, the anti-toxin component of the module is degraded, resulting in the release of the toxin. The overall result of this is the inhibition of numerous cellular processes, such as DNA production and translation.[62] In short, persister cells can utilise this relatively simplistic mechanism to exhibit multi-drug tolerance through cleavage of mRNA and subsequent induction of cell stasis.[63]

One of the first mechanisms validated for bacterial persistence involves the SOS-induced TisB toxin. In the laboratory, much of our understanding of the SOS stress response has come from exposure of bacterial cultures to fluoroquinolones. These antibiotics specifically target both DNA gyrase and topoisomerase, leading to breakages in the bacterial DNA. In response to the breakages, bacteria up-regulate DNA repair functions through induction of the SOS stress response.[64] In *E. coli*, ciprofloxacin dosing has been shown to decrease adenosine triphosphate (ATP) by the direct disruption of proton motive force. This is achieved through the SOS-linked production of a membrane-associated peptide called TisB.[65] The TisB response is intriguing as it involves the simultaneous repair of bacterial DNA (through the SOS response) coupled with the down-regulation of ATP production (through the TisB peptide) and the generation of dormant cells.

Lewis argued that the mechanism by which TisB peptides lead to dormancy suggests a link between an array of environmental stresses and the formation of persistence.[66] As well as antibiotics and biocides targeted to bacterial DNA, he suggests that persistence can be induced by exposure to changes in pH and temperature.[66]

THE CLINICAL SIGNIFICANCE OF BACTERIAL PERSISTENCE

The routine investigation of antimicrobial resistance involves the screening of clinical isolates under nutrient-rich planktonic conditions and thus persister cells have often been overlooked as a cause of biofilm recalcitrance. By definition, persister cells are able to survive sustained periods of antibiotic exposure, reverting to a susceptible, actively growing phenotype in the absence of environmental stress. The clinical relevance of this is as follows:

- Persister cells may be associated with chronic or recurrent infections, notably cystic fibrosis,[67] candidiasis[68] and tuberculosis.[69]
- By tolerating therapy with antimicrobials, persisters may be a reservoir for antibiotic-resistant mutants.

Even though there is no direct evidence for the latter, it is recognised that bacterial pure cultures can become less susceptible to antimicrobials through 'training' experiments, whereby the bacterium is exposed to subinhibitory concentrations of antimicrobial using gradient plates or disk diffusion assays. Over time, this can potentially result in increases in minimum inhibitory concentrations to the antimicrobial.[70]

Although antibiotics alone are often ineffective against persisters, there is considerable interest in identifying persistence-specific targets against which novel antimicrobials or 'persisticides' can be developed. Alternatively, antibiotics can be combined with agents that maximise effectiveness against persisters. For example, persister cells containing knockouts of the global regulator PhoU were found to be markedly more sensitive to antibiotic dosing, including ampicillin and gentamicin.[71] More recently, the killing of persister cells has been demonstrated through the combination of aminoglycosides and specific metabolites in an *in vivo* mouse model, leading to the conclusion that these dormant cells are 'primed for metabolite uptake and respiration'.[72] Through such approaches, antimicrobials capable of inactivating with even the most recalcitrant of bacterial biofilms may be within reach.

CONCLUSION

Biofilms are populations of microbial cells encased in a self-produced layer of polymeric substances. Overall, the lack of any single unifying theory for biofilm recalcitrance may suggest a multi-faceted model of biofilm resistance, as illustrated in Figure 14.2. Physiological gradients and matrix-mediated diffusion limitations offer a degree of resistance to sublethal or short-term antimicrobial exposure. During prolonged antimicrobial exposure, these mechanisms can temporarily retard diffusion into deeper underlying layers, where resident cells exhibit altered metabolic states and therefore reduced sensitivities to antimicrobial dosing. Despite this, mathematical modelling suggests that these growth-dependent phenotypes provide only a delayed eradication of the biofilm community. A key concept in recalcitrance, therefore, is bacterial persistence. It is these cells that represent the least susceptible genotype; thus, they are likely able to recolonise a surface in the absence of further environmental stresses.

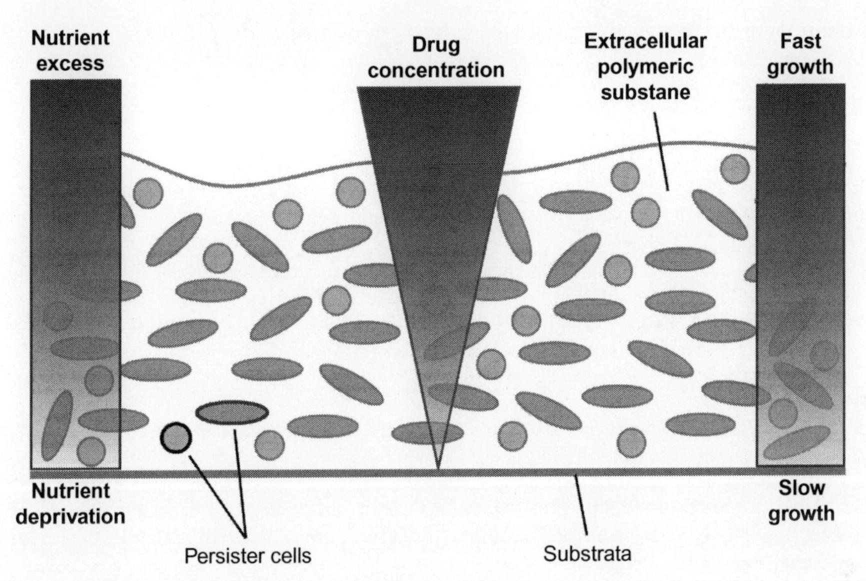

Figure 14.2 ***A multifactorial model for biofilm recalcitrance.*** Physiological gradients arise through nutrient limitation at the depths of the biofilm, resulting in significantly reduced growth rates. Extracellular polymeric substances can innately mediate reaction diffusion limitation, resulting in reduced drug penetration. Biofilm-specific phenotypes, in particular the persister cells, are also illustrated because of their ability to recolonise the substrata following antimicrobial intervention.

REFERENCES

1. Allison DG. The biofilm matrix. *Biofouling* 2003;**19**(2):139–50.
2. Gilbert P, Maira-Litran T, McBain AJ, Rickard AH, Whyte FW. The physiology and collective recalcitrance of microbial biofilm communities. *Adv Microb Physiol* 2002;**46**:202–56.
3. Hall-Stoodley L, Costerton JW, Stoodley P. Bacterial biofilms: from the natural environment to infectious diseases. *Nat Rev Microbiol* 2004;**2**(2):95–108.
4. Reysenbach AL, Cady SL. Microbiology of ancient and modern hydrothermal systems. *Trends Microbiol* 2001;**9**(2):79–86.
5. Gill HS, Rutherfurd KJ, Cross ML, Gopal PK. Enhancement of immunity in the elderly by dietary supplementation with the probiotic *Bifidobacterium lactis* HN019. *Am J Clin Nutr* 2001;**74**(6):833–9.
6. Lu L, Walker WA. Pathologic and physiologic interactions of bacteria with the gastrointestinal epithelium. *Am J Clin Nutr* 2001;**73**(6):1124S–1130SS.
7. Gilbert P, McBain AJ. Biofilms: their impact on health and their recalcitrance toward biocides. *Am J Infect Control* 2001;**29**(4):252–5.
8. Ashby MJ, Neale JE, Knott SJ, Critchley IA. Effect of antibiotics on non-growing planktonic cells and biofilms of *Escherichia coli*. *J Antimicrob Chemother* 1994;**33**(3):443–52.
9. Tambyah PA, Halvorson KT, Maki DG. A prospective study of pathogenesis of catheter-associated urinary tract infections. *Mayo Clin Proc* 1999;**74**(2):131–6.
10. Stickler DJ. Bacterial biofilms in patients with indwelling urinary catheters. *Nat Clin Pract Urol* 2008; **5**(11):598–608.
11. Mah TF, O'Toole GA. Mechanisms of biofilm resistance to antimicrobial agents. *Trends Microbiol* 2001; **9**(1):34–9.

12. Porte L, Soto A, Andrighetti D, Dabanch J, Braun S, Saldivia A, et al. Catheter-associated bloodstream infection caused by *Leifsonia aquatica* in a haemodialysis patient: a case report. *J Med Microbiol* 2012;**61**(Pt 6):868–73.
13. Stoodley P, Braxton Jr. EE, Nistico L, Hall-Stoodley L, Johnson S, Quigley M, et al. Direct demonstration of Staphylococcus biofilm in an external ventricular drain in a patient with a history of recurrent ventriculoperitoneal shunt failure. *Pediatr Neurosurg* 2010;**46**(2):127–32.
14. Allan JM, Jacombs AS, Hu H, Merten SL, Deva AK. Detection of bacterial biofilm in double capsule surrounding mammary implants: findings in human and porcine breast augmentation. *Plast Reconstr Surg* 2012;**129**(3):578e–580ee.
15. Pawlowski KS, Wawro D, Roland PS. Bacterial biofilm formation on a human cochlear implant. *Otol Neurotol* 2005;**26**(5):972–5.
16. Kathju S, Nistico L, Lasko LA, Stoodley P. Bacterial biofilm on monofilament suture and porcine xenograft after inguinal herniorrhaphy. *FEMS Immunol Med Microbiol* 2010;**59**(3):405–9.
17. Darouiche RO. Treatment of infections associated with surgical implants. *N Engl J Med* 2004;**350**(14):1422–9.
18. Donlan RM. Biofilms and device-associated infections. *Emerg Infect Dis* 2001;**7**(2):277–81.
19. Donlan RM. Biofilms: microbial life on surfaces. *Emerg Infect Dis* 2002;**8**(9):881–90.
20. Slack MP, Nichols WW. Antibiotic penetration through bacterial capsules and exopolysaccharides. *J Antimicrob Chemother* 1982;**10**(5):368–72.
21. Stewart PS. Theoretical aspects of antibiotic diffusion into microbial biofilms. *Antimicrob Agents Chemother* 1996;**40**(11):2517–22.
22. Campanac C, Pineau L, Payard A, Baziard-Mouysset G, Roques C. Interactions between biocide cationic agents and bacterial biofilms. *Antimicrob Agents Chemother* 2002;**46**(5):1469–74.
23. Chan C, Burrows LL, Deber CM. Alginate as an auxiliary bacterial membrane: binding of membrane-active peptides by polysaccharides. *J Pept Res* 2005;**65**(3):343–51.
24. Stewart PS, Grab L, Diemer JA. Analysis of biocide transport limitation in an artificial biofilm system. *J Appl Microbiol* 1998;**85**(3):495–500.
25. Walters 3rd MC, Roe F, Bugnicourt A, Franklin MJ, Stewart PS. Contributions of antibiotic penetration, oxygen limitation, and low metabolic activity to tolerance of *Pseudomonas aeruginosa* biofilms to ciprofloxacin and tobramycin. *Antimicrob Agents Chemother* 2003;**47**(1):317–23.
26. Sutherland I. Biofilm exopolysaccharides: a strong and sticky framework. *Microbiology* 2001;**147**(Pt 1):3–9.
27. Giwercman B, Jensen ET, Hoiby N, Kharazmi A, Costerton JW. Induction of beta-lactamase production in *Pseudomonas aeruginosa* biofilm. *Antimicrob Agents Chemother* 1991;**35**(5):1008–10.
28. Sondossi M, Rossmore HW, Wireman JW. Induction and selection of formaldehyde based resistance in *Pseudomonas aeruginosa*. *J Ind Microbiol* 1986;**1**(2):97–103.
29. Huang CT, Yu FP, McFeters GA, Stewart PS. Nonuniform spatial patterns of respiratory activity within biofilms during disinfection. *Appl Environ Microbiol* 1995;**61**(6):2252–6.
30. Huang CT, Xu KD, McFeters GA, Stewart PS. Spatial patterns of alkaline phosphatase expression within bacterial colonies and biofilms in response to phosphate starvation. *Appl Environ Microbiol* 1998;**64**(4):1526–31.
31. Xu KD, Stewart PS, Xia F, Huang CT, McFeters GA. Spatial physiological heterogeneity in *Pseudomonas aeruginosa* biofilm is determined by oxygen availability. *Appl Environ Microbiol* 1998;**64**(10):4035–9.
32. Borriello G, Werner E, Roe F, Kim AM, Ehrlich GD, Stewart PS. Oxygen limitation contributes to antibiotic tolerance of *Pseudomonas aeruginosa* in biofilms. *Antimicrob Agents Chemother* 2004;**48**(7):2659–64.
33. Field TR, White A, Elborn JS, Tunney MM. Effect of oxygen limitation on the in vitro antimicrobial susceptibility of clinical isolates of *Pseudomonas aeruginosa* grown planktonically and as biofilms. *Eur J Clin Microbiol Infect Dis* 2005;**24**(10):677–87.
34. Anderl JN, Zahller J, Roe F, Stewart PS. Role of nutrient limitation and stationary-phase existence in *Klebsiella pneumoniae* biofilm resistance to ampicillin and ciprofloxacin. *Antimicrob Agents Chemother* 2003;**47**(4):1251–6.

35. Evans DJ, Allison DG, Brown MR, Gilbert P. Effect of growth-rate on resistance of gram-negative biofilms to cetrimide. *J Antimicrob Chemother* 1990;**26**(4):473–8.
36. Evans DJ, Brown MR, Allison DG, Gilbert P. Susceptibility of bacterial biofilms to tobramycin: role of specific growth rate and phase in the division cycle. *J Antimicrob Chemother* 1990;**25**(4):585–91.
37. Brown MR, Collier PJ, Gilbert P. Influence of growth rate on susceptibility to antimicrobial agents: modification of the cell envelope and batch and continuous culture studies. *Antimicrob Agents Chemother* 1990;**34**(9):1623–8.
38. McBain AJ, Allison D, Gilbert P. Emerging strategies for the chemical treatment of microbial biofilms. *Biotechnol Genet Eng Rev* 2000;**17**:267–79.
39. Roberts ME, Stewart PS. Modeling antibiotic tolerance in biofilms by accounting for nutrient limitation. *Antimicrob Agents Chemother* 2004;**48**(1):48–52.
40. Anwar H, Dasgupta M, Lam K, Costerton JW. Tobramycin resistance of mucoid *Pseudomonas aeruginosa* biofilm grown under iron limitation. *J Antimicrob Chemother* 1989;**24**(5):647–55.
41. Binkhathlan Z, Lavasanifar A. P-glycoprotein inhibition as a therapeutic approach for overcoming multidrug resistance in cancer: current status and future perspectives. *Curr Cancer Drug Targets* 2013;**13**(3):326–46.
42. Rodriguez MR, Nuevo R, Chatterji S, Ayuso-Mateos JL. Definitions and factors associated with subthreshold depressive conditions: a systematic review. *BMC Psychiatry* 2012;**12**:181.
43. Li XZ, Nikaido H. Efflux-mediated drug resistance in bacteria: an update. *Drugs* 2009; **69**(12):1555–623.
44. Takacs D, Cerca P, Martins A, Riedl Z, Hajos G, Molnar J, et al. Evaluation of forty new phenothiazine derivatives for activity against intrinsic efflux pump systems of reference *Escherichia coli, Salmonella enteritidis, Enterococcus faecalis* and *Staphylococcus aureus* strains. *In Vivo* 2011;**25**(5):719–24.
45. DeMarco CE, Cushing LA, Frempong-Manso E, Seo SM, Jaravaza TA, Kaatz GW. Efflux-related resistance to norfloxacin, dyes, and biocides in bloodstream isolates of *Staphylococcus aureus*. *Antimicrob Agents Chemother* 2007;**51**(9):3235–9.
46. Papadopoulos CJ, Carson CF, Chang BJ, Riley TV. Role of the MexAB-OprM efflux pump of *Pseudomonas aeruginosa* in tolerance to tea tree (*Melaleuca alternifolia*) oil and its monoterpene components terpinen-4-ol, 1,8-cineole, and alpha-terpineol. *Appl Environ Microbiol* 2008;**74**(6):1932–5.
47. Pages JM, Masi M, Barbe J. Inhibitors of efflux pumps in Gram-negative bacteria. *Trends Mol Med* 2005;**11**(8):382–9.
48. De Kievit TR, Parkins MD, Gillis RJ, Srikumar R, Ceri H, Poole K, et al. Multidrug efflux pumps: expression patterns and contribution to antibiotic resistance in *Pseudomonas aeruginosa* biofilms. *Antimicrob Agents Chemother* 2001;**45**(6):1761–70.
49. Gillis RJ, White KG, Choi KH, Wagner VE, Schweizer HP, Iglewski BH. Molecular basis of azithromycin-resistant *Pseudomonas aeruginosa* biofilms. *Antimicrob Agents Chemother* 2012;**49**(9):3858–67.
50. Zhang L, Mah TF. Involvement of a novel efflux system in biofilm-specific resistance to antibiotics. *J Bacteriol* 2008;**190**(13):4447–52.
51. Schaible B, Taylor CT, Schaffer K. Hypoxia increases antibiotic resistance in *Pseudomonas aeruginosa* through altering the composition of multidrug efflux pumps. *Antimicrob Agents Chemother* 2012; **56**(4):2114–8.
52. Humphreys GJ, McBain AJ. Continuous culture of sessile human oropharyngeal microbiotas. *J Med Microbiol Feb* 2013:28.
53. Kvist M, Hancock V, Klemm P. Inactivation of efflux pumps abolishes bacterial biofilm formation. *Appl Environ Microbiol* 2008;**74**(23):7376–82.
54. Baugh S, Ekanayaka AS, Piddock LJ, Webber MA. Loss of or inhibition of all multidrug resistance efflux pumps of *Salmonella enterica* serovar Typhimurium results in impaired ability to form a biofilm. *J Antimicrob Chemother* 2012;**67**(10):2409–17.
55. Stewart PS, Costerton JW. Antibiotic resistance of bacteria in biofilms. *Lancet* 2001;**358**(9276):135–8.
56. Bigger JW. Treatment of staphylococcal infections with penicillin by intermittent sterilisation. *Lancet* 1944;**244**(6320):497–500.
57. Kaldalu N, Mei R, Lewis K. Killing by ampicillin and ofloxacin induces overlapping changes in *Escherichia coli* transcription profile. *Antimicrob Agents Chemother* 2004;**48**(3):890–6.

58. Brooun A, Liu S, Lewis K. A dose-response study of antibiotic resistance in *Pseudomonas aeruginosa* biofilms. *Antimicrob Agents Chemother* 2000;**44**(3):640–6.
59. Balaban NQ, Merrin J, Chait R, Kowalik L, Leibler S. Bacterial persistence as a phenotypic switch. *Science* 2004;**305**(5690):1622–5.
60. Lewis K. Programmed death in bacteria. *Microbiol Mol Biol Rev* 2000;**64**(3):503–14.
61. Schumacher MA, Piro KM, Xu W, Hansen S, Lewis K, Brennan RG. Molecular mechanisms of HipA-mediated multidrug tolerance and its neutralization by HipB. *Science* 2009;**323**(5912):396–401.
62. Schuster CF, Bertram R. Toxin-antitoxin systems are ubiquitous and versatile modulators of prokaryotic cell fate. *FEMS Microbiol Lett* 2013;**340**(2):73–85.
63. Keren I, Shah D, Spoering A, Kaldalu N, Lewis K. Specialized persister cells and the mechanism of multidrug tolerance in *Escherichia coli. J Bacteriol* 2004;**186**(24):8172–80.
64. Courcelle J, Khodursky A, Peter B, Brown PO, Hanawalt PC. Comparative gene expression profiles following UV exposure in wild-type and SOS-deficient *Escherichia coli. Genetics* 2001;**158**(1):41–64.
65. Dorr T, Vulic M, Lewis K. Ciprofloxacin causes persister formation by inducing the TisB toxin in *Escherichia coli. PLoS Biol* 2010;**8**(2):e1000317.
66. Lewis K. Persister cells. *Annu Rev Microbiol* 2010;**64**:357–72.
67. Mulcahy LR, Burns JL, Lory S, Lewis K. Emergence of *Pseudomonas aeruginosa* strains producing high levels of persister cells in patients with cystic fibrosis. *J Bacteriol* 2010;**2192**(23):6191–9.
68. Lafleur MD, Qi Q, Lewis K. Patients with long-term oral carriage harbor high-persister mutants of *Candida albicans. Antimicrob Agents Chemother* 2010;**54**(1):39–44.
69. Fauvart M, De Groote VN, Michiels J. Role of persister cells in chronic infections: clinical relevance and perspectives on anti-persister therapies. *J Med Microbiol* 2011;**60**(Pt 6):699–709.
70. Latimer J, Forbes S, McBain AJ. Attenuated virulence and biofilm formation in *Staphylococcus aureus* following sublethal exposure to triclosan. *Antimicrob Agents Chemother* 2012;**56**(6):3092–100.
71. Li Y, Zhang Y. PhoU is a persistence switch involved in persister formation and tolerance to multiple antibiotics and stresses in *Escherichia coli. Antimicrob Agents Chemother* 2007;**51**(6):2092–9.
72. Allison KR, Brynildsen MP, Collins JJ. Metabolite-enabled eradication of bacterial persisters by aminoglycosides. *Nature* 2011;**473**(7346):216–20.

CHAPTER FIFTEEN

Microbial Resistance and Superbugs

Lim S. Jones and Robin A. Howe

INTRODUCTION

The availability of safe and effective antimicrobials is relatively recent. Many of the infections that we now can cure once caused considerable morbidity and mortality. In his Nobel Prize acceptance speech in 1945, Alexander Fleming stated,

> *It is not difficult to make microbes resistant to penicillin in the laboratory by exposing them to concentrations not sufficient to kill them.... There is the danger that the ignorant man may easily underdose himself and by exposing his microbes to non-lethal quantities of the drug make them resistant.*[1]

By 1947, clinicians had observed penicillin resistance to *Staphylococcus aureus*, and the subsequent antimicrobial introductions have frequently been closely followed by acquired resistance in clinical practice.[2] Many bacteria have adapted in the face of the widespread use and abuse of antimicrobials; multi-drug-resistant (MDR) phenotypes are thus frequently encountered. Meticillin-resistant *Staphylococcus aureus* (MRSA) has been the exemplar of the MDR virulent pathogen. It has spread successfully in hospitals in many parts of the world and is now a growing cause of community-acquired infections.[3,4]

During the past decade, multi-drug-resistant Gram-negative organisms have become much more common, largely as a result of the spread of broad-spectrum β-lactamases which have compromised the use of extended-spectrum penicillins and cephalosporins.[5,6] Enzymes capable of hydrolysing carbapenems, formerly the drugs of last resort, have been increasingly identified. In a few cases, strains have been identified which are resistant to all clinically useful antimicrobials.[7,8]

Some experts warn that we may be heading towards a post-antibiotic era.[5,8] Such dire warnings may be premature, but the control of antimicrobial resistance and the development of novel therapies need to be addressed as top priorities for medicine during the twenty-first century.

In this chapter, we discuss how antimicrobial resistance is defined and the mechanisms and associated genetic changes that underlie it. We then introduce some of the

Biofilms in Infection Prevention and Control
Doi: http://dx.doi.org/10.1016/B978-0-12-397043-5.00015-3

MDR bacterial "superbugs" which are currently considered to be of clinical importance. The chapter does not consider antibiotic resistance in mycobacteria or the challenges of drug resistance of fungi or viruses.

DEFINITIONS OF ANTIMICROBIAL RESISTANCE

The concept of antimicrobial resistance arose from our need to predict whether an antimicrobial is effective in the treatment of specific infections/pathogens. Hence, clinical definitions, such as those from the European Committee on Antimicrobial Susceptibility Testing (EUCAST), have been developed (Table 15.1). The EUCAST definition specifies an organism as susceptible, intermediate or resistant to an antimicrobial according to whether the minimum inhibitory concentration (MIC) of the antimicrobial for that organism is above or below published breakpoints.

Such breakpoints are derived from the study of pharmacokinetic/pharmacodynamic determinants of clinical efficacy and are published according to standard methods by different groups around the world, including EUCAST, the British Society for Antimicrobial Chemotherapy (BSAC) and the Clinical and Laboratory Standards Institute (CLSI). It should be noted that the clinical breakpoints may be altered if, for example, new dosing regimens are developed or new information emerges on the relationship between drug levels and clinical efficacy. Indeed, the breakpoints are not always consistent across various standards, because the various groups may interpret data differently. Nevertheless, it has been shown that susceptibility testing is useful in predicting clinical outcome.[9]

Some organisms may be inherently susceptible or resistant to a given antimicrobial, usually because the antimicrobial target is missing or because of impermeability. For example, Gram-positive bacteria are resistant to colistin because they lack its target (i.e., an outer cell membrane), and Gram-negative bacteria are resistant to glycopeptides, such as vancomycin, because these large molecules cannot penetrate the outer cell membrane that Gram-negatives have. Organisms that are inherently susceptible to an antimicrobial may acquire a resistance mechanism, and this gives rise to the concept of microbiological resistance (see Table 15.1).

Wild-type organisms can be defined through examination of the population distribution of MICs and the application of an appropriate cut-off, termed the epidemiological cut-off (ECOFF). They may be inherently susceptible or resistant, but do not have an acquired mechanism of resistance. If an organism has such a mechanism, it should have a raised MIC compared to the wild-type population. However, it is important to recognise that the presence of an acquired mechanism of resistance does not necessarily confer clinical resistance. For example, wild-type *Streptococcus pneumoniae* have penicillin MICs of ≤0.06 mg/L.

Table 15.1 Clinical and Microbiological Definitions of Antimicrobial Susceptibility

Clinical Definition	
Clinically susceptible	A micro-organism is defined as susceptible by a level of antimicrobial activity associated with a high likelihood of therapeutic success. A susceptible micro-organism is categorized by applying the appropriate breakpoint in a defined phenotypic test system. This breakpoint may be altered with legitimate changes in circumstances.
Clinically intermediate	A micro-organism is defined as intermediate by a level of antimicrobial agent activity associated with an uncertain therapeutic effect. This implies that an infection due to the isolate may be appropriately treated in body sites where drugs are physically concentrated or when a high dose of the drug can be used; it also indicates a buffer zone that should prevent small, uncontrolled, technical factors from causing major discrepancies in interpretations. A micro-organism is categorised as intermediate by applying the appropriate breakpoints in a defined phenotypic test system. These breakpoints may be altered with legitimate changes in circumstances.
Crlinically resistant	A micro-organism is categorised as resistant by applying the appropriate breakpoint in a defined phenotypic test system. This breakpoint may be altered with legitimate changes in circumstances.
Microbiological Definition	
Wild type	A micro-organism is categorised as wild type (WT) for a species by applying the appropriate cut-off value in a defined phenotypic test system. The value is not altered by changing circumstances. Wild-type micro-organisms may or may not respond clinically to antimicrobial treatment.
Non-wild type	A micro-organism for a species is defined as non-wild type (NWT) by the presence of an acquired or mutational mechanism of resistance to the drug in question. A NWT micro-organism is categorised as non-wild type for a species by applying the appropriate cut-off value in a defined phenotypic test system. This value is not altered by changing circumstances. Non-wild-type micro-organisms may or may not respond clinically to antimicrobial treatment.

Source: Taken from the EUCAST website: www.srga.org/Eucastwt/eucastdefinitions.htm.

Strains with altered penicillin-binding proteins (PBPs) may have MICs of 0.12 mg/L or more, but it has been shown that pneumonia caused by strains with MICs of 0.12 to 2 mg/L still respond to treatment with penicillin.[10] These organisms are currently categorised as clinically intermediate (Figure 15.1). Further complexity occurs when there is variable expression of a resistance mechanism or multiple mechanisms (see the following section).

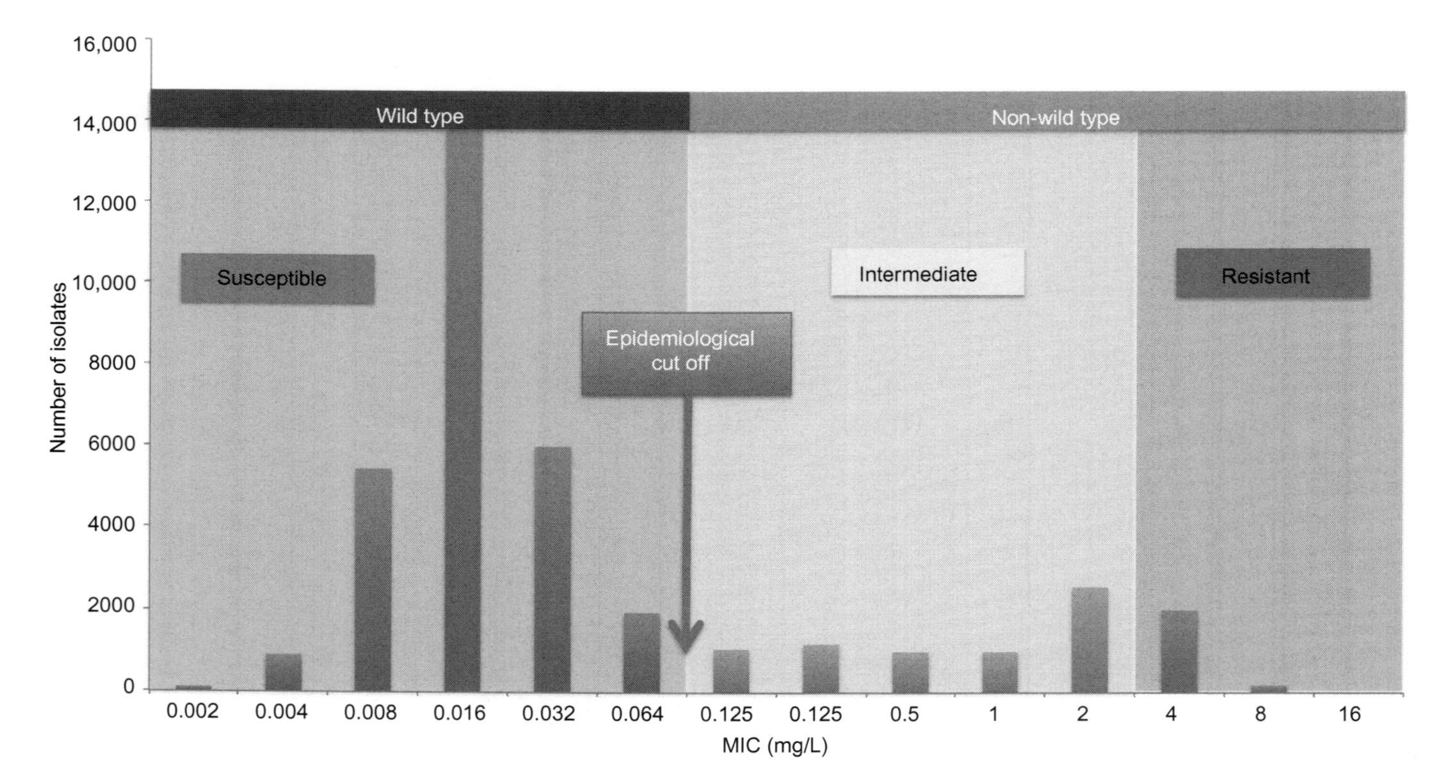

Figure 15.1 ***Population distribution of MIC of benzylpenicillin for* S. pneumoniae *(n = 37,742).*** *Source: Data taken from the EUCAST website: http://mic.eucast.org/Eucast2/SearchController/search.jsp?action*=performSearch&BeginIndex=0&Micdif=mic&NumberIndex=50&Antib=-1&Specium=12*).*

MECHANISMS OF ANTIMICROBIAL RESISTANCE

There are four main mechanisms by which bacteria can overcome the effects of antibiotics and become resistant; they are shown in Table 15.2 with examples. Other mechanisms exist but are much less common; these include, among others, target site protection (e.g., qnr proteins that protect the fluorquinolone-binding site) and failure to form active drug. For example, reduction of metronidazole by pyruvate dehydrogenase is required for activity; reduced pyruvate dehydrogenase expression leads to metronidazole resistance.

Table 15.2 Principle Mechanisms of Antimicrobial Resistance

Mechanism		**Example**
Target site modification	Target is modified so that the drug binds less efficiently.	Mutations in the RNA polymerase target of rifampicin confer resistance. The *erm* gene (acquired in *Staphylococci* and *Streptococci*) encodes an enzyme that methylates a specific adenosine residue located in the binding site of macrolides and lincosamides on the 23S rRNA. The *mecA* gene (acquired in *Staphylococci*) encodes an altered PBP2a. β-lactams cannot bind, leading to the MRSA phenotype.
Antibiotic-modifying enzymes	Aminoglycoside-modifying enzymes inactivate or modify the antibiotic.	β-lactamases produced by Gram-positive or Gram-negative bacteria inactivate β-lactam antibiotcs. Different β-lactamases have different substrate specificities that determine which antibiotics they are active against. There are many types of acetyltransferases, posphotransferases and nucleotidyltransferases with different activities against different aminoglycosides.
Impermeability	Bacterial envelope is modified so that the antibiotic cannot penetrate to its site of action.	Reduced expression of the outer membrane porin protein, OprD, in *P aeruginosa* confers reduced susceptibility to carbapenems, especially imipenem.
Efflux	Antibiotic is actively removed from the bacterial cell through an efflux pump.	Overexpression of the MexAB-OprM tripartite efflux pump in *P. aeruginosa* confers reduced susceptibility to β-lactams, fluoroquinolones, tetracyclines, chloramphenicol, macrolides and trimethoprim. Overexpression of the NorA efflux pump in *S. aureus* confers reduced susceptibility to fluoroquinolones.

In certain cases, the presence of a resistance mechanism correlates directly with a predictable level of resistance. For example, mutations in the gene for the RNA polymerase target of rifampicin give predictable resistance. However, in most cases the situation is much more complicated and the resistance phenotype results from a complex interplay of factors.

Resistance mechanisms may be expressed to a greater or lesser extent, perhaps because of coordinated control of expression in response to the environment, such as for efflux pumps, or because of loss of control due to mutations within control systems, leading, for example, to overexpression of efflux pumps or underexpression of porins. β-lactamases may be expressed at different levels depending on the number of gene copies present or the presence or otherwise of upstream gene promoters (see Table 15.3).

In addition, bacteria may possess and express a multitude of different resistance mechanisms directed against the same agent. For Gram-negative bacteria such as *E. coli*, then, the phenotypic level of resistance to a β-lactam antibiotic is a summation of the number, expression and specificity of β-lactamases; the expression or otherwise of porins; and the expression of efflux pumps. For example, for a collection of *E. coli* that possess the CTX-M gene that encodes a β-lactamase active against cefotaxime, the MICs of cefotaxime range from 1 mg/L (clinically susceptible) to >256 mg/L (clinically resistant) (Figure 15.2).

The complex interplay of resistance mechanisms to determine resistance phenotype is particularly pertinent in the case of biofilms. In the classic model of biofilm infection, chronic pseudomonal lung infection in patients with cystic fibrosis, hyper-expression of chromosomal β-lactamase and hyper-expression of a range of efflux pumps combine with the mucoid alginate biofilm matrix (which provides a barrier to antimicrobial diffusion) and the mutator phenotype (which leads to more profuse biofilm growth) to cause significant resistance to multiple classes of antimicrobials.[11] This phenotypic resistance not only compromises the ability to treat the infection but also reduces the ability to predict antimicrobial efficacy through susceptibility testing, as the biofilm conditions cannot be readily reproduced in the diagnostic laboratory.

GENETICS OF RESISTANCE

As can be seen from some of the examples already given, much of the genetics of antimicrobial resistance is determined by chromosomal genes and their regulatory networks. These may, like constitutively expressed efflux pumps and chromosomal β-lactams, contribute to the intrinsic resistance of the organism.[12] Resistance can increase because intrinsic genes mutate, gene expression is altered or genes are disrupted.

However, much of the increase in resistance in recent years is attributed to resistant genes acquired on mobile genetic elements from other bacteria.[5,13,14] For example, the class of extended spectrum-β-lactamases, CTX-M, which results in resistance to most

Table 15.3 Classification of β-Lactamases

Functional Classification	Molecular Class (Subclass)	Inhibited by		Defining Characteristic(s)	Representative Enzyme(s)
		CA or TZBa	EDTA		
1	C	No	No	Increased hydrolysis of cephalosporins than benzylpenicillin; hydrolyzes cephamycins	*E. coli* AmpC, P99, ACT-1, CMY-2, FOX-1, MIR-1
1e		No	No	Increased hydrolysis of ceftazidime and often other oxyimino-β-lactams	GC1, CMY-37
2a	A	Yes	No	Increased hydrolysis of benzylpenicillin than cephalosporins; predominant enzymes in Gram-positive cocci, including *S. Aureus*	PC1
2b		Yes	No	Similar hydrolysis of benzylpenicillin and cephalosporins	TEM-1, TEM-2, SHV-1
2be		Yes	No	Increased hydrolysis of oxyimino-β-lactams (cefotaxime, ceftazidime, ceftriaxone, cefepime, aztreonam)	TEM-3, SHV-2, CTX-M-15, PER-1, VEB-1
2br		No	No	Resistance to clavulanic acid, sulbactam, and tazobactam	TEM-30, SHV-10
2ber		No	No	Increased hydrolysis of oxyimino-β-lactams combined with resistance to clavulanic acid, sulbactam and tazobactam	TEM-50
2c		Yes	No	Increased hydrolysis of carbenicillin.	PSE-1, CARB-3
2ce		Yes	No	Increased hydrolysis of carbenicillin, cefepime and cefpirome	RTG-4
2d	D	V	No	Increased hydrolysis of cloxacillin or oxacillin	OXA-1, OXA-10
2de		V	No	Hydrolyzes cloxacillin or oxacillin and oxyimino-β-lactams	OXA-11, OXA-15
2df		V	No	Hydrolyzes cloxacillin or oxacillin and carbapenems	OXA-23, OXA-48
2e	A	Yes	No	Hydrolyzes cephalosporins;inhibited by clavulanic acid but not aztreonam	CepA
2f		V	No	Increased hydrolysis of carbapenems, oxyimino-β-lactams, cephamycins	KPC-2, IMI-1, SME-1
3a	B (B1)	No	Yes	Broad-spectrum hydrolysis, including carbapenems but not monobactams	IMP-1, VIM-1, NDM-1
	B (B3)				L1, CAU-1, GOB-1, FEZ-1
3b	B (B2)	No	Yes	Preferential hydrolysis of carbapenems	CphA, Sfh-1

Source: Adapted from Bush K, Jacoby GA. Updated functional classification of beta-lactamases. Antimicrob Agents Chemother. 2010. 54(3):969–76.
Note: V–Variable; CA–Clavulanic acid; TZBa–Tazobactam.

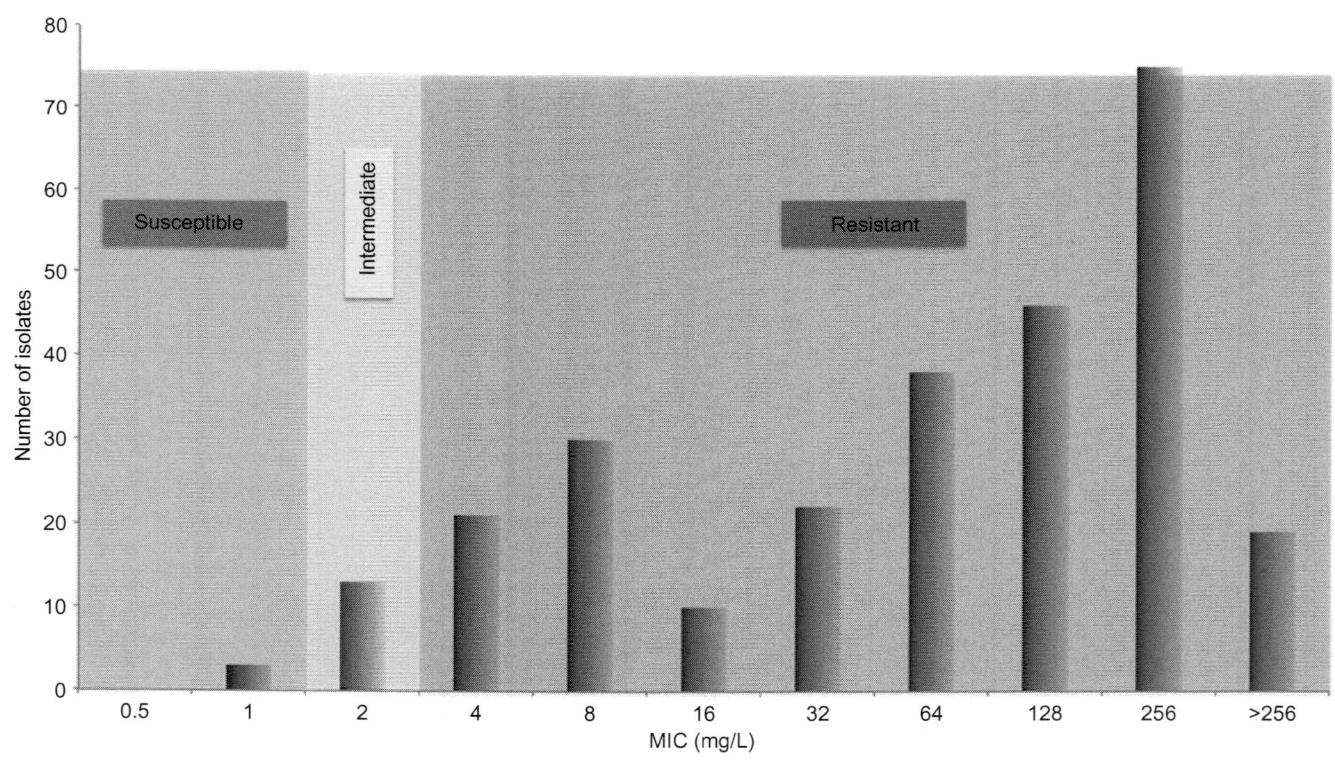

Figure 15.2 ***Distribution of MICs of cefotaxime for 277*** **E. coli** ***positive for the CTX-M gene.*** *Source: Data on file.*

β-lactams other than carbapenems, has become widespread internationally in *E. coli* and *Klebsiella* spp.; this is true although these genes are likely to have originated in environmental organisms of the genus *Kluyvera*.[15] This may have important implications for infection control because of the potential for the spread of resistance mechanisms among organisms.

Although most bacteria replicate by binary fission so that daughter cells are almost genetically identical to the parent cell, a number of strategies exist by which bacteria can acquire new genetic information. Some bacteria can take up and incorporate naked DNA from the environment—*transformation*.[16] Certain bacteriophages (i.e., viruses which target bacteria) can incorporate their DNA into the chromosome of a bacterial host and with it introduce foreign DNA acquired from previous hosts—*transduction*.[17] Bacteria can also directly share genetic information by *conjugation*, in which conjugative elements mediate the formation of a channel between bacterial cells through which the cells can pass to the new host.

Probably the most important of these elements in terms of the spread of resistance genes are plasmids—genetic elements which replicate independently of bacterial chromosomes.[18] Integrative and conjugative elements (ICEs) can also transfer resistance genes by conjugation, but differ in that they replicate while integrated in the bacterial chromosome.[19]

Insertion sequences (ISs) are elements that can move themselves from one genetic location to another. In so doing they increase resistance by moving strong promoter sequences upstream of some genes and directly disrupting others by inserting themselves into them. Transposons are formed by two IS elements close to each other which can move together and transfer the intervening DNA with them to a new genetic location. In this way, genes normally found on the chromosome can integrate into other elements, such as plasmids or ICEs, and transfer to a new host.

Integrons are also important to the evolution of antimicrobial resistance. These are genetic elements that code for a site-specific recombinase (integrase) and an integration site in which one or more mobile gene cassettes can be incorporated. Gene cassettes consist of a single gene and a recombination site. Although integrons themselves are not mobile, they can acquire and express multiple resistance genes as cassettes. If the transposons and/or plasmids incorporate the entire integron structure, they can act as potent vehicles for the rapid spread of resistance.[14,20]

Many of the resistance determinants currently causing the greatest concern are associated with several of these mobile elements. For example, the gene coding for the NDM-1 enzyme, which leads to resistance to almost all β-lactams except aztreonam, is usually found on plasmids and is associated with many different insertion sequence elements/transposons.[18] These plasmids often also have other resistance genes, including some on integrons. The combination of genetic elements and resistance genes has almost certainly contributed to the broad range of Gram-negative species found to produce NDM-1 and to a situation in which most of these bacteria are resistant not simply to β-lactams but to most classes of antibiotic currently available.[21,22]

SUPERBUGS

The term *superbug* has become common in the media to reflect the concern expressed by the medical community regarding certain bacteria. It is usually applied to bacteria that are resistant to multiple classes of antimicrobials (or, more specifically, susceptible to very few) and those that display evidence of transmissibility and so cause outbreaks in healthcare facilities or those that have a particular pathogenicity. Over time, as more antimicrobials have been developed, some threats have diminished. For example, epidemic gentamicin-resistant *S. aureus* was a concern during the 1970s,[11] but it is not a prominent problem now because more alternative antimicrobials are available.

The key superbugs of current concern are described in the following subsections. They include bacteria of relatively low pathogenicity, such as *Enterococci*, which may be resistant to almost all antimicrobials, and bacteria, such as *E. coli*, which are major causes of human morbidity and mortality and are developing an extremely worrying level of resistance. The particular concern at present is the lack of novel antimicrobials in the development pipeline; this is causing real apprehension that we may be entering a post-antibiotic era.

More informative terms for describing antimicrobial resistance in bacteria have recently been defined by expert consensus for the key resistant pathogens *S. aureus*, *Enterococci*, Enterobacteriaceae, *Pseudomonas aeruginosa* and *A. baumannii*.[23] Definitions are based only on agents to which these organisms are not intrinsically resistant.

The term *multi-drug-resistant* is applied to bacteria not susceptible to at least one antibiotic out of three or more of the antimicrobial categories used for treating that organism. *Extensively drug-resistant* (XDR) is the term proposed for organisms not susceptible to at least one antibiotic of all but two or fewer antimicrobial categories used for treating that organism. Finally, *pan-drug-resistant* (PDR) bacteria are resistant to all agents available to treat the organism. The preceding terms generally are preferred for communicating information regarding antibiotic resistance to healthcare professionals and for use in the scientific literature. They are adopted throughout the rest of this chapter.

S. aureus: MRSA, VISA, VRSA

S. aureus causes a wide range of infections, most commonly those involving simple skin and soft tissue, known as SSTIs (e.g., cellulitus, impetigo and furuncles). It can also cause a range of invasive infections, including osteomyelitis, septic arthritis, peritonitis, pneumonia and endocarditis.[24] In the pre-antibiotic era, the majority of *S. aureus* bacteraemia cases resulted in death,[4] and mortality remains in the range of 20 to 30% in modern times.[24,25]

S. aureus is inherently susceptible to many antimicrobials, such as β-lactams (e.g., penicillins, cephalosporins and carbapenems), glycopeptides (e.g., vancomycin), macrolides (e.g., erythromycin), clindamycin, tetracycline and aminoglycosides (e.g., gentamicin). However, the mainstay of therapy has been β-lactams and, for patients with β-lactam allergy, macrolides. When penicillin was first introduced, it was effective against *S. aureus* infections, but resistance rapidly emerged. Most clinical isolates are now penicillin-resistant. This is secondary to the production of a β-lactamase, which hydrolyses the β-lactam ring and renders the drug inactive.[2]

Many of the β-lactams that have been developed since pencillin are stable to the *S. aureus* β-lactamase. They include the cephalosporins, the carbapenems, β-lactam/ β-lactamase inhibitor combinations and the penicillinase-resistant isoxazoylpenicillins such as flucloxacillin.[26] Because of their potency, relatively narrow spectrum of activity and favourable side effect profile,[27] the latter drugs are the drugs of choice for treating mild to severe invasive infections caused by *S. aureus*.

Unfortunately, shortly after the introduction of meticillin in 1960, resistant strains of *S. aureus* were isolated *in vitro*[28] and subsequently from clinical infections.[29] Since meticillin was the first of the isoxazoyl penicillins to enter clinical practice, resistance to it and to related agents has been termed *meticillin-resistant S. aureus* (MRSA), although *β-lactam-resistant S. aureus* would be a more informative name. In MRSA an altered penicillin-binding protein, PBP2a, for which most β-lactam antibiotics have poor affinity, is the mechanism of resistance.

PBP2a is encoded by the *mecA* gene, thought to have been horizontally acquired from another bacterial species, on a genetic element called the staphylococcal cassette chromosome mec (SCC*mec*).[4,30] Until the recent development of cephalosporins with some MRSA activity, this meant that no available β-lactams were considered effective. In August 2012, the European Medicines Agency approved ceftaroline fosamil; it is the first β-lactam agent with activity against MRSA approved in Europe.[31]

Prevalence rates for healthcare-acquired MRSA (HA-MRSA) vary, from greater than 50% in the United States and parts of South America to less than 1% in some Northern European countries.[4] The United Kingdom, the same as much of the rest of Europe, has seen a relatively high prevalence, which has fallen in recent years. Results for invasive *S. aureus* infections submitted to the European Antimicrobial Resistance Surveillance System (EARSS) show that the percentage of meticillin-resistant isolates fell from a high of 47.3% in 2001 to 13.6% in 2011.[32]

The success of MRSA in spreading widely around the world is due to the epidemiology of *S. aureus* colonisation, imprudent use of antimicrobials in hospitals and lapses in basic infection control procedures. About 25 to 35% of normal healthy individuals are colonised with *S. aureus*, providing a reservoir of infection for the colonised individual and his or her contacts.[33] Sites of colonisation are the nose, the skin (especially if broken or damaged), the throat and the groin.[33]

S. aureus is a prominent cause of both community-acquired and healthcare-acquired infections, with MRSA being one of the most frequently identified resistant pathogens in hospitals in many parts of the world.[4] The organism can survive for significant periods of time in the environment and on the skin and hands of patients, visitors and healthcare workers, so it can readily spread in the hospital environment.[34]

Conventional cleaning regimens do not always eradicate the organism from the environment.[34,35] The success of some MRSA strains in causing hospital infections is a likely result of selection pressure from the large number of hospitalised patients on antibiotics, which reduces the growth of or kills the competing bacterial flora to a greater extent than the more resistant strains.[36,37]

S. aureus can form biofilms on prosthetic devices such as cannulas, long lines, prosthetic joints and prosthetic heart valves. In the biofilm, the organism is protected from both the immune system and antimicrobials,[38] making treatment a challenge if it is not straightforward to remove the device.

In the 1990s cases of community-acquired or community-associated MRSA (CA-MRSA) began to emerge in the United States. CA-MRSA cases are not linked to hospitalisation and are more likely to affect younger, previously well patients, especially those from certain groups such as participants in team sports or intravenous drug users.[39] The strains responsible are more sensitive to antimicrobials than HA-MRSA, and typing methods show them to be distinct from HA-MRSA strains.[39,40]

CA-MRSA has since become endemic in the United States and has been reported in many countries, including the United Kingdom,[39] where the prevalence of the CA-MRSA strains has been estimated at less than 1% of all MRSA isolates. However, for various reasons this prevalence is likely to have been significantly underestimated.[40] The boundaries between HA- and CA-MRSA have become blurred as strains previously associated with community infections have displaced the traditional HA-MRSA strains and as antibiotic resistance to other agents has increased in CA-MRSA strains.[4,39–41]

Effective and safe treatments for MRSA infections are quite limited, although the proportion of isolates sensitive to alternative antibiotics varies considerably according to strain (i.e., HA-MRSA or CA-MRSA) geographic area and time. Ideally empiric therapy decisions take into account up-to-date local resistance trends.[27] Earlier in Figure 15.1 the 2001–2011 resistance rates for MRSA bacteraemia isolates in the United Kingdom are shown, as published by the British Society of Antimicrobial Chemotherapy.[42]

For severe infections with MRSA, vancomycin has been the treatment of choice,[25] although its efficacy has been questioned. For MSSA bacteraemia, most studies suggest that β-lactams are more effective.[43–45] Several studies have also shown that outcomes with vancomycin are worse, with relatively modest MIC increases against it. This is despite the fact that the MICs are below resistance breakpoints and that vancomycin dosing achieves optimal trough blood levels.[25,43,46] Some studies, mainly from the United States, suggest that there has been a steady 'creep' upward in MICs to vancomycin.[47–49] In response, treatment guidelines have advised aiming for higher vancomycin levels to maintain the drug's effectiveness.[50]

A discussion of vancomycin resistance in *S. aureus* is complicated by the changing definitions and breakpoints applied. The term vancomycin-intermediate *S. aureus* (VISA) has been commonly used to refer to an organism with MICs that are just above the breakpoint. A phenotype with lower MICs but with sub-populations able to grow at much higher vancomycin concentrations has been termed heteroresistant-*VISA* (hVISA).[48] The mechanism of resistance for these phenotypes is not fully understood, but there is cell wall thickening which is thought to limit penetration of the large vancomycin molecule to the site of action.[48,51] Both hVISA and VISA were first isolated in Japan in 1996 and 1997, respectively,[52,53] and have since been encountered in many countries but remain uncommon.[48,51]

In 2002 the first case of high-level vancomycin-resistant *S. aureus* (VRSA) was identified in the United States, in this instance resulting from the *vanA* gene cluster previously responsible for vancomycin resistance in some *Enterococci*.[54] The effect of *vanA* is that the affinity of vancomycin for its target is reduced and so it is less effective at blocking the cross-linking which gives the bacterial cell wall its strength.[48,55]

Most reports of *vanA*-mediated resistance have been from the United States, but reports from India and Iran have also been published and an isolate from Portugal has been reported.[48,51,54,56,57] Many of the cases in these studies have been of questionable clinical significance, and colonisation has often been successfully eradicated with other antimicrobials.[54]

Of the more recently available drugs used to treat severe MRSA infections, both linezolid and daptomycin have been shown to be non-inferior to vancomycin for the management of some infections.[25,58] Clinical trials suggest that linezolid is a good alternative to vancomycin in regimens for SSTIs and pneumonia, but the evidence that outcomes are superior is at present unconvincing.[58] Daptomycin is licensed for both complicated skin and soft tissue infections; it is also licenced for bloodstream infections, including right-sided endocarditis.

In a controlled comparative study between daptomycin and standard therapy for staphylococcal bacteraemia and endocarditis, daptomycin met non-inferiority criteria but outcomes for left-sided endocarditis were disappointing.[27,25] Resistance has been encountered occasionally for both linezolid and daptomycin. In both cases this occurs through mutational changes, which have been observed to arise during therapy.[58,59] The hVISA/VISA-resistance phenotypes also seem to be correlated with increased resistance to daptomycin.[60] Other recently developed agents active against MRSA are discussed in the section on newer antimicrobials.

Enterococcus spp.

Enterococci are part of the normal flora of the lower gastrointestinal tract of most healthy adults; the common species found are *E. faecalis* and *E. faecium*, and indeed these are the typical species causing infections. It is often suggested that enterococci are not very virulent. However, they are an important cause of endocarditis, being responsible for 10 to 15% of native-valve cases. Non-endocarditis enterococcal bacteraemia usually arises from urinary tract infections (UTIs) or intra-abdominal, biliary or pelvic infections as part of a polymicrobial infection; it is associated with significant mortality often due to an underlying condition.[61] Enterococci have emerged as important nosocomial pathogens largely because of their resistance to many antimicrobials (to be discussed later) and subsequent selection during antimicrobial therapy.

Enterococci are inherently resistant to multiple antimicrobial classes, and they have acquired resistance to many more. It is important to note that they are resistant to all cephalosporins and have reduced susceptibility to penicillin, amoxicillin and carbapenems, compared with other streptococci because of the production of low-affinity penicillin-binding proteins such as PBP5.[62] MICs of these penicillins and carbapenems are 4-fold higher for *E. faecium* than for *E. faecalis* and are higher than the breakpoint; therefore, *E. faecium* is considered resistant.

Enterococci also have low-level inherent resistance to aminoglycosides so that these agents are not active as monotherapy. However, synergy is observed between active β-lactams (e.g., amoxicillin and aminoglycosides), and this combination is a mainstay of therapy for enterococcal endocarditis. High-level resistance is present in some strains because of the acquisition of an aminoglycoside-modifying enzyme. At this level of resistance, synergy with β-lactams does not occur.

Inherently, enterococci are resistant to clindamycin, but also have very commonly acquired *erm* genes that encode a ribosomal methylase and confer resistance to macrolides (e.g., erythromycin), lincosamides (e.g., clindamycin) and streptogramin B (i.e., the MLS_B phenotype). They are also usually resistant to chloramphenicol because of an acquired chloramphenicol acetyl transferase.[63]

Given the inherent and acquired resistance of enterococci, the mainstays of therapy were at one time amoxicillin for *E. faecalis* and vancomycin for *E. faecium*; they were also used for patients with penicillin allergy. Therefore, it was a significant concern when, in 1988, enterococci's resistance to vancomycin was reported.[64] Vancomycin usually acts by binding to the terminal D-alanyl-D-alanine (D-ala-D-ala) amino acids of the pentapeptide responsible for cross-linking of the bacterial cell wall.

In the archetypal resistance phenotype (VanA), there is a change in the terminal amino acids of the pentapeptide from D-ala-D-ala to D-alanyl-D-lactate, to which vancomycin cannot bind. This is achieved not only by genes encoding acquired ligases (e.g., *VanA*) that produce the different terminal amino acids but also by acquired genes encoding enzymes to convert pyruvate to D-lactate (e.g., *vanH*) and dipeptidases (e.g., *VanX*) that cleave any D-ala-D-ala dipeptide that is formed.

These genes, together with regulatory genes, are located on transposable elements: *Tn1546* in the case of VanA, which may be plasmid or chromosomally located.[65] There are at least six types of acquired vancomycin resistance in enterococci (i.e., VanA, VanB, VanD, VanE, VanG and VanL) as well a similar inherent resistance mechanism seen in *E. gallinarum* and *E. casseliflavus* (VanC); in the latter the terminal amino acids are altered to D-alanyl-D-serine.[66] These mechanisms all confer full resistance to vancomycin but variable resistance to teicoplanin.

Since 1988, vancomycin resistance has spread widely. In 2011, the prevalence of resistance in *E. faecalis* across European countries was 0 to 1.8% in the United Kingdom and 6.2% in Greece; for *E. faecium* it was 0 to 8.9% in the United Kingdom and 34.9% in Ireland (see www.ecdc.europa.eu/en/activities/surveillance/EARS-Net/database/Pages/tables_report.aspx).

In the United States much higher rates of resistance have been reported: 9.6% for *E. faecalis* and 72.4% for *E. faecium*.[67] Vancomycin-resistant enterococci (VRE) are primarily nosocomially acquired, and there have been numerous reports of hospitals outbreaks. It is now clear that spread within and between hospitals is clonal and that there are particular strains that have the ability to spread and cause infection outbreaks. Multi-locus sequence typing (MLST) has revealed that isolates from a particular group of related sequence types (i.e., clonal complex-17, or CC17) is hospital-adapted and multiply-resistant, plus it has a putative pathogenicity island.[68]

During the 1990s, the options for treatment of VRE were extremely limited. Quinupristin-Dalfopristin, a combination of streptogramins, had some activity against *E. faecium* but was ineffective against *E. faecalis*. Treatment options have been significantly improved by the introduction in the last few years of linezolid, tigecycline and daptomycin, and rates of resistance remain low—US isolates of *enterococci* collected in 2011 show susceptibility of 99.7%, 99.7%, and 99.9%, respectively.[69] However, resistance to all of these agents has been described in enterococci and therefore is likely to spread in the future.

Enterobacteriaceae

The Enterobacteriaceae are a large group of Gram-negative bacilli which are part of the commensal flora of human and animal guts. Several species are opportunistic pathogens in humans. *Escherichia coli* is the most common cause of UTIs and has also become the most frequently identified and significant organism from blood cultures in many parts of the world, including the United Kingdom.[5,70]

Other species frequently associated with infections are *Klebsiella* spp., *Enterobacter cloacae*, *Serratia marcesens*, *Citrobacter* spp. and *Proteus mirabilis*. All enterobacteriaceae can cause a wide range of infections other than urinary tract ones, including wound infections, pneumonia, peritonitis and infections of prosthetic medical devices. However, most require a profound breach in normal host defences to become established.

Being Gram-negative, Enterobacteriaceae have an outer membrane that prevents the penetration of many agents such as vancomycin, macrolides, linezolid, and daptomycin. The degree of drug susceptibility is thus determined by permeability via porin channels and extent of clearence by efflux pumps. In general, Enterobacteriaceae outer membranes are more permeable and their drug clearance is less efficient than other potentially MDR Gram-negative pathogens (e.g., *Pseudomonas aeruginosa*).

Nevertheless, all Enterobacteriaceae are resistant to antimicrobials which are too large to cross the outer membrane (e.g., vancomycin, daptomycin) and several of the penicillins (e.g., benzylpenicillin, flucloxacillin, penicillin V).[5,12] *E. coli* are intrinsically sensitive to most other β-lactams, including amoxicillin, piperacillin, β-lactam/β-lactamase inhibitor combinations (BL/BLI), cephalosporins and carbapenems. *Klebsiella pneumoniae* intrinsically contains a β-lactamase, SHV-1, which confers resistance to amoxicillin.

The situation with β-lactams is more complicated for *Enterobacter cloacae*, *Citrobacter freundii*, *S. marcesens* and other species. All of these organisms have constitutively expressed AmpC-type β-lactamases for which expression is inducible following β-lactam exposure. These inducible AmpC enzymes confer resistance to other penicillins and BL/BLIs because they are not substrates for clinically available inhibitors and first-generation cephalosporins. Extended-spectrum cephalosporins remain active; however, they are considered unreliable for treating Enterobacteriaceae with inducible chromosomal AmpC enzymes, since mutations can readily arise. This can lead to the expression becoming permanently elevated and causing resistance to these agents.[71]

Many other classes of antibiotics are potentially active against Enterobacteriaceae, including aminoglycosides, fluorquinolones, tetracyclines, sulphonamides, trimethoprim and nitrofurantoin. Resistance to all of these antibiotics has been encountered in Enterobacteriaceae as acquired traits.[72–74]

Resistance to fluoroquinolones has risen substantially in most Enterobacteriaceae, mainly through mutations in the genes *gyrA*, *gyrB* and *parC*, which code for the enzymatic targets of fluorquinolones, but more recently are secondary to the spread by plasmids of resistance genes *qnr* (target protection), *AAC(6′)-Ib-cr* (enzymatic inactivation) and various efflux pumps.[75] The recent association of fluoroquinolone use with overgrowth of *Clostridium difficile* in the gut has also prompted many hospitals in Europe and the United States to limit its use as much as possible. Resistance to nitrofurantoin, which should only be used for treating UTIs, is less common in most Enterobacteriaceae, although *Proteus* spp. are intrinisically resistant to this agent.[73]

The major shift in antimicrobial resistance in the past two decades in Enterobacteriaceae has been the rapid rise of resistance to β-lactams, which has been paralleled by increasing resistance to aminoglycosides. In both cases resistance is largely a result of horizontally acquired resistance genes, often identified on plasmids.[5,18] A classification of β-lactamases are given in Table 15.2 earlier in the chapter.

Early examples of horizontally acquired β-lactamases were SHV-1 and TEM-1, found on plasmids in *E. coli* and other Enterobacteriaceae, which resulted in relatively narrow-spectrum resistance to β-lactams, including amoxicillin and first-generation cephalosporins.[5] Variants of both enzymes—mainly found in *E. coli* and *K. pneumoniae*—that increase MICs to extended-spectrum cephalosporins and aztreonam as well as penicillins are referred to as extended-spectrum β-lactamases (ESBLs); these are inhibited by β-lactamase inhibitors (e.g., clavulanate), although MICs against the amoxicillin-clavulanate combination are usually significantly raised.

Other BL/BLIs (e.g., piperacillin-tazobactam) are generally active *in vitro* unless additional resistance mechanisms are present even though there are concerns about the clinical effectiveness of these agents for treating invasive infections with ESBL producers. TEM and SHV ESBLs, however, remained fairly infrequent.

From around the turn of the century onward, a new group of ESBLs, CTX-M enzymes, became prevalent globally, but there are large geographic differences.[76] For example, data from intra-abdominal infections collected for the worldwide Study for Monitoring Antimicrobial Resistance Trends (SMART) reported rates of 17.9% and 11.6% ESBL positivity in *K. pneumoniae* and *E. coli*, respectively.[77] The equivalent rates reported by SMART for India were 55% and 67%.[78] It is common for strains with ESBLs to be resistant to many other classes of antibiotics by other mechanisms, including aminoglycosides and fluoroquinolones; thus, many are MDR.[5]

Over the same period, a number of horizontally acquired AmpC-type β-lactamases became increasingly identified in Enterobacteriaceae, including organisms which either lack or usually express chromosomal AmpCs at a low rate. These enzymes are closely related to the chromosomal AmpCs described previously; however, they are often found on plasmids and their expression is often sufficient to result in a resistance phenotype similar to that seen with ESBLs. Because AmpCs are not susceptible to inhibitors (e.g., clavulanate) BL/BLIs are also compromised.[71]

Although several treatments are available for treating ESBLs or acquired AmpC producers, it is often the case that carbapenems (e.g., meropenem and imipenem) are the most clinically reliable and safe agents available.[79] The recent emergence in Enterobacteriaceae of various β-lactamases capable of hydrolysing carbapenems is therefore a substantial cause for concern. These include metallo-β-lactams (MBLs) mainly of the vimentin (VIM) and NDM groups, KPC enzymes and the carbapenem-hydrolysing oxacillinase, known as OXA-48.[80]

The epidemiology and phenotypes associated with each of the enzymes just mentioned differ somewhat. In Northern Europe and most of the United States, carbapenem resistance remains rare in Enterobacteriaceae, the exception being KPC enzymes, which first emerged in *K. pneumoniae* in the United States. Figure 15.3 summarises the current global spread of the most common carbapenemases in Enterobacteriaceae. These are found most often in *K. pneumoniae* and *E. coli*; although, particularly in the case of NDM, they have been identified in a wide range of GNBs, including many Enterobacteriaceae.[80,81]

KPC producers are probably the most clinically relevant at present, with the ST258 strain of *K. pneumoniae* having demonstrated its potential to spread globally and cause significant local outbreaks.[5,81] During 2009–2010 ST258 was estimated to be present in ~40% of *K. pneumoniae* bloodstream isolates in Greece.[82]

According to several studies, NDM appears to have emerged recently in South Asia, where its prevalence is already worryingly high. However, the studies have been hotly contested by some scientists in India, clinicians and politicians. VIM has been most prevalent in Greece, although rates are now reported to be declining.[81] OXA-48 was first identified in Turkey, from where it has spread to North Africa, the Middle East and parts of Europe.

Carbapenem-resistant Enterobacteriaceae (CRE) are usually multi-drug-resistant by virtue of their presence in strains with other resistance determinants, and many examples are XDR.[5,80] There is some variation in resistance profiles, however. In the absence of additional resistance mechanisms, carbapenem MICs may not be above resistance breakpoints; moreover, the fact that they are carbapenemase producers may be missed in routine microbiology laboratories. This is especially true for OXA-48, which does not significantly hydrolyse carbapenems and has very little activity against cephalosporins.[81]

It is important to note that the production of several of these β-lactamases has been associated with a number of adverse outcomes for patients with bloodstream infections. This has been shown for ESBL, VIM and KPC producers.[79,83,84] For example, in a cohort study performed in the United States, mortality associated with *K. pneumoniae* bacteraemia was 32.1% for those infected with KPC producers compared to 8.9% for those infected with non-KPC producers.[83]

In all cases, this effect may partly derive from the fact that MDR Enterobacteriaciae infections are more likely to occur in the most severely ill patients; however, it is also the case that it takes longer on average for these patients to receive appropriate antibiotic therapy. Although the clinical impact of producers of some other β-lactamases has not been studied to the same extent, we can hypothesise that their impact is likely to be similar if they become established in strains with significant pathogenic potential.

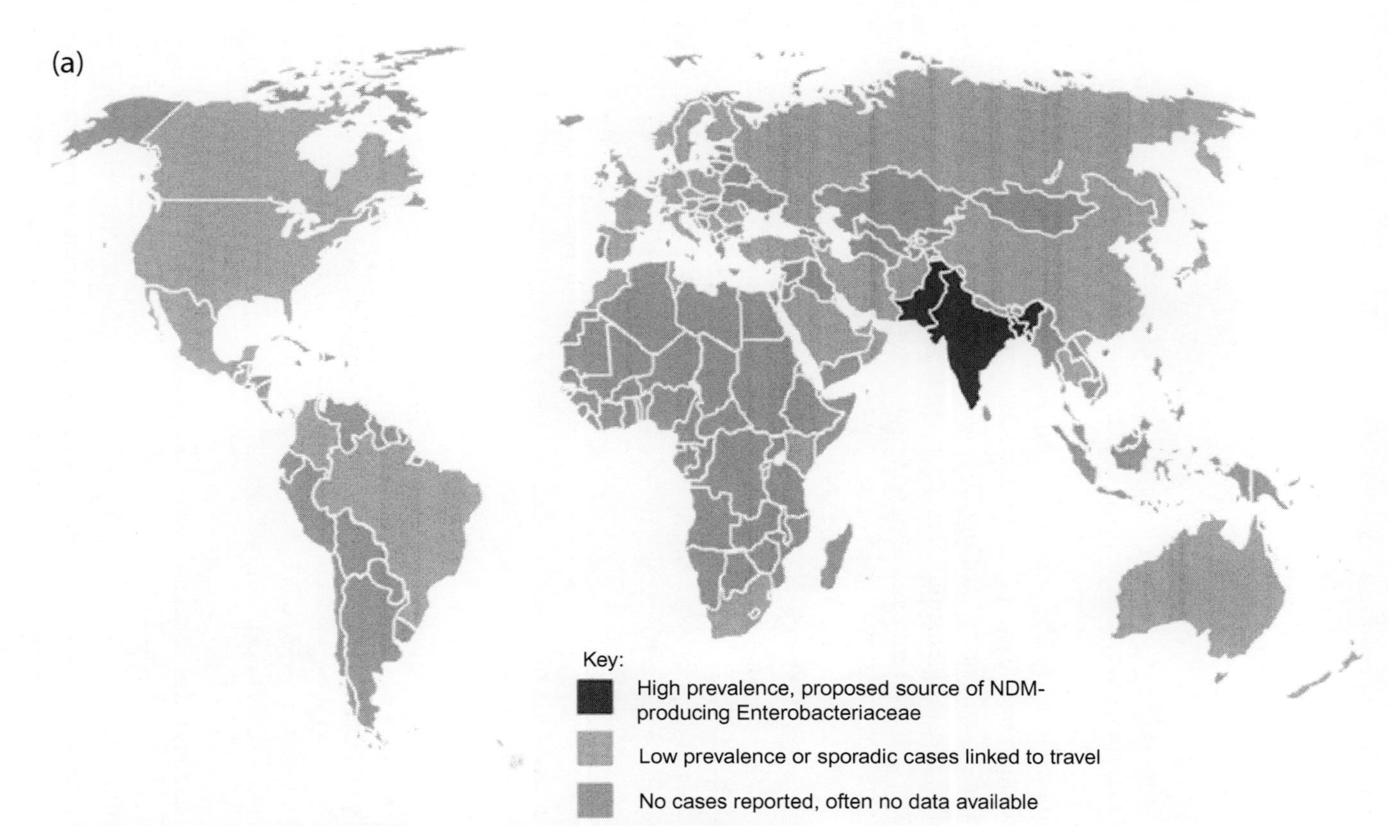

Figure 15.3 ***Global spread of Enterobacteriaceae with (a) NDM and (b) KPC.***

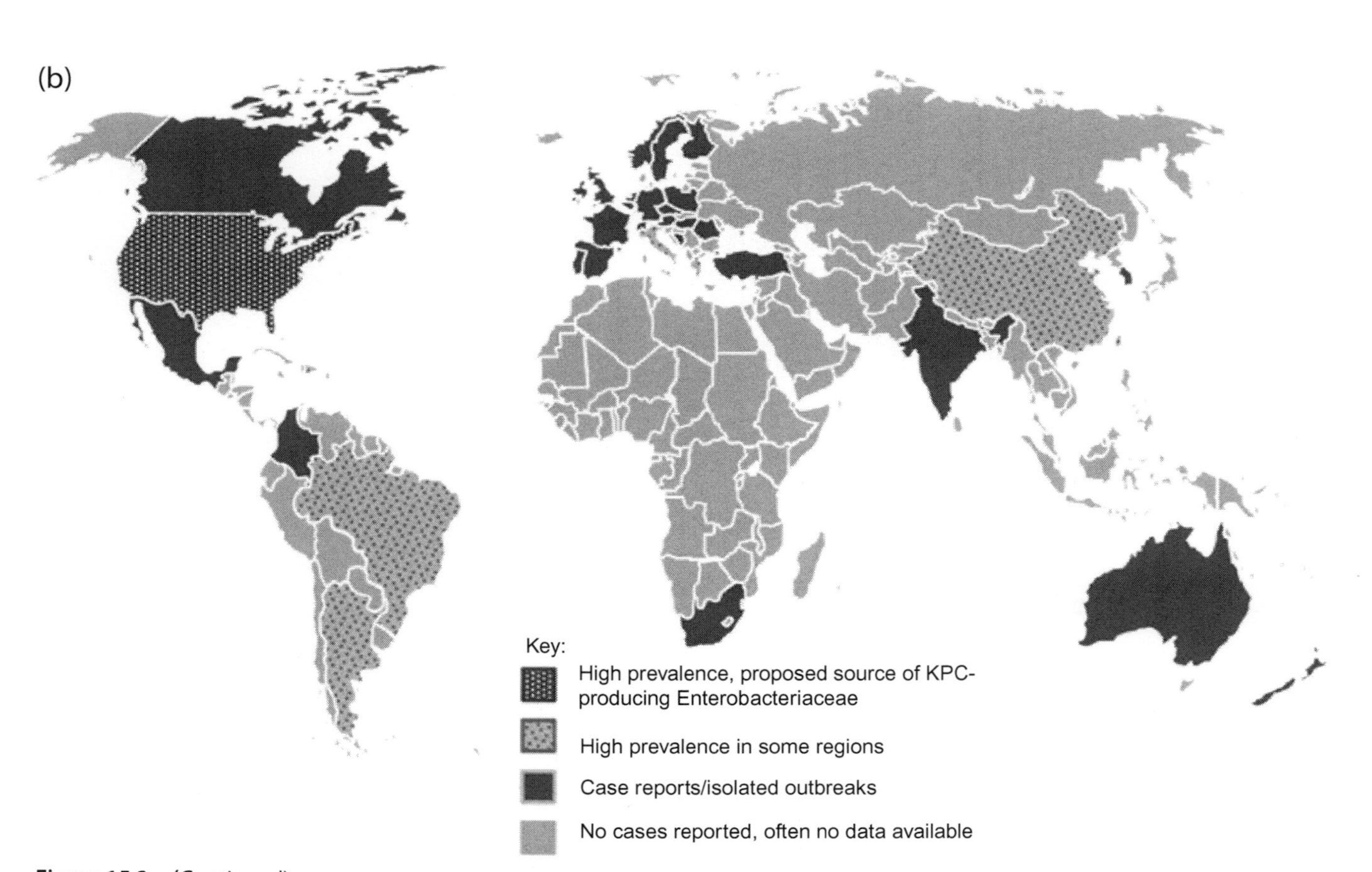

Figure 15.3 (Continued)

Therapeutic options in XDR Enterobacteriaceae–producing carbapenemases are often quite limited. Although phenotypes are variable, the agents most likely to remain active are aminoglycosides, tigecycline, colistin and fosfomycin.[5,84] As mentioned previously, aminoglycoside resistance genes are commonly found in these bacteria, often on the same plasmids harbouring the β-lactamase genes.[5,18] However, aminoglycoside-modifying enzymes have varying activity against different agents. Enterobacteriaceae producing 16s RNA methylases are resistant to all currently available aminoglycosides but are most often identified in Asia.

Gentamicin often retains activity against KPC and VIM producers, while most NDM producers are resistant to all aminoglycosides, probably because of co-production of RNA methylases.[5,81] Colistin resistance is infrequent in many XDR Enterobacteriaceae, but some of these organisms (e.g., *S. marcesens* and *Proteae* spp.) are intrinsically resistant to this agent. The mechanisms of colistin resistance are not fully understood, but probably involve multiple genetic and regulatory changes. Use of both aminoglycosides and colistin is associated with a significant nephrotoxicity.[81,85]

Many XDR Enterobacteriaceae remain susceptible to tigecycline, but resistance is encountered at variable rates depending on the species.[74,86] Susceptibility in the Proteae is reduced, at least in part, because of their resistance-nodulation-division (RND) efflux pumps. Cases of resistance in other species are thought to be related to increased expression of efflux pumps, especially pumps in the RND family. Low levels of tigecylcine achieved in the bloodstream relative to resistance breakpoints have raised concerns that these agents may not be terribly effective for treating bacteraemia.[72,86]

As renal excretion of tigecycline is limited, it is also unlikely to be useful as a treatment for UTIs.[72] Many strains are sensitive to fosfomycin, but this antibiotic has been predominantly used for uncomplicated UTIs in most countries, and there are concerns that resistance may arise readily during treatment. The clinical utility of fosfomycin for treating XDR Enterobacteraiceae, either in UTIs or systemic infections, is uncertain.[87]

Pseudomonas aeruginosa

The opportunistic pathogen *P. aeruginosa* is widespread in the environment, particularly where there is moisture. Its ability to survive and grow in the presence of limited nutrients and over a wide range of environmental conditions, added to its inherent resistance to many antimicrobials, makes it ideal as a nosocomial pathogen. It is also one of the first organisms for which the importance of biofilm growth was established with respect to both infections, as in the lung in cystic fibrosis, and the environment, as in colonisation of water pipes.[88]

P. aeruginosa possesses mechanisms for resistance to many antimicrobials. As noted earlier, its outer membrane provides a barrier to the penetration of many agents. In the case of β-lactams, *P. aeruginosa* is intrinsically resistant to most penicillins and cephalosporins because of a combination of reduced permeability,[89] low-affinity PBPs, basal production of an AmpC chromosomal β-lactamase, intrinsic β-lactamase PBP5[90] activity and efflux. Thus, it is susceptible only to a limited number of anti-pseudomonal penicillins (e.g., piperacillin), third- and fourth-generation cephalosporins (e.g., ceftazidime, cefepime), and carbapenems (e.g., meropenem). While *P. aeruginosa* is intrinsically susceptible to aminoglycosides, fluoroquinolones and colistin, acquired resistance to the these agents has become commonplace.

Although acquisition of resistance mechanisms through horizontal gene transfer is important in some situations, particularly the acquisition of potent β-lactamases, resistance to many antimicrobials in *P. aeruginosa* is mediated through chromosomal mutations that alter the expression of intrinsic resistance mechanisms. Notable among these are efflux pumps, the most important of which, in terms of antimicrobial resistance, are the RND (resistance nodulation division) pumps. These tripartite structures typically have an inner membrane portion, a periplasmic fusion protein (MFP) and an outer membrane protein (known as the outer membrane factor, or OMF); also, they accommodate a broad range of unrelated molecules, including most classes of antimicrobials, biocides, dyes, toxic fatty acids, bile salt and homoserine lactones (associated with quorum sensing).[91]

The first of the RND pumps to be characterised was MexAB-OprM, which has β-lactams, fluoroquinolones, sulphonamides, chloramphenicol, trimethoprim and tigecycline among its substrates. Other pumps (e.g., MexEF-OprN) can efflux aminoglycosides. *P. aeruginosa* is notable in having many efflux systems. From analysis of the genome, 10 putative RND pumps (compared with 4 in *E. coli*) and 20 in the Major Facilitator Superfamily (MFS) group have been identified.[92] Expression of the RND pumps is regulated and inducible in the presence of their substrates, contributing to resistance to many antimicrobials. Mutations in their regulatory controls lead to hyper-expression and thereby multi-resistance.

The multi-factorial nature of resistance arising in the genome of *P. aeruginosa* is exemplified by a study that screened a comprehensive library for mutants with increased or decreased susceptibility to ceftazidime, imipenem and meropenem. This study identified 41 gene mutations with reduced susceptibility and 37 with increased susceptibility.[89]

In addition to resistance that is mediated through intrinsic mechanisms that are constitutive, inducible or hyper-expressed and through mutational target site changes, *P. aeruginosa* achieves resistance by its acquisition of exogenous genes. Notable are the aminoglycoside-modifying enzymes, of which the most common are AAC(6')-I and APH(3')-II.[93] *P. aeruginosa* is also able to acquire β-lactamases, including TEM, SHV, OXA and a number of carbapenemases, including VIM, IMP, SPM and GIM.[94,95]

Both intrinsic and acquired-resistance mechanisms contribute to the increasing number of MDR and XDR strains of *P. aeruginosa*. The susceptibility of any given strain is unpredictable without formal testing, and even when testing is performed *in vitro* the results may not reflect the resistance phenotype expressed *in vivo*. The only agent that maintains fairly consistent activity is colistin (polymyxin E), which acts primarily on the outer membrane. Unfortunately, *P. aeruginosa* can develop resistance because of lipid alterations, reduced magnesium and calcium contents and alterations in specific proteins in the outer cell membrane.[96–98]

Acinetobacter baumannii

Acinetobacter baumannii is an opportunistic pathogen or coloniser of hospitalised patients, especially severely unwell patients on intensive care units. Its ability to survive in the hospital environment likely contributes to its ability to spread nosocomially and to its association with hospital outbreaks. *A. baumannii* is implicated in a range of infections, the most common of which is ventilator-associated pneumonia.[99] MICs to many commonly used antibiotics, including *β*-lactams, are high relative to, for example, *E. coli*.[89]

Intrinsic resistance mechanisms have not been well studied, but the outer membrane of *A. baumannii* is much less permeable than that of *E. coli*. This is likely connected to porins which do not allow antibiotics to readily cross the outer membrane, together with constitutive expression of efflux pumps.[100] Several different classes of these pumps have been found to contribute to resistance to antimicrobials including *β*-lactams, aminoglycosides, chloramphenicol, quinolones, tetracyclines and tigecycline. In addition, all *A. baumannii* contain chromosomal genes coding for *β*-lactamases of the ADC (*acinetobacter*-derived cephalosporinase) and OXA-51-like types, but the level of gene expression is usually low.[101]

A. baumannii has been particularly successful at accumulating acquired-resistance mechanisms. Intrinsic resistance mechanisms, outlined previously, can be increased by various genetic modifications. IS elements can move upstream of both of the chromosomal *β*-lactamses genes and provide a strong promoter, resulting in raised MICs to broad-spectrum *β*-lactams, in the case of OXA-51-like enzymes, including the carbapenems.[102,103] Down-regulation or disruption by IS elements of several outer membrane proteins has been associated in several reports of *β*-lactam resistance. Acquired OXA-type enzymes can be plasmid-associated or chromosomal and can lead to raised MICs to *β*-lactams, including carbapenems.

Other *β*-lactamases are fairly infrequent, although *A. baumannii*–producing metallo-*β*-lactamases of the NDM family are increasingly recognised in some parts of the world.[101] Some efflux pumps (e.g., those specifically targeting tetracycline) are likely to be acquired mechanisms of resistance.[104] Many types of aminoglycoside-modifying enzymes are common in drug-resistant isolates.[105] Quinolone resistance arises readily, predominantly by point mutations in *gyrA* and *parC* genes, as in other GNBs. Most isolates remain susceptible to colistin, although resistance has been described occasionally and, as in other GNBs, is not completely understood.[101]

PREVENTION AND CONTROL OF MDR ORGANISMS

The prevention and control of MDR 'superbugs' requires controlling the transmission of resistant organisms and reducing the selective pressure provided by inappropriate antimicrobial use in the environment. Control of transmission is a major element in infection control procedures in hospitals. It is discussed in detail elsewhere in this book; however, it should be remembered that, while the prevalence of most superbugs has been 'amplified' by the conditions of high selective pressure seen in hospitals, superbugs have in some cases 'escaped' into the community. When this happens, infection control procedures should be tailored to the particular situation (e.g., nursing homes), where ESBL-producing Enterobacteriaceae are a problem,[106] or other institutions, where PVL-producing CA-MRSA may be more prevalent.[107]

Antimicrobial stewardship is the term that has been globally adopted to describe activities to reduce inappropriate antimicrobial use. It is estimated that physicians use antimicrobials inappropriately about 50% of the time, and many strategies have been suggested to improve this situation.[108] At the governmental level, WHO has recognised the importance of antimicrobial resistance as a global threat to public health; more locally, the European Commission has urged member states to promote appropriate antimicrobial use.

CONCLUSION

The antibiotic 'golden age' was during the 1970s and 1980s. Since that time significant resistance has emerged, first in Gram-positive organisms (e.g., *S. aureus* and *enterococci*) and then in Gram-negative organisms. We are now faced with organisms, such as *E. coli*, *P. aeruginosa* and *A. baumanii*, that are resistant to most or all available agents. The development pipeline of new agents includes a few members of existing antimicrobial classes but very few are really novel. New strategies that target virulence expression rather than growth inhibition are being explored,[109] but it is likely to be many years before these enter clinical practice. In the meantime, the best defence against superbugs is good infection control and effective antimicrobial stewardship.

REFERENCES

1. Fleming A. *Sir Alexander Fleming—Nobel Lecture: Penicillin.* <www.nobelprize.org/nobel_prizes/medicine/laureates/1945/fleming-lecture.html>; 1945 [accessed 16.02.13].
2. McDonald LC. Trends in antimicrobial resistance in health care-associated pathogens and effect on treatment. *Clin Infect Dis* 2006;**42**(Suppl. 2):S65–71.
3. Mediavilla JR, Chen L, Mathema B, Kreiswirth BN. Global epidemiology of community-associated methicillin resistant *Staphylococcus aureus* (CA-MRSA). *Curr Opin Microbiol* 2012;**15**(5):588–95.
4. Stefani S, Chung DR, Lindsay JA, et al. Meticillin-resistant *Staphylococcus aureus* (MRSA): global epidemiology and harmonisation of typing methods. *Int J Antimicrob Agents* 2012;**39**(4):273–82.

5. Livermore DM. Current epidemiology and growing resistance of Gram-negative pathogens. *Korean J Internal Med* 2012;**27**(2):128–42.
6. Perez F, Endimiani A, Hujer KM, Bonomo RA. The continuing challenge of ESBLs. *Curr Opin Pharmacol* 2007;**7**(5):459–69.
7. Nordmann P, Naas T, Poirel L. Global spread of Carbapenemase-producing Enterobacteriaceae. *Emerg Infect Dis* 2011;**17**(10):1791–8.
8. Nordmann P, Poirel L, Toleman MA, Walsh TR. Does broad-spectrum beta-lactam resistance due to NDM-1 herald the end of the antibiotic era for treatment of infections caused by Gram-negative bacteria? *J Antimicrob Chemother* 2011;**66**(4):689–92.
9. Lorian V, Burns L. Predictive value of susceptibility tests for the outcome of antibacterial therapy. *J Antimicrob Chemother* 1990;**25**(1):175–81.
10. Pallares R, Liñares J, Vadillo M, et al. Resistance to penicillin and cephalosporin and mortality from severe pneumococcal pneumonia in Barcelona, Spain. *N Engl J Med* 1995;**333**(8):474–80.
11. Ciofu O, Mandsberg LF, Wang H, Høiby N. Phenotypes selected during chronic lung infection in cystic fibrosis patients: implications for the treatment of *Pseudomonas aeruginosa* biofilm infections. *FEMS Immunol Med Microbiol* 2012;**65**(2):215–25.
12. Cox G, Wright GD. Intrinsic antibiotic resistance: mechanisms, origins, challenges and solutions. *Int J Med Microbiol* 2013;**303**(6–7):287–92.
13. Palmer KL, Kos VN, Gilmore MS. Horizontal gene transfer and the genomics of enterococcal antibiotic resistance. *Curr Opin Microbiol* 2010;**13**(5):632–9.
14. Toleman MA, Walsh TR. Combinatorial events of insertion sequences and ICE in Gram-negative bacteria. *FEMS Microbiol Rev* 2011;**35**(5):912–35.
15. Nordmann P, Lartigue MF, Poirel L. Beta-lactam induction of ISEcp1B-mediated mobilization of the naturally occurring bla(CTX-M) beta-lactamase gene of *Kluyvera ascorbata. FEMS Microbiol Lett* 2008;**288**(2):247–9.
16. Charpentier X, Polard P, Claverys JP. Induction of competence for genetic transformation by antibiotics: convergent evolution of stress responses in distant bacterial species lacking SOS? *Curr Opin Microbiol* 2012;**15**(5):570–6.
17. Novick RP, Christie GE, Penades JR. The phage-related chromosomal islands of Gram-positive bacteria. *Nat Rev Microbiol* 2010;**8**(8):541–51.
18. Carattoli A. Plasmids and the spread of resistance. *Int J Med Microbiol* 2013:2013.
19. Wozniak RAF, Waldor MK. Integrative and conjugative elements: mosaic mobile genetic elements enabling dynamic lateral gene flow. *Nat Rev Microbiol* 2010;**8**(8):552–63.
20. Domingues S, da Silva GJ, Nielsen KM. Integrons: vehicles and pathways for horizontal dissemination in bacteria. *Mob Genet Elem* 2012;**2**(5):211–23.
21. Walsh TR, Weeks J, Livermore DM, Toleman MA. Dissemination of NDM-1 positive bacteria in the New Delhi environment and its implications for human health: an environmental point prevalence study. *Lancet Infect Dis* 2011;**11**(5):355–62.
22. Nordmann P, Poirel L, Walsh TR, Livermore DM. The emerging NDM carbapenemases. *Trends Microbiol* 2011;**19**(12):588–95.
23. Magiorakos AP, Srinivasan A, Carey RB, et al. Multidrug-resistant, extensively drug-resistant and pandrug-resistant bacteria: an international expert proposal for interim standard definitions for acquired resistance. *Clin Microbiol Infect* 2012;**18**(3):268–81.
24. Que YA, Moreillon P. *Staphylococcus aureus* (including Staphylococcal toxic shock). In: Mandell GL, Bennett JE, Dolin R, editors. *Mandell, Douglas and Bennett's principles and practice of infectious diseases.* Elsevier; 2010.
25. Corey GR. *Staphylococcus aureus* bloodstream infections: definitions and treatment. *Clin Infect Dis* 2009;**48**(Suppl. 4):S254–9.
26. Chambers HF. Penicillins and β-lactam inhibitors. In: Mandell GL, Bennett JE, Dolin R, editors. *Mandell, Douglas and Bennett's principles and practice of infectious diseases*, 7th ed. Elsevier; 2010.
27. Gould FK, Brindle R, Chadwick PR, et al. Guidelines (2008) for the prophylaxis and treatment of methicillin-resistant *Staphylococcus aureus* (MRSA) infections in the United Kingdom. *J Antimicrob Chemother* 2009;**63**(5):849–61.
28. Barber M. Methicillin-resistant staphylococci. *J Clin Pathol* 1961;**14**:385–93.

29. Harding JW. Infections due to methicillin-resistant strains of *Staphylococcus pyogenes*. *J Clin Pathol* 1963;**16**:268–70.
30. Chambers HF, Hartman BJ, Tomasz A. Increased amounts of a novel penicillin-binding protein in a strain of methicillin-resistant *Staphylococcus aureus* exposed to nafcillin. *J Clin Invest* 1985;**76**(1):325–31.
31. European Medicines Agency, Committee for Medicinal Products for Human Use (CHMP). *Zinforo: European Public Assessment Report,* September 18, 2012. <www.ema.europa.eu/ema/> [accessed 07.03.13].
32. European Antimicrobial Surveilance Network. *Susceptibility of Staphylococcus aureus isolates to methicillin in United Kingdom, 1998–2011.* <http://ecdc.europa.eu/en/activities/surveillance/EARS-Net/database/>; 2013 [accessed 03.07.13].
33. Wertheim HF, Melles DC, Vos MC, et al. The role of nasal carriage in *Staphylococcus aureus* infections. *Lancet Infect Dis* 2005;**5**(12):751–62.
34. Dancer SJ. Importance of the environment in methicillin-resistant *Staphylococcus aureus* acquisition: the case for hospital cleaning. *Lancet Infect Dis* 2008;**8**(2):101–13.
35. French GL, Otter JA, Shannon KP, Adams NM, Watling D, Parks MJ. Tackling contamination of the hospital environment by methicillin-resistant *Staphylococcus aureus* (MRSA): a comparison between conventional terminal cleaning and hydrogen peroxide vapour decontamination. *J Hosp Infect* 2004;**57**(1):31–7.
36. Henderson DK. Managing methicillin-resistant staphylococci: a paradigm for preventing nosocomial transmission of resistant organisms. *Am J Infect Cont* 2006;**34**(5 Suppl. 1):S46–54. discussion S64–73.
37. Coia JE, Duckworth GJ, Edwards DI, et al. Guidelines for the control and prevention of methicillin-resistant *Staphylococcus aureus* (MRSA) in healthcare facilities. *J Hosp Infect* 2006;**63**(Suppl. 1):S1–44.
38. Gordon RJ, Lowy FD. Pathogenesis of methicillin-resistant *Staphylococcus aureus* infection. *Clin Infect Dis* 2008;**46**(Suppl. 5):S350–9.
39. Gorwitz RJ. Community-associated methicillin-resistant *Staphylococcus aureus*: epidemiology and update. *Pediatric Infect Dis J* 2008;**27**(10):925–6.
40. Elston JW, Barlow GD. Community-associated MRSA in the United Kingdom. *J Infect* 2009;**59**(3):149–55.
41. Green SM, Marsh P, Ahmad N, Jefferies JM, Clarke SC. Characterization of community and hospital *Staphylococcus aureus* isolates in Southampton, UK. *J Med Microbiol* 2010;**59**(Pt 9):1084–8.
42. British Society of Antimicrobial Chemotherapy. *Bacteraemia data: SIR summary MRSA from UK and Ireland, 2001–2011.* <www.bsacsurv.org/> [accessed 08.03.13].
43. Holmes NE, Turnidge JD, Munckhof WJ, et al. Antibiotic choice may not explain poorer outcomes in patients with *Staphylococcus aureus* bacteremia and high vancomycin minimum inhibitory concentrations. *J Infect Dis* 2011;**204**(3):340–7.
44. Chang FY, Peacock Jr. JE, Musher DM, et al. *Staphylococcus aureus* bacteremia: recurrence and the impact of antibiotic treatment in a prospective multicenter study. *Medicine* 2003;**82**(5):333–9.
45. Kim SH, Kim KH, Kim HB, et al. Outcome of vancomycin treatment in patients with methicillin-susceptible *Staphylococcus aureus* bacteremia. *Antimicrob Agents Chemother* 2008;**52**(1):192–7.
46. Hidayat LK, Hsu DI, Quist R, Shriner KA, Wong-Beringer A. High-dose vancomycin therapy for methicillin-resistant *Staphylococcus aureus* infections: efficacy and toxicity. *Arch Intern Med* 2006;**166**(19):2138–44.
47. Wang G, Hindler JF, Ward KW, Bruckner DA. Increased vancomycin MICs for *Staphylococcus aureus* clinical isolates from a university hospital during a 5-year period. *J Clin Microbiol* 2006;**44**(11):3883–6.
48. Howden BP, Davies JK, Johnson PD, Stinear TP, Grayson ML. Reduced vancomycin susceptibility in *Staphylococcus aureus*, including vancomycin-intermediate and heterogeneous vancomycin-intermediate strains: resistance mechanisms, laboratory detection, and clinical implications. *Clin Microbiol Rev* 2010;**23**(1):99–139.
49. Steinkraus G, White R, Friedrich L. Vancomycin MIC creep in non-vancomycin-intermediate *Staphylococcus aureus* (VISA), vancomycin-susceptible clinical methicillin-resistant *S. aureus* (MRSA) blood isolates from 2001–05. *J Antimicrob Chemother* 2007;**60**(4):788–94.
50. Rybak M, Lomaestro B, Rotschafer JC, et al. Therapeutic monitoring of vancomycin in adult patients: a consensus review of the American Society of Health-System Pharmacists, the Infectious

Diseases Society of America, and the Society of Infectious Diseases Pharmacists. *Am J Health Syst Pharm* 2009;**66**(1):82–98.

51. Sujatha S, Praharaj I. Glycopeptide resistance in Gram-positive cocci: a review. *Interdiscip Perspect Infect Dis* 2012;**2012**:781679.
52. Hiramatsu K, Hanaki H, Ino T, Yabuta K, Oguri T, Tenover FC. Methicillin-resistant *Staphylococcus aureus* clinical strain with reduced vancomycin susceptibility. *J Antimicrob Chemother* 1997;**40**(1):135–6.
53. Hiramatsu K, Aritaka N, Hanaki H, et al. Dissemination in Japanese hospitals of strains of *Staphylococcus aureus* heterogeneously resistant to vancomycin. *Lancet* 1997;**350**(9092):1670–3.
54. Sievert DM, Rudrik JT, Patel JB, McDonald LC, Wilkins MJ, Hageman JC. Vancomycin-resistant *Staphylococcus aureus* in the United States, 2002–2006. *Clin Infect Dis* 2008;**46**(5):668–74.
55. Leclercq R, Derlot E, Duval J, Courvalin P. Plasmid-mediated resistance to vancomycin and teicoplanin in *Enterococcus faecium*. *N Engl J Med* 1988;**319**(3):157–61.
56. Azimian A, Havaei SA, Fazeli H, et al. Genetic characterization of a vancomycin-resistant *Staphylococcus aureus* isolate from the respiratory tract of a patient in a university hospital in northeastern Iran. *J Clin Microbiol* 2012;**50**(11):3581–5.
57. Melo-Cristino J, Resina C, Manuel V, Lito L, Ramirez M. First case of infection with vancomycin-resistant *Staphylococcus aureus* in Europe. *Lancet* 2013;**382**(9888):205.
58. Watkins RR, Lemonovich TL, File Jr. TM. An evidence-based review of linezolid for the treatment of methicillin-resistant *Staphylococcus aureus* (MRSA): place in therapy. *Core Evid* 2012;**7**:131–43.
59. Bayer AS, Schneider T, Sahl HG. Mechanisms of daptomycin resistance in *Staphylococcus aureus*: role of the cell membrane and cell wall. *Ann NY Acad Sci* 2013;**1277**:139–58.
60. Cui L, Tominaga E, Neoh HM, Hiramatsu K. Correlation between reduced daptomycin susceptibility and vancomycin resistance in vancomycin-intermediate *Staphylococcus aureus*. *Antimicrob Agents Chemother* 2006;**50**(3):1079–82.
61. Murray BE. The life and times of the Enterococcus. *Clin Microbiol Rev* 1990;**3**(1):46–65.
62. Williamson R, le Bouguénec C, Gutmann L, Horaud T. One or two low affinity penicillin-binding proteins may be responsible for the range of susceptibility of *Enterococcus faecium* to benzylpenicillin. *J Gen Microbiol* 1985;**131**(8):1933–40.
63. Brunton J. Antibiotic resistance in streptococci. In: Bryan LT, editor. *Antimicrobial drug resistance*. Academic Press; 1984. p. 530–65.
64. Uttley AH, Collins CH, Naidoo J, George RC. Vancomycin-resistant enterococci. *Lancet* 1988; **1**(8575–6):57–8.
65. Woodford N. Glycopeptide-resistant *enterococci*: a decade of experience. *J Med Microbiol* 1998; **47**(10):849–62.
66. Woodford N, Livermore DM. Infections caused by Gram-positive bacteria: a review of the global challenge. *J Infect* 2009;**59**(Suppl. 1):S4–16.
67. Draghi DC, Sheehan DJ, Hogan P, Sahm DF. *In vitro* activity of linezolid against key gram-positive organisms isolated in the United States: results of the LEADER 2004 surveillance program. *Antimicrob Agents Chemother* 2005;**49**(12):5024–32.
68. Willems RJ, Top J, van Santen M, et al. Global spread of vancomycin-resistant *Enterococcus faecium* from distinct nosocomial genetic complex. *Emerg Infect Dis*. 11(6): 821–28.
69. Flamm RK, Mendes RE, Ross JE, Sader HS, Jones RN. Linezolid surveillance results for the United States: LEADER surveillance program 2011. *Antimicrob Agents Chemother* 2013;**57**(2):1077–81.
70. Public Health England. *Escherichia coli bacteraemia in England, Wales and Northern Ireland, 2008–2012*. <www.hpa.org.uk/Topics/InfectiousDiseases/InfectionsAZ/EscherichiaColi/VoluntarySurveillance/>; 2013.
71. Jacoby GA. AmpC beta-lactamases. *Clin Microbiol Rev* 2009;**22**(1):161–82.
72. Seputiene V, Povilonis J, Armalyte J, Suziedelis K, Pavilonis A, Suziedeliene E. Tigecycline—how powerful is it in the fight against antibiotic-resistant. *Medicina (Kaunas)* 2010;**46**(4):240–8.
73. Guay DR. Contemporary management of uncomplicated urinary tract infections. *Drugs* 2008; **68**(9):1169–205.
74. Livermore DM, Warner M, Mushtaq S, Doumith M, Zhang J, Woodford N. What remains against carbapenem-resistant Enterobacteriaceae? Evaluation of chloramphenicol, ciprofloxacin, colistin, fosfomycin, minocycline, nitrofurantoin, temocillin and tigecycline. *Int J Antimicrob Agents* 2011; **37**(5):415–9.

75. Ruiz J, Pons MJ, Gomes C. Transferable mechanisms of quinolone resistance. *Int J Antimicrob Agents* 2012;**40**(3):196–203.
76. Canton R, Coque TM. The CTX-M beta-lactamase pandemic. *Curr Opin Microbiol* 2006;**9**(5): 466–75.
77. Hawser SP, Bouchillon SK, Lascols C, et al. Susceptibility of European *Escherichia coli* clinical isolates from intra-abdominal infections, extended-spectrum beta-lactamase occurrence, resistance distribution, and molecular characterization of ertapenem-resistant isolates (SMART 2008–2009). *Clin Microbiol Infect* 2011.
78. Hawser SP, Badal RE, Bouchillon SK, Hoban DJ. Antibiotic susceptibility of intra-abdominal infection isolates from India. *J Med Microbiol* 2010;**59**(Pt 9):1050–4.
79. Vardakas KZ, Tansarli GS, Rafailidis PI, Falagas ME. Carbapenems versus alternative antibiotics for the treatment of bacteraemia due. *J Antimicrob Chemother* 2012;**67**(12):2793–803.
80. Walsh TR. Emerging carbapenemases: a global perspective. *Int J Antimicrob Agents* 2010;**36**(Suppl. 3): S8–S14.
81. Tzouvelekis LS, Markogiannakis A, Psichogiou M, Tassios PT, Daikos GL. Carbapenemases in *Klebsiella pneumoniae* and other Enterobacteriaceae: an evolving. *Clin Microbiol Rev* 2012;**25**(4):682–707.
82. Giakkoupi P, Papagiannitsis CC, Miriagou V, et al. An update of the evolving epidemic of blaKPC-2-carrying *Klebsiella pneumoniae*. *J Antimicrob Chemother* 2011;**66**(7):1510–3.
83. Gasink LB, Edelstein PH, Lautenbach E, Synnestvedt M, Fishman NO. Risk factors and clinical impact of *Klebsiella pneumoniae* carbapenemase-producing. *Infect Control Hosp Epidemiol* 2009; **30**(12):1180–5.
84. Daikos GL, Petrikkos P, Psichogiou M, et al. Prospective observational study of the impact of VIM-1 metallo-beta-lactamase on. *Antimicrob Agents Chemother* 2009;**53**(5):1868–73.
85. Nation RL, Li J. Colistin in the 21st century. *Curr Opin Infect Dis* 2009;**22**(6):535–43.
86. Kelesidis T, Karageorgopoulos DE, Kelesidis I, Falagas ME. Tigecycline for the treatment of multidrug-resistant Enterobacteriaceae: a. *J Antimicrob Chemother* 2008;**62**(5):895–904.
87. Falagas ME, Kastoris AC, Karageorgopoulos DE, Rafailidis PI. Fosfomycin for the treatment of infections caused by multidrug-resistant. *Int J Antimicrob Agents* 2009;**34**(2):111–20.
88. Costerton JW. The etiology and persistence of cryptic bacterial infections: a hypothesis. *Rev Infect Dis* 1984;**6**(Suppl. 3):S608–16.
89. Hancock RE. Resistance mechanisms in *Pseudomonas aeruginosa* and other nonfermentative gram-negative bacteria. *Clin Infect Dis* 1998;**27**(Suppl. 1):S93–9.
90. Smith JD, Kumarasiri M, Zhang W, et al. Structural analysis of the role of *Pseudomonas aeruginosa* penicillin-binding protein 5 in β-lactam resistance. *Antimicrob Agents Chemother* 2013;**57**(7):3137–46.
91. Poole K. Efflux-mediated multiresistance in Gram-negative bacteria. *Clin Microbiol Infect* 2004; **10**(1):12–26.
92. Stover CK, Pham XQ, Erwin AL, et al. Complete genome sequence of *Pseudomonas aeruginosa* PAO1, an opportunistic pathogen. *Nature* 2000;**406**(6799):959–64.
93. Bonomo RA, Szabo D. Mechanisms of multidrug resistance in *Acinetobacter* species and *Pseudomonas aeruginosa*. *Clin Infect Dis* 2006;**43**(Suppl. 2):S49–56.
94. Nordmann P, Poirel L. Emerging carbapenemases in Gram-negative aerobes. *Clin Microbiol Infect* 2002;**8**(6):321–31.
95. Toleman MA, Simm AM, Murphy TA, et al. Molecular characterization of SPM-1, a novel metallo-beta-lactamase isolated in Latin America: report from the SENTRY antimicrobial surveillance programme. *J Antimicrob Chemother* 2002;**50**(5):673–9.
96. Gunn JS, Lim KB, Krueger J, et al. PmrA-PmrB-regulated genes necessary for 4-aminoarabinose lipid A modification and polymyxin resistance. *Mol Microbiol* 1998;**27**(6):1171–82.
97. Denton M, Kerr K, Mooney L, et al. Transmission of colistin-resistant *Pseudomonas aeruginosa* between patients attending a pediatric cystic fibrosis center. *Pediatr Pulmonol* 2002;**34**(4):257–61.
98. Moore RA, Chan L, Hancock RE. Evidence for two distinct mechanisms of resistance to polymyxin B in *Pseudomonas aeruginosa*. *Antimicrob Agents Chemother* 1984;**26**(4):539–45.
99. Visca P, Seifert H, Towner KJ. *Acinetobacter* infection—an emerging threat to human health. *IUBMB Life* 2011;**63**(12):1048–54.

100. Vila J, Martí S, Sánchez-Céspedes J. Porins, efflux pumps and multidrug resistance in *Acinetobacter baumannii*. *J Antimicrob Chemother* 2007;**59**(6):1210–5.
101. Roca I, Espinal P, Vila-Farrés X, Vila J. The *Acinetobacter baumannii* oxymoron: commensal hospital dweller turned pan-drug-resistant menace. *Front Microbiol* 2012;**3**:148.
102. Hujer KM, Hamza NS, Hujer AM, et al. Identification of a new allelic variant of the *Acinetobacter baumannii* cephalosporinase, ADC-7 beta-lactamase: defining a unique family of class C enzymes. *Antimicrob Agents Chemother* 2005;**49**(7):2941–8.
103. Turton JF, Woodford N, Glover J, Yarde S, Kaufmann ME, Pitt TL. Identification of *Acinetobacter baumannii* by detection of the blaOXA-51-like carbapenemase gene intrinsic to this species. *J Clin Microbiol* 2006;**44**(8):2974–6.
104. Huys G, Cnockaert M, Vaneechoutte M, et al. Distribution of tetracycline resistance genes in genotypically related and unrelated multiresistant *Acinetobacter baumannii* strains from different European hospitals. *Res Microbiol* 2005;**156**(3):348–55.
105. Nemec A, Dolzani L, Brisse S, van den Broek P, Dijkshoorn L. Diversity of aminoglycoside-resistance genes and their association with class 1 integrons among strains of pan-European *Acinetobacter baumannii* clones. *J Med Microbiol* 2004;**53**(Pt 12):1233–40.
106. Rooney PJ, O'Leary MC, Loughrey AC, et al. Nursing homes as a reservoir of extended-spectrum beta-lactamase (ESBL)-producing ciprofloxacin-resistant *Escherichia coli*. *J Antimicrob Chemother* 2009;**64**(3):635–41.
107. Lowy FD, Aiello AE, Bhat M, et al. *Staphylococcus aureus* colonization and infection in New York State prisons. *J Infect Dis* 2007;**196**(6):911–8.
108. Davey P, Brown E, Charani E, et al. Interventions to improve antibiotic prescribing practices for hospital inpatients. *Cochrane Database Syst Rev* 2013;**4**: CD003543.
109. Hurley MN, Cámara M, Smyth AR. Novel approaches to the treatment of Pseudomonas aeruginosa infections in cystic fibrosis. *Eur Respir J* 2012;**40**(4):1014–23.

CHAPTER SIXTEEN

Preventing Infection Associated with Urethral Catheter Biofilms

Rachael P.C. Jordan and Lindsay E. Nicolle

INTRODUCTION

Urinary tract infection (UTI) is one of the most important healthcare-acquired infections (HCAI), and it is estimated that 70 to 80% of these infections can be attributed to use of an indwelling urethral catheter.[1,2] A small number of catheter-associated urinary tract infections (CA-UTIs) occur following introduction of organisms into the bladder at the time of catheter insertion. However, most healthcare-acquired UTIs result from biofilm formation along an indwelling urethral catheter.[3,4] Biofilm development is consistent for all indwelling urinary devices—that is, urethral and suprapubic catheters, ureteric stents and nephrostomy tubes. This chapter addresses only the indwelling urethral catheter, but much of the discussion is also relevant to these other devices.

Indwelling urethral catheters remain in place for variable durations, from a few minutes to years. A catheter *in situ* for less than 30 days is considered a short-term indwelling catheter and if present for over 30 days is a long-term (chronic) indwelling catheter. The term 'catheter acquired urinary tract infection' refers to a symptomatic UTI occurring in an individual with a current or recently removed indwelling urethral catheter.[1] When significant bacteriuria is present in the urine without symptoms, this is referred to as catheter-associated asymptomatic bacteriuria (CAASB). Asymptomatic bacteriuria is far more common than symptomatic urinary infection. Many early studies addressing catheter-acquired infection did not, however, differentiate between symptomatic UTI and bacteriuria, and the term CA-UTI was used to encompass any positive culture, irrespective of symptoms.[1]

EPIDEMIOLOGY

An estimated 12 to 16% of acute care hospital in-patients have a urinary catheter placed at some time during their hospital stay.[4] The daily risk of acquisition of

Biofilms in Infection Prevention and Control
Doi: http://dx.doi.org/10.1016/B978-0-12-397043-5.00016-5

bacteriuria varies from 3 to 7% while the catheter remains *in situ*.[1] The duration of catheterisation is the most important determinant of CAASB. Being female and of an increased age are also consistent associations of acquisition of bacteriuria.

Clinical

The determinants of symptomatic infection are not well described.[5] While CAASB is common in patients with indwelling urethral catheters, CA-UTI is relatively uncommon. However, given the large number of patients for whom catheters are used, the impact of CA-UTI on the healthcare system is substantial.[1,6]

The frequency of CA-UTI in recent reports is 4.32/1000 device days for acute care hospitals.[2] In 2007, CA-UTI occurred in 2 to 3/1000 urine catheter days in US adult intensive care units (ICUs) participating in the CDC National Healthcare Safety Network.[7] At a Veterans hospital, 0.3% of catheter days involved a symptomatic UTI.[8] From 1990 to 2007, the incidence of CA-UTI in US critical care units declined, with a proportional decrease varying from 18.6% in cardiothoracic units to 67% in medical/surgical units. The CA-UTI infection rate in 398 intensive care units in Shanghai over a five-year period was 6.4/1000 catheter days, and varied from 0 in burn ICUs to 12.8/1000 in coronary care units.[8]

CA-UTI may be a source for bacteremia, but this complication occurs in less than 1% of patients with short-term catheters.[5] Catheter obstruction or bladder mucosal trauma with bleeding increases the risk of bacteremia. During a three-year period, 21% of all bloodstream infections, acquired 48 hours or longer after hospitalisation in 61 Quebec acute care hospitals, were from a urinary source: 1.4 episodes/10,000 patient days.[9] Of these, 71% were associated with urinary catheters. The all-cause mortality at 30 days was 15% for a urinary source of bacteremia, but attributable mortality was not reported. CA-UTI also has been associated with increased mortality and prolonged length of hospital stay,[1] although recent studies of critical care unit patients suggest the reported associations are largely attributable to confounding by unmeasured variables.[10]

Between 3 to 5% of residents of long-term care facilities are managed with chronic indwelling catheters.[1,4] The incidence of CA-UTI varied from 0 to 7.3/1000 catheter days (mean 3.2/1000) in Idaho long-term care facilities, compared to 0 to 2.28/1000 resident days for all urinary infections (mean 0.57/1000). Episodes of fever from a presumed urinary source are three times more common in residents with a chronic urethral catheter, with an estimated incidence of 0.7 to 1.1/100 catheter days.[4,11] An autopsy study reported that histological evidence of acute pyelonephritis was eight times more frequent for residents with chronic indwelling catheters compared to bacteriuric residents without catheters. CA-UTI is the source of more than 50% of episodes of bacteremia in long-term care facilities.[1,4,12]

The risk of bacteremia is 3 to 36 times higher for residents with an indwelling catheter compared with those with a positive urine culture but without a catheter.[13]

Residents with chronic indwelling catheters are also reported to have increased mortality, but this is likely due to confounding by patient characteristics.[14] When chronic indwelling urethral catheters were routinely used for voiding management of patients with spinal cord injuries, urosepsis, renal failure and other complications of infection, they were the major cause of death. Mortality from UTI and its complications are now uncommon following the introduction and widespread use of intermittent catheterisation for voiding management of spinal cord injured patients. Symptomatic UTI is reported as 2.72/100 person days for patients with spinal cord injuries managed with a chronic indwelling catheter, compared with 0.41/100 days for those with intermittent catheterisation.[4]

Local infections attributed to use of a chronic catheter include urethritis, Bartholin's gland and urethral gland abscesses, and prostatitis in men. A chronic indwelling catheter is a risk factor for urolithiasis. The indwelling urethral catheter also is associated with non-infectious adverse outcomes. Genito-urinary trauma is the most important of these and is reported to occur in 1.5% of catheter days.[15] Bacteriuria is frequently treated with antimicrobials, inappropriately, as symptomatic infection. This contributes to the development of antimicrobial resistance and *Clostridium difficile* colitis.[16] The catheter drainage bag is often colonised with a high concentration of organisms and may be a reservoir for resistant organisms that are transmitted to other catheterised patients in the facility.[1]

Microbiology

The organisms most frequently associated with CA-UTI in acute care facilities are *Escherichia coli*, *Candida* species, *Enterococcus* species, *Pseudomonas aeruginosa* and *Klebsiella* species. In long-term care facility residents with chronic catheters, the most common bacteria are *E. coli*, *Klebsiella pneumoniae*, *Proteus mirabilis*, *P. aeruginosa*, *Providencia* species and *Enterococcus* species (Table 16.1). For bacteremic CA-UTI in these residents *E. coli*, *P. mirabilis* and *Enterococcus* species are most frequently isolated.[12]

The initial episode of bacteriuria or CA-UTI in a patient with a short-term catheter is often with a single organism. *Escherichia coli* is most common, but a broad range of other organisms, including *Staphylococcus epidermidis* and *Enterococcus faecalis* may be involved depending on host colonisation, prior healthcare and antimicrobial exposures (see Table 16.1).

As duration of catheter use increases, and for all chronic indwelling catheters, multiple organisms, usually 3 to 5 species at any time, can be isolated.[19] These may include *P. aeruginosa*, *P. mirabilis*, *Providencia stuartii*, *Morganella morganii*, *K. pneumoniae* and other Gram-negative bacteria, as well as less pathogenic organisms such as coagulase-negative staphylococci or enterococci.[21,22] Micro-organisms are difficult to eradicate with antibiotics when the catheter is in place.[22,23] When a patient receives antimicrobial therapy, organisms of increased resistance, including yeast, are more likely to be present.[24]

Table 16.1 Range of Micro-organisms Isolated from Patients with Indwelling Urethral Catheters

Catheter	Short term	Long term (chronic)
Source references	17,18	19, 20
Escherichia coli	24–52%	10–37%
Klebsiella species	8.1–24%	3–21%
Enterobacter species	4.5%	8.4%
Serratia species	3%	–
Citrobacter species	3–7.2% (Citrobacter species and Proteus mirabilis combined)	10–60%
Proteus mirabilis		9–36%
Morganella morganii	9.9–24%	5–61%
Providencia stuartii	–	50%
Pseudomonas aeruginosa	9.9–24%	5–30%
Other Gram-negative bacteria	1–2%	7–16%
Enterococcus species	21%	1–28%
Other *Streptococcus* species	8.1%	8.5%
Staphylococcus aureus	–	4.7%
Other Gram-positive bacteria	26–23%	1–20%
Candida species	32–17%	NS

Note: Shown in % of total isolates.

The polymicrobial microbiology of long-term catheters is dynamic, with continuous acquisition and loss of organisms in the biofilm. Different species appear to persist for variable time periods and new organisms can be isolated at a rate of 3 to 7% per day.[12] Urease-producing organisms, particularly *P. mirabilis*, persist for a longer time compared with *E. coli*, *K. pneumoniae* and *P. stuartii*,[19] and the adherence and persistence of both *K. pneumoniae* and *P. stuartii* is associated with the presence of the MR/K adhesins.[25] *Proteus mirabilis* swarms rapidly over all catheter types.[26] *Enterococcus* species tend to persist only for a few weeks and coagulase-negative staphylococci may persist for only a few days.

BIOFILM ON THE URINARY CATHETER

Biofilm growth is a basic survival strategy used by bacteria in a wide range of environmental, industrial and clinical settings.[27] Biofilms are defined as a microbial-derived sessile community, characterised by cells attached to a substratum or interface, or to each other; they are enclosed in an extracellular polysaccharide, which the micro-organisms themselves produce.[28] Non-cellular material may also be found in the biofilm, depending on the environment in which it was formed.[29]

Biofilms exhibit an altered phenotype with respect to growth rate and gene transcription compared to planktonic cells,[30] and cells in biofilm communities are protected from environmental stresses by their extracellular polysaccharide matrix. This protection has particular advantages for the bacteria in biofilms that develop *in vivo.*

Micro-organisms sensitive to antibiotics and antiseptics using conventional laboratory testing methods may exhibit resistance or tolerance when present as an *in vivo* biofilm.[31]

Biofilm formation consists of several steps: the first involves the deposition of the micro-organisms on a surface, followed by their attachment by microbial adhesion and anchorage to the surface by exopolymer production.[32] Each bacterial cell has an array of adhesins within their cell walls, which allow colonisation of many types of substrate. The subsequent production of exopolysaccharides secures the attachment of the biofilm. Following production of exopolysaccharides, bacteria multiply to form micro-colonies, which subsequently spread over the colonised surface forming populations embedded in a gel-like polysaccharide matrix.[31]

Biofilm Formation

Biofilms formed on indwelling urinary catheters and other urinary devices differ from biofilms on non-urinary devices (e.g., indwelling vascular lines or endotracheal tubes) because of the incorporation of urine components.[13,32] These include Tamm-Horsfall protein, as well as magnesium and calcium ions. Urease-producing organisms, particularly *P. mirabilis*, may produce higher amounts of exopolysaccharide.

In patients who develop bacteriuria during short-term catheterisation, bacterial colonisation of the catheter occurs but the biofilms are generally sparse and, because the catheter is removed after a few days, these biofilms cause few problems.[33] Mature biofilms develop on all chronic catheters; they contain mixed bacterial communities of up to five species, the most common of which are *E. faecalis*, *P. aeruginosa*, *E. coli* and *P. mirabilis*.[31]

Biofilms form on both the interior and exterior surfaces of the indwelling catheter[1,3,4] following ascension of organisms colonising the urethral meatus along the catheter-mucosa interface or with intraluminal spread to the bladder when the collecting tube or drainage bag is contaminated. The risk of bacteriuria is related to the length of time a catheter is in place. Because catheterisation allows external bacteria easy access to the bladder, the longer a patient is catheterised the greater the likelihood that bacteriuria will occur.[18,34] The higher rate of acquisition of bacteriuria in women is attributed to shorter urethral length relative to men. Patients with pre-existing bacteriuria presumably can have biofilm initiated anywhere along the catheter. Biofilm development begins immediately following catheter insertion and continues as long as the device remains in place. Extensive biofilm formation may be visible to the naked eye.

Immediately following insertion of the catheter, a conditioning layer of host proteins is laid down on the catheter surface.[3,34,35] Bacterial or yeast adhesins recognise host cell receptors on the mucosa or catheter and initiate attachment. Following attachment, organisms transform phenotypically with subsequent production of exopolysaccharides. The bacteria or yeast grow within this mucoid glycocalyx substance, forming micro-colonies and mature biofilm. Sessile colonies in the biofilm are in an environment where there is limited access to host defences such as leukocytes, or antimicrobials. Biofilm-originating cells typically migrate to the bladder within a few days of catheter insertion.[3] Some organisms within the biofilm may break away in small clumps or as single organisms and establish bacteriuria or colonisation and thus biofilm further along the device.

Crystalline Biofilms

Crystalline biofilms within urinary catheters are formed by bacterial species, which have the ability to produce the enzyme urease.[36,37] Urease hydrolyses urea leading to the formation of ammonium and carbonate ions and this leads to an increase in urinary pH.[38,39] As the pH of the urine becomes more alkaline, magnesium and calcium phosphate crystals are precipitated and aggregates of this crystalline material accumulate in the urine and biofilm on the catheter surface, leading to catheter blockage.[26,35,40,41] This is similar to the material found in urolithiasis, and forms crusts along the catheter. It is the major cause of obstruction of indwelling catheters.

Crystalline biofilms can form on the outside of the catheter, around the retention balloon and catheter tip and can cause trauma to the bladder and urethral epithelium. After deflation of the retention balloon, crystalline debris from the biofilm can shed from the catheter and enter the bladder, possibly initiating stone formation. The main complication, however, is blockage of the flow of urine through the catheter as a result of the accumulation of crystalline material in the catheter lumen. This can lead to incontinence, retention of urine in the bladder, reflux of infected urine, pyelonephritis and septicaemia if the blockage is not detected.[42,43] Approximately half of all patients who undergo long-term catheterisation will suffer catheter blockage at some time and some patients are prone to very frequent and early blockage.[44–46] *Proteus mirabilis* is isolated from more than 80% of catheters obstructed by encrustations.[35]

The crystalline deposits on catheters have a similar composition to infection-inducing bladder and kidney stones. Struvite and a poorly crystalline form of apatite are the principle crystalline components.[47,48] Microscopy has shown that large numbers of bacilli are associated with the crystals and culture techniques have confirmed the persistence of a range of bacteria, notably species capable of producing urease. Species that may be able to produce urease and have been isolated from catheter biofilms include *P. aeruginosa*, *K. pneumoniae*, *M. morganii*, *Proteus* species, some *Providencia* species and some strains of *Staphylococcus aureus* and coagulase-negative staphylococci.

Proteus mirabilis is the most common species isolated from urine of patients suffering from recurrent catheter encrustation and blockage.[49,50] This species is also the most common species recovered from patients' encrusted catheters.[39] Urease-producing bacteria, which do not produce alkaline urine (e.g., *M. morganii*, *K. pneumoniae* and *P. aeruginosa*), also do not produce appreciable encrustation on catheters.[51] Based on epidemiological and experimental data, *P. mirabilis* is primarily responsible for the formation of crystalline biofilms.[52]

Physical forces can also initiate the development of a crystalline biofilm. Latex-based catheters have uneven surfaces and irregularities are common around the eye-holes of silicone catheters.[53] Within two hours, bacterial cells become trapped in these uneven surfaces and micro-colonies develop.[53] Once the urinary pH rises sufficiently, crystals begin to form and spread down the catheter lumen, initiating blockage.[54] The chemical environment also has a role to play as colonisation is increased in alkaline urine.[55]

Proteus mirabilis

Proteus mirabilis is not usually found during early colonisation of the catheterised urinary tract, so is uncommon in patients undergoing short-term catheterisation.[56] The longer a catheter is in place, however, the more likely it is that *P. mirabilis* will be isolated from the urine. In patients with long-term catheterisation, *P. mirabilis* has been isolated from 40% of urine samples.[57] Genotyping of *P. mirabilis* isolated from catheterised urinary tracts suggests that strains are remarkably stable, with the same genotype persisting in the patient's urinary tract despite many catheter changes, courses of antibiotics and even catheter-free periods.[58]

Genotyping also determined that the *P. mirabilis* strains isolated from bladder stones and encrusted catheters of the same patient were identical, and that the majority of patients were infected with genetically distinct strains.[59] *Proteus mirabilis* is an enteric organism and subsequent analysis showed that bacteria isolated from faecal material and catheter biofilms of the same patient were identical, indicating that catheter encrustation of long-term catheterised patients occurs from *P. mirabilis* originating from their own faecal flora.[60]

All types of Foley catheters are vulnerable to *P. mirabilis* biofilms. Several different adhesins have been identified on *P. mirabilis* which can facilitate adherence to catheter surfaces.[35,61] Although the ability of *P. mirabilis* to bind to the catheter is an important factor in the development of crystalline biofilms, the most important factor seems to be synthesis of a potent urease. Urease from *P. mirabilis* can hydrolyse urea several times faster than urease produced by other species.[62] Experimental work has shown that urease-negative *P. mirabilis* mutants fail to form crystalline biofilms; whereas urease-positive strains, which lack flagella or the ability to swarm, are still able to encrust and block catheters, similar to wild-type parent strains.[55,63]

Proteus mirabilis also exhibits a type of motility referred to as 'swarming', in which multi-cellular rafts of elongated, hyperflagellated swarmer cells form with the ability to move rapidly over solid surfaces. It has been suggested that swarming is important in the pathogenesis of CA-UTI and is essential for migration of *P. mirabilis* over all-silicone catheters. Swarming-deficient mutants were attenuated for migration over hydrogel-coated latex catheters, but those capable of swimming motility were able to move over and colonise these surfaces. The flagellar filaments of *P. mirabilis* are highly organised during raft migration and are interwoven to form helical connections between adjacent swarmer cells. Mutants lacking organised flagellar structures fail to swarm successfully, suggesting these structures are important for migration and formation of multi-cellular rafts.

In addition, the highly organised structure of multi-cellular rafts enables *P. mirabilis* to initiate CA-UTI by migration over catheter surfaces from the urethral meatus into the bladder.[64] The transformation of *P. mirabilis* from small swimming bacilli into elongated, highly flagellated swarmer cells is accompanied by a substantial increase in the production of urease. This may accelerate the generation of alkaline urine and increase the deposition of crystalline material into the bladder and onto urinary catheters.[65,66] Additional studies show that *P. mirabilis* has the ability to migrate successfully across sections of hydrogel-coated latex, hydrogel/silver-coated latex, silicone-coated latex and all-silicone catheters.[67]

DIAGNOSIS

A positive urine culture is necessary to confirm a diagnosis of urinary infection in every patient.

Urine Culture

The microbiological diagnosis of bacteriuria or urinary infection requires a urine specimen with bacteria or yeast isolated with a quantitative count $\geq 10^5$ colony forming units (cfu)/ml. It has been suggested that lower threshold counts of $\geq 10^3$ cfu/ml are sufficient for indwelling catheter specimens for diagnosis of CA-UTI.[1] However, the validity of lower counts has not been critically evaluated. When a lower count is obtained, critical review of the clinical circumstances is necessary before diagnosing CA-UTI. Lower counts could be relevant in a few selected patients, such as individuals undergoing diuresis or receiving antimicrobials.

For most patients, however, $\geq 10^5$ cfu/ml should be isolated as a small pool of urine remains undrained in the bladder due to the catheter balloon, allowing bacteria in this pool to achieve high quantitative counts.[1] Following insertion of an indwelling catheter in subjects without pre-existing bacteriuria, organisms are initially isolated in low numbers of $\geq 10^2$ cfu/ml.[17] The quantitative count increases to $\geq 10^5$ cfu/ml within 48 to 72 hours of an initial positive culture if antimicrobial treatment is not initiated and the catheter remains in place. The initial low count likely reflects early biofilm formation along the catheter rather than bladder bacteriuria.

For chronic indwelling catheters, the biofilm incorporates a greater number of organisms than are present in bladder urine.[68] Organisms in the biofilm contaminate a urine specimen obtained through the biofilm of the catheter lumen, which may not be present in the bladder. *Escherichia coli* and *K. pneumoniae*, if isolated, are more likely to be present in the bladder, while urease-producing and Gram-positive bacteria are more likely to be associated only with the biofilm.[68] When a clinical specimen is required, a catheter that has been *in situ* for two weeks or longer should be removed and replaced by a fresh one, and a urine specimen obtained for culture through the new catheter to sample only bladder urine and not biofilm.[1,68] Organisms isolated at $<10^5$ cfu/ml following catheter replacement tend not to persist.[69]

Other Laboratory Tests

CA-UTI may be accompanied by bacteremia.[9] A positive blood culture is helpful in determining which organisms are most likely clinically relevant if multiple organisms are isolated from the urine specimen. This is useful for optimal antimicrobial selection, particularly when resistant organisms are present.

Pyuria is common in patients with an indwelling urinary catheter irrespective of the presence or absence of bacteriuria.[5] Pyuria is universal in patients with a chronic indwelling catheter and does not differentiate CAASB from CA-UTI.[4] Thus, pyuria is not a useful diagnostic test for CA-UTI in bacteriuric patients.

Clinical Considerations

CA-UTI appears to be substantially overdiagnosed and overtreated. In a prospective cohort study of patients following insertion of a short-term indwelling catheter, the prevalence of symptoms potentially attributable to UTI and the mean daily temperature were similar for patients with and without bacteriuria.[9] The most common clinical presentation of CA-UTI is fever without localising findings.[4,11,70]

Acute hematuria, catheter obstruction or costovertebral angle pain or tenderness, when present, supports a urinary source. However, there is no definitive way to diagnose CA-UTI in a catheterised patient with bacteriuria and non-localising fever unless a blood culture is positive with an organism also isolated from the urine culture. Thus, the clinical diagnosis of CA-UTI is usually a diagnosis of exclusion; the urine specimen is positive and no alternate source is apparent. For residents with an indwelling catheter in long-term care facilities, consensus guidelines suggest initiation of antimicrobial therapy for CA-UTI is appropriate if there is new costovertebral tenderness or, if no other source is apparent, rigours or new onset delirium.[71]

CA-UTI PREVENTION GUIDELINES AND RECOMMENDATIONS

Evidence-based guidelines have provided recommendations for prevention of CA-UTI in healthcare facilities. Guidelines for preventing infections associated with the insertion and maintenance of short-term indwelling urinary catheters in acute care were published by the Department of Health in Great Britain in 2001 and updated in 2006.[72] The US Health and Human Services, Health Care Infection Control and Prevention Advisory Committee (HICPAC), published guidelines in 2009 to update the 1981 CDC guidelines.[73] In 2009, the Infectious Diseases Society of America, in collaboration with other international organisations, published international clinical practice guidelines addressing the diagnosis, prevention and treatment of CA-UTI in adults.[1]

Recommendations in all guidelines are generally similar, with some variation whenever new evidence has become available. The HICPAC document is the most comprehensive and addresses strategies for avoidance of catheter use, catheter choice, policies for catheter insertion and maintenance, surveillance of CA-UTI and catheter use, program implementation and recommended quality indicators. A 2008 compendium published by The Society for Healthcare Epidemiology of America (SHEA) and other US healthcare organisations provides a concise summary of recommendations to assist acute healthcare facilities in implementation of programs to limit CA-UTI.[74]

Avoidance of Catheter Use

The most effective intervention to prevent CA-UTI is to avoid use of an indwelling catheter and prompt removal of necessary catheters once the indication for use is no longer present.[1,73] Approaches to limit catheter use include restricted criteria for use and alternate approaches to voiding management, such as a condom[75] or intermittent[76] catheter, whenever possible. The limited indications for an indwelling urinary catheter include the presence of urinary obstruction; a requirement for output monitoring in critically ill patients; for selected surgical procedures; and, rarely, end-of-life care for patient comfort (Exhibit 16.1).[73] The indication of assisting with healing of sacral pressure ulcers in incontinent patients should be individualised depending on the patient's characteristics.

EXHIBIT 16.1 Indications for Indwelling Urethral Catheter Use

- Acute urinary retention or obstruction
- Accurate measurement of urinary output in critically ill patients
- Selected peri-operative use
  - Surgery on contiguous structures of the genito-urinary tract
  - Anticipated prolonged duration of surgery
  - Large volume infusions or diuretics during surgery
  - Intra-operative monitoring of urine output
- To assist in healing open sacral or perineal wounds in incontinent patients
- When prolonged immobilisation is necessary (e.g., unstable thoracic or lumbar spine, pelvic fractures)
- To improve comfort for end-of-life care, if needed

Source: Adapted from Gould et al.[73]

Indwelling urinary catheters should not be used as a substitute for nursing care of the patient or resident with incontinence, as a means of obtaining urine for culture or other diagnostic tests when the patient can voluntarily void, for prolonged post-operative duration without appropriate indications, or routinely for patients receiving epidural anaesthesia/analgesia.[73] Previously, insertion of an indwelling urethral catheter was routine for many surgical procedures, with the catheter often remaining in place for a prolonged period post-surgery. However, there are only a few surgical procedures for which an indwelling catheter is essential and maintaining the catheter following surgery is seldom appropriate.[1,73] For instance, a routine indwelling urinary catheter is not required for Caesarean section[77] or when epidural analgesia for pain control is prolonged in thoracic surgery patients.[78] Early post-surgical removal for most procedures has not resulted in excess recatheterisations or other morbidity. Following removal of the urethral catheter post-surgery, monitoring of bladder volume by ultrasound can identify the few patients who may require further assistance with bladder emptying.

Studies consistently report that healthcare providers are often unaware that a patient has a urinary catheter in place.[1,73,79] This results in unnecessarily prolonged catheterisation for many patients. Strategies to limit the duration of catheterisation include daily review of all catheterised patients to confirm a continued need for the catheter and automatic nursing stop orders when criteria for discontinuation are met. A meta-analysis of the effectiveness of reminder systems prompting catheter reassessment and removal, if no longer required, reported a 52% reduction in CA-UTI with use of a reminder or stop order and a 37% decrease in the mean duration of catheter days.[79]

Selection of a Urinary Catheter

Urinary catheters are made of latex or silicone; silicone catheters should be used for individuals with latex allergies. There is, however, no evidence to date of a decreased frequency of bacteriuria or CA-UTI with silicone compared to latex catheters.[1] The decreased frequency of catheter obstruction observed with a silicone catheter is attributed to the larger bore size of silicone compared with latex catheters. Thus, silicone may be preferable to reduce the frequency of obstruction from catheter encrustation in long-term catheterised patients where frequent obstruction has been a problem.[73]

Hydrogel-coated (hydrophilic) catheters have a surface which decreases the likelihood of mucosal trauma from their use.[1] These catheters have not been shown to decrease bacteriuria or CA-UTI compared to catheters without hydrogel. Antibacterial coatings also have been developed to prevent CA-UTI. Previously developed silver oxide-coated catheters did not prevent bacteriuria or CA-UTI.[1] Subsequently, silver alloy-coated catheters were reported to decrease bacteriuria, but not CA-UTI, during the first week of use.[1,80]

A meta-analysis reported that while a short-term benefit of silver alloy silicone catheters compared to latex catheters was reported in early trials, later studies, which compared silicone catheters with silver alloy silicone catheters, showed little or no benefit.

It was suggested that silicone, rather than the silver alloy, might have contributed to reported differences in the acquisition of bacteriuria in the earlier studies.[81]

A large, prospective randomised US trial of hydrogel-coated silver alloy silicone catheters compared with hydrogel-coated silicone catheters showed no differences between the two catheters for bacteriuria or CA-UTI.[82] Another recent large, randomised British study reported no clinical benefit with silver alloy latex compared with plain latex catheters for CA-UTI.[83] Thus, evidence does not support the use of silver alloy catheters to decrease bacteriuria or CA-UTI.

Nitrofurazone, an antimicrobial related to nitrofurantoin, has also been used as an antimicrobial coating for urinary catheters. *In vitro* studies report a delay in the onset of *E. coli* adherence and biofilm formation of up to five days with these catheters, but less impact with other organisms.[4] Clinical trials with trauma patients in critical care[84] and Korean medical and surgical patients[85] reported that the nitrofurazone catheter delayed acquisition of bacteriuria or funguria for catheters in place longer than five days; however, these trials did not report the frequency of CA-UTI. The previously mentioned British study included a nitrofurazone silicone catheter arm and reported a small but significant decrease in CA-UTI with this catheter compared to the plain latex or silver alloy latex catheters.[83] It is not clear, however, whether the reported benefit was attributable to the silicone or the nitrofurazone, and the nitrofurazone catheter was associated with an increased frequency of adverse effects. The benefit observed with the nitrofurazone silicone catheter was not attractive from a cost-effectiveness perspective for uniform use of the catheter. None of the current antimicrobial catheters have a role for use in patients using chronic indwelling catheters.[1]

Other technical modifications of urinary catheter systems, referred to as 'complex urinary catheter systems' have been developed with the aim of decreasing infection.[73] These include systems with filters placed in the line to prevent ascension of biofilm and antiseptic-releasing cartridges at the urine collection bag drain port. Evidence from clinical trials does not suggest a benefit of any of these systems over simple closed-drainage systems.[73] Clinical trials of systems with the catheter and tubing junction sealed so that they cannot be disconnected have reported contradictory results with decreased bacteriuria in one study and no benefit in another.[1] Catheter valves have also not been shown to decrease bacteriuria or CA-UTI.[69] Thus, none of these catheter system modifications are currently recommended for routine use to prevent CA-UTI.

Catheter Insertion and Maintenance

Evidence-based practices recommended for catheter insertion and maintenance are summarised in Exhibits 16.2 and 16.3. Catheter insertion should use aseptic technique and sterile equipment. Drapes, gloves, catheter and meatal cleaning and lubrication solutions for catheter insertion should all be sterile. Following insertion, the catheter must be anchored to the skin to prevent urethral trauma.

EXHIBIT 16.2 Proper Techniques for Urinary Catheter Insertion

- Only properly trained persons who know the correct technique of aseptic catheter insertion should perform this procedure.
- Perform hand hygiene immediately before and after insertion or any manipulation of the catheter device or site.
- In the acute care hospital setting, insert urinary catheters using aseptic techniques and sterile equipment :
    - Sterile gloves, drape, sponges and appropriate antiseptic or sterile solution for periurethral cleaning and a single-use packet of sterile lubricant jelly for insertion
- Unless otherwise clinically indicated, use the smallest bore catheter possible, consistent with good drainage, to minimise urethral trauma.
- Properly secure indwelling catheter after insertion to prevent movement and urethral traction.

Source: Adapted from Hooton et al.[1] *and Gould et al.*[73]

EXHIBIT 16.3 Proper Techniques for Catheter Maintenance

- Maintain a sterile continuously closed drainage system:
    - If breaks in aseptic technique, disconnection or leakage occur replace the catheter and collecting system using aseptic technique and sterile equipment
    - Consider using urinary catheter systems with pre-connected sealed catheter tubing junctions
- Maintain unobstructed urine flow:
    - Keep the catheter and collecting tube free from kinking
    - Keep the collecting bag below the level of the bladder at all times
    - Do not rest the bag on the floor
    - Empty the collecting bag regularly using a separate container for each patient and avoid contact with drainage spigotte with a non-sterile collecting container
- Use standard precautions, including the use of gloves and gowns, as appropriate, during any manipulation of the catheter or collecting system.
- Unless obstruction is anticipated, bladder irrigation is not recommended.

Source: Adapted from Hooton et al.[1] *and Gould et al.*[73]

The benefits of a closed-drainage system for decreasing the incidence of bacteriuria and CA-UTI have been repeatedly documented.[1,73] Urine specimens should be collected by aspiration through the sampling port or catheter tubing so that the junction is not disconnected. Unobstructed urine flow should be ensured and the drainage bag must be below the level of the bladder at all times. Urine collecting containers should be dedicated to individual patients to limit cross-transmission of organisms among catheterised patients.

Some practices evaluated in clinical trials are not effective and not recommended.[1,73] Daily meatal cleaning with antiseptic solutions or sterile saline in patients with long-term catheters does not prevent bacteriuria and may, in fact, increase its risk. Catheter irrigation with saline or antiseptics should be avoided unless necessary to prevent obstruction for post-surgical patients with bleeding. Routine catheter irrigation was associated with an increased bacteriuria, presumed to be due to interruption of the closed system.

Installation of antiseptic or antimicrobial solutions into urinary drainage bags is not effective in decreasing bacteriuria or CA-UTI. Systemic antimicrobial therapy delays the onset of bacteriuria for short-term catheters and decreases the prevalence of bacteriuria at the time of catheter removal, but when infections occur the organisms isolated are of increased resistance. Thus, antimicrobial therapy should not be given to prevent CA-UTI in patients requiring catheterisation.

CA-UTI Prevention in Long-Term Care Facilities

In long-term care facilities, programs should focus on limiting the use of chronic indwelling catheters.[1,73] For patients managed with chronic indwelling catheters, practices must limit catheter trauma and support early identification of catheter obstruction or other malfunction. Other recommendations such as maintaining a closed drainage system as much as possible and avoiding kinking or other obstruction of urine flow are also appropriate. Residents with chronic catheters may use a leg bag for drainage. The facility should have policies in place addressing reuse and cleaning or replacement of these leg bags. Catheters should not be changed routinely.[1,73] Chronic indwelling catheters should be changed only when they are obstructed or malfunctioning, or prior to initiating antimicrobial therapy for urinary tract infection to allow collection of a specimen of bladder urine without biofilm contamination. Antimicrobial therapy does not decrease the frequency of bacteriuria or CA-UTI for residents using chronic catheters but will increase numbers of resistant organisms.

CA-UTI Prevention in Acute Care Facilities

All acute care facilities require programs to limit CA-UTI. Worldwide experience reports improved outcomes following implementation of suggested programs (see[86–90]). Further investigation is required to characterise the optimal institutional program for preventing CA-UTI, including specific components.

To be effective, any program should be individualised to be relevant to local experience, population characteristics and resources. Essential components of an effective program are leadership at the senior management level; collaboration among professional groups; and continuing surveillance of catheter use, catheter indications, and CA-UTI together with timely feedback of this information to relevant professionals.[91]

Implementation

There must be adequate infrastructure to support a prevention program.[73] This should include policies for catheter indications, catheter selection, insertion and management. Adequate staffing and education, and appropriate supplies must be provided. There should be appropriate guidelines addressing perioperative catheter management to minimise catheter use wherever possible. These should include recommendations for early removal and monitoring with ultrasound bladder scanners, where available, to identify retention of urine. There should be documentation of urinary catheter use, including dates of insertion and removal. Electronic patient records, when available, can facilitate data collection and notification. This may also be useful for automatic reminders.[73,74] There should be a means to ensure that patients with indwelling catheters are identified and reviewed on a continual basis.

A qualitative study identified four main themes relevant to use of preventive practices in US hospitals.[91] These were, recognition of the importance of early catheter removal; a focus on non-infectious complications, together with a clinical 'champion' on the ward area who promotes the prevention program; the availability of hospital-specific pilot studies for devices; and external forces such as public reporting. The development of such programs has been advanced by the institution of 'bundles' for prevention of device-associated HCAI.

This approach implements a number of simultaneous interventions, and requires explicit documentation of adherence through the use of checklists.[92] A US statewide initiative in Michigan describes a program to introduce a bladder bundle using a collaborative model with strategies to facilitate implementation described under concepts of 'engage and educate', 'execute' and 'evaluate' (Table 16.2).[92]

Monitoring

Effective programs for prevention of HCAI require continuous monitoring and timely reporting of outcomes. Surveillance of catheter use, including indications and duration, as well as incidence of CA-UTI and catheter-associated bacteremia, are recommended outcomes.[73] Obtaining screening urine cultures is not recommended, except prior to an invasive procedure. Monitoring laboratory results for bacteriuria, and pharmacy for antimicrobial prescriptions, facilitates surveillance by focusing on patients likely to have symptomatic UTI. Electronic patient records which document catheter insertion and removal are helpful to facilitate surveillance, if available, and as reliable as bedside

Table 16.2 Components of Catheter-Associated UTI 'Bundle' Implemented in Michigan Critical Care Units

Concise Summary of Guideline Recommendations (Bundle Components)	Adherence to general infection control principles (e.g., hand hygiene, surveillance and feedback, aseptic insertion, proper maintenance, education) is important. Bladder ultrasound may avoid indwelling catheterisation. Condom catheters or other alternatives to an indwelling catheter (e.g., intermittent catheterisation) should be considered in appropriate patients. Do not use the indwelling catheter unless you must. Early removal of the catheter using a reminder or nurse-initiated removal protocol appears warranted.

Bladder Bundle Practices and Outcome Measures

Key practices	Measures (all pre- and post-interventions)
Nurse-initiated urine catheter discontinuation protocol	Prevalence rate of urinary catheter utilisation (urinary catheter days/total number of patient days during a period of time)
Urinary catheter reminders and removal prompts	Indication for each insertion
Alternatives to indwelling urinary catheterisation	Prevalence rate of unnecessary urinary catheter use (number unnecessary urinary catheter days/total number of patient days)
Portable bladder ultrasound monitoring	Rate of discontinuation of unnecessary urinary catheters (number of unnecessary catheters discontinued/number of urinary catheters evaluated for which no indication was found)
Insertion care and maintenance	

Source: Adapted from Saint et al. (2009).[92]

review.[93] There should be a means to ensure that indications for catheterisation are included in the patient record.

Facilitywide surveillance is not usually appropriate. Programs should allow intermittent review and identification of groups at high risk. This would normally include some continuing measurement of the use of catheters, including whether appropriate criteria for catheter use and duration are met. Each facility should use surveillance data to identify local rates and to compare these to external rates. Should the rates be high in any area in the facility, intensification of programs to decrease CA-UTI should be undertaken.

Surveillance requires the use of standardised definitions. The National Healthcare Safety Network (NHSN) definitions for CA-UTI are provided in Table 16.3.

Table 16.3 NHSN Criteria for UTI for Patients with Indwelling Catheters

Criterion	
1a	Patient has an indwelling catheter in place and at least one of fever (>38 °C), suprapubic tenderness, or costovertebral angle pain or tenderness and no other recognised source and urine culture $\geq 10^5$ cfu/ml with ≤ 2 species.
2a	Patient has an indwelling urinary catheter in place with symptoms as in 1a and a positive urinalysis based on: a: positive dipstick for leukocyte esterase and/or nitrite b: pyuria (urine specimen with ≥ 10 white blood cells/cubic millimetre or ≥ 3 white blood cells per high-power field of unspun urine) c: micro-organism seen on gram stain of unspun urine and a positive urine culture of $\geq 10^3$ and $<10^5$ cfu/ml with no more than two species of micro-organisms
3	Patient ≤ 1 years of age (with or without an indwelling catheter) at least one of fever >38 °C, hypothermia (<38 °C core), apnea, bradycardia, dysuria, lethargy or vomiting and a urine culture $\geq 10^5$ cfu/ml with no more than two species of micro-organisms.
4	Clinical criteria as in 3, and a positive urinalysis demonstrated by one of the criterion in 2a.

Source: Adapted from Gould et al. (2010).[73]

Bacteriuria is not an outcome, and criteria for identifying symptomatic urinary tract infection are restrictive. Despite this, the definitions lack specificity as they allow over-diagnosis of symptomatic UTI, where CA-UTI is diagnosed in the absence of localising signs or symptoms (i.e., fever and positive cultures).

Each facility should identify specific performance measures to monitor processes and outcomes relevant to CA-UTI (Exhibit 16.4). CA-UTI is reported with a denominator of /1000 catheter days.[1,73] However, the most effective means of preventing infection is to avoid use of a catheter or limit duration of use. If there is optimal use of indwelling urethral catheters, patients who require a catheter are likely to be more ill and at greater risk of infection than patients for whom catheters are avoided. Using catheter days rather than patient days for the denominator of CA-UTI or bacteremia may then underestimate or obscure an effective program as this metric does not incorporate reductions in catheter use.[94,95] Thus, a denominator of patient days may be considered in addition to catheter days. The most useful surveillance outcome for long-term care facilities is bacteremia-attributed CA-UTI in patients with chronic catheters. Continued surveillance of antimicrobial use may also identify problems with inappropriate use of antibiotics for bacteriuria in these patients.

EXHIBIT 16.4 HICPAC Proposed Quality Indicators (Performance Measures) for Programs to Limit Catheter-Acquired UTIs

1. **Process Measures**
   a. Compliance with educational program: calculate percent of personnel who have proper training
      - Numerator: number of personnel who insert urinary catheters and who have proper training
      - Denominator: number of personnel who insert urinary catheters
   b. Compliance with documentation of catheter insertion and removal dates: conduct random audits of selected units and calculate percent compliance rate
      - Numerator: number of patients on unit with catheters with proper documentation of insertion and removal dates
      - Denominator: number of patients with catheters in place at some point during admission
   c. Compliance with documentation of indication for catheter placement: conduct random audits of selected units and calculate percent compliance rate
      - Numerator: number of patients on unit with catheters with proper documentation of indication
      - Denominator: number of patients on the unit with catheter in place
2. **Outcome Measures**
   a. Rates of CA-UTI/1000 catheter days: use NHSN definitions
      - Numerator: number of CA-UTIs in each location monitored
      - Denominator: total number of urinary catheter-days for all patients who have an indwelling urinary catheter in each location monitored—that is, with bloodstream infections
   b. Rate of bloodstream infection secondary to CA-UTI/1000 catheter days: Use NHSN definitions
      - Numerator: number of episodes of bloodstream infections secondary to CA-UTI
      - Denominator: total number of urinary catheter-days for all patients who have an indwelling urinary catheter in each location monitored

Source: Adapted from Gould et al.[73]

FUTURE DIRECTIONS

Development of biofilm-resistant catheters would largely obviate the problem of CA-UTI. Despite substantial interest and investigations into different catheter materials, including antimicrobial catheters, nothing developed to date has provided a substantial breakthrough. Continuing development and evaluation of potential biofilm-resistant biomaterials for urinary catheters is important. In addition, the unique role of

P. mirabilis in chronic catheters and catheter obstruction suggests interventions targeted specifically to this organism should be pursued.

The guidelines in this chapter identify a number of specific practice questions which require further critical evaluation.[1,73] Continued evaluation of alternatives to catheters is needed. This includes optimal use of condom catheters in male long-term care facility residents, and characterisation of the optimal role for supra-pubic catheters and urethral stents.

Other prevention methods that require study include the role of irrigation with antiseptic solutions or oral urease inhibitors to prevent catheter encrustation. Studies evaluating the utility and reliability of portable ultrasound for bladder volume assessment are also necessary for selected clinical settings. Finally, use of spatial separation of patients with urinary catheters to prevent transmission of organisms and CA-UTI needs to be explored. Other prevention strategies, such as bacterial interference to induce bacteriuria with avirulent organisms, are in preliminary evaluation and the potential role is not yet clear.

REFERENCES

1. Hooton TM, Bradley SF, Cardenas DD, Colgan R, Geerlings SE, Rice JC, et al. Diagnosis, prevention and treatment of catheter-associated urinary tract infection in adults: 2009 International Clinical Practice Guidelines from the Infectious Diseases Society of America. *Clin Infect Dis* 2010;**50**:625–63.
2. Weber DJ, Sickbert-Bennett EE, Gould CV, Brown VM, Huslage K, Rutala WA. Incidence of catheter-associated and non-catheter-associated urinary tract infections in a healthcare system. *Infect Control Hosp Epidemiol* 2011;**32**:822–3.
3. Saint S, Chenoweth CE. Biofilms and catheter-associated urinary tract infections. *Infect Dis Clin North Am* 2003;**17**:411–32.
4. Nicolle LE. Urinary catheter-associated infections. *Infect Dis Clin North Am* 2012;**26**:13–28.
5. Tambyah PA, Maki DG. Catheter-associated urinary tract infection is rarely symptomatic: a prospective study of 1,497 catheterized patients. *Arch Intern Med* 2000;**160**:678–87.
6. Tambyah PA, Knasinski V, Maki DG. The direct costs of nosocomial catheter-associated urinary tract infection in the era of managed care. *Infect Control Hosp Epidemiol* 2002;**23**:27–31.
7. Burton DC, Edwards JR, Srinivasan A, Fridkin SK, Gould CV. Trends in catheter-associated urinary tract infections in adult intensive care units—United States, 1990-2007. *Infect Control Hosp Epidemiol* 2011;**32**:748–56.
8. Tao L, Hu B, Rosenthal VD, Gao X, He L. Device-associated infection rates in 398 intensive care units in Shanghai, China: International Nosocomial Infection Control Consortium (INICC) findings. *Int. J. Infect. Dis.* 2011;**15**:e774–80.
9. Fortin E, Rocher I, Frenette C, Tremblay C, Quach C. Healthcare-associated bloodstream infections secondary to a urinary focus: the Québec provincial surveillance results. *Infect Control Hosp Epidemiol* 2012;**33**:456–62.
10. Chant C, Smith DM, Marshall JC, Friedrich JO. Relationship of catheter-associated urinary tract infection to mortality and length of stay in critically ill patients: a systematic review and meta-analysis of observational studies. *Crit Care Med* 2011;**39**:1167–73.
11. Warren JW, Damron D, Tenney JH, Hoopes JM, Deforge B, Muncie Jr. HL. Fever, bacteremia, and death as complications of bacteriuria in women with long-term urethral catheters. *J Infect Dis* 1987;**155**:1151–8.
12. Mylotte JM. Nursing home-acquired bloodstream infection. *Infect Control Hosp Epidemiol* 2005;**26**:837–8.

13. Nicolle LE. Urinary tract infections in the elderly. *Clin Geriatr Med* 2009;**25**:423–36.
14. Kunin CM, Chin QF, Chambers S. Morbidity and mortality associated with indwelling urinary catheters in elderly patients in a nursing home—confounding due to the presence of associated diseases. *J Am Geriatr Soc* 1987;**35**:1001–6.
15. Leuck A-M, Wright D, Ellingson L, Kraemer L, Kuskowski MA, Johnson JR. Complications of Foley catheters—is infection the greatest risk? *J Urol* 2012;**187**:1662–6.
16. Cope M, Cevallos ME, Cadle RM, Darouiche RO, Musher DM, Trautner BW. Inappropriate treatment of catheter-associated asymptomatic bacteriuria in a tertiary care hospital. *Clin Infect Dis* 2009;**48**:1182–8.
17. Stark RP, Maki DG. Bacteriuria in the catheterized patient. What quantitative level of bacteriuria is relevant? *N Engl J Med* 1984;**311**:560–4.
18. Garibaldi RA, Burke JP, Dickman ML, Smith CB. Factors predisposing to bacteriuria during indwelling urethral catheterization. *N Engl J Med* 1974;**291**:215–9.
19. Warren JW, Tenney JH, Hoopes JM, Muncie HL, Anthony WC. A prospective microbiologic study of bacteriuria in patients with chronic indwelling urethral catheters. *J Infect Dis* 1982;**146**:719–23.
20. Nicolle LE. The chronic indwelling catheter and urinary infection in long-term care facility residents. *Infect Control Hosp Epidemiol* 2001;**22**:316–21.
21. Warren JW. The catheter and urinary tract infection. *Med Clin North Am.* 1991;**75**:481–93.
22. Clayton CL, Chawla JC, Stickler DJ. Some observations on urinary tract infections in patients undergoing long-term bladder catheterization. *J Hosp Infect* 1982;**3**:39–47.
23. Warren JW, Anthony WC, Hoopes JM, Muncie Jr HL. Cephalexin for susceptible bacteriuria in afebrile, long-term catheterized patients. *JAMA* 1982;**248**:454–8.
24. Nicolle LE, Bradley S, Colgan R, Rice JC, Schaeffer A, Hooton TM. Infectious Diseases Society of America guidelines for the diagnosis and treatment of asymptomatic bacteriuria in adults. *Clin Infect Dis* 2005;**40**:643–54.
25. Mobley H, Chippendale GR, Tenney JH, Mayrer AR, Crisp LJ, Penner JL, et al. MR/K hemagglutination of *Providencia stuartii* correlates with adherence to catheters and with persistence in catheter associated bacteriuria. *J Infect Dis* 1988;**157**:264–71.
26. Stickler DJ, Feneley RCL. The encrustation and blockage of long-term indwelling bladder catheters: a way forward in prevention and control. *Spinal Cord* 2010;**48**:784–90.
27. Costerton JW, Cheng KJ, Geesey GG, Ladd TI, Nickel JC, Dasgupta M, et al. Bacterial biofilms in nature and disease. *Annu Rev Microbiol* 1987;**41**:435–64.
28. Hojo K, Nagaoka S, Ohshima T, Maeda N. Bacterial interactions in dental biofilm development. *J Dent Re* 2009;**88**:982–90.
29. Donlan RM. Biofilms: microbial life on surfaces. *Emerg Infect Dis* 2002;**8**:881–90.
30. Donlan RM, Costerton JW. Biofilms: survival mechanisms of clinically relevant microorganisms. *Clin Microbiol Rev* 2002;**15**:167–93.
31. Stickler DJ. Bacterial biofilms in patients with indwelling urinary catheters. *Nat Clin Pract Urol* 2008;**5**:598–608.
32. Tenke P, Riedl CR, Jones GL, Williams GJ, Stickler D, Nagy E. Bacterial biofilm formation on urologic devices and heparin coating as preventive strategy. *Int J Antimicrob Agents* 2004;**23**(Suppl. 1):S67–74.
33. Ohkawa M, Sugata T, Sawaki M, Nakashima T, Fuse H, Hisazumi H. Bacterial and crystal adherence to the surfaces of indwelling urethral catheters. *J Urol* 1990;**143**:717–21.
34. Kunin CM. Care of the urinary catheter. In: Kunin CM, editor. *Urinary tract infections: Detection, prevention and management*, 5th ed. Williams & Wilkins; 1997. p. 226–78.
35. Jacobsen SM, Stickler DJ, Mobley HL, Shirtliff ME. Complicated catheter-associated urinary tract infections due to *Escherichia coli* and *Proteus mirabilis*. *Clin Microbiol Rev* 2008;**21**:26–59.
36. Getliffe KA, Mulhall AB. The encrustation of indwelling catheters. *Br J Urol* 1991;**67**:337–41.
37. Stickler DJ, Zimakoff J. Complications of urinary tract infections associated with devices used for long-term bladder management. *J Hosp Infect* 1994;**28**:177–94.
38. Cox AJ, Hukins DW, Sutton TM. Infection of catheterised patients: bacterial colonisation of encrusted Foley catheters shown by scanning electron microscopy. *Urol Res* 1989;**17**:349–52.
39. Stickler D, Ganderton L, King J, Nettleton J, Winters C. *Proteus mirabilis* biofilms and the encrustation of urethral catheters. *Urol Res* 1993;**21**:407–11.

40. Morris NS, Stickler DJ, McLean RJ. The development of bacterial biofilms on indwelling urethral catheters. *World J Urol* 1999;**17**:345–50.
41. McLean RJC, Stickler DJ, Nickel JC. Biofilm mediated calculus formation in the urinary tract. *Cells Mater* 1996;**6**:165–74.
42. Liedl B. Catheter-associated urinary tract infections. *Curr Opin Urol* 2001;**11**:75–9.
43. Kunin CM. Care of the urinary catheter. In: Kunin CM, editor. *Detection, prevention and management of urinary tract infections*, 4th Ed. Lea & Febiger; 1987. p. 245–98.
44. Cools HJ, Van der Meer JW. Restriction of long-term indwelling urethral catheterisation in the elderly. *Br J Urol* 1986;**58**:683–8.
45. Getliffe KA. The characteristics and management of patients with recurrent blockage of long-term urinary catheters. *J Adv Nurs* 1994;**20**:140–9.
46. Kohler-Ockmore J, Feneley RC. Long-term catheterization of the bladder: prevalence and morbidity. *Br J Urol* 1996;**77**:347–51.
47. Hedelin H, Eddeland A, Larsson L, Pettersson S, Ohman S. The composition of catheter encrustations, including the effects of allopurinol treatment. *Br J Urol* 1984;**56**:250–4.
48. Cox AJ, Hukins DW. Morphology of mineral deposits on encrusted urinary catheters investigated by scanning electron microscopy. *J Urol* 1989;**142**:1347–50.
49. Mobley HL, Warren JW. Urease-positive bacteriuria and obstruction of long-term urinary catheters. *J Clin Microbiol* 1987;**25**:2216–7.
50. Kunin CM. Blockage of urinary catheters: role of microorganisms and constituents of the urine on formation of encrustations. *J Clin Epidemiol* 1989;**42**:835–42.
51. Stickler D, Morris N, Moreno MC, Sabbaba N. Studies on the formation of crystalline bacterial biofilms on urethral catheters. *Eur J Clin Microbiol Infect Dis* 1998;**17**:649–52.
52. Macleod SM, Stickler DJ. Species interactions in mixed-community crystalline biofilms on urinary catheters. *J Med Microbiol* 2007;**56**:1549–57.
53. Stickler D, Young R, Jones G, Sabbuba N, Morris N. Why are Foley catheters so vulnerable to encrustation and blockage by crystalline bacterial biofilm? *Urol Res* 2003;**31**:306–11.
54. Stickler DJ, Morgan SD. Observations on the development of the crystalline bacterial biofilms that encrust and block Foley catheters. *J Hosp Infect* 2008;**69**:350–60.
55. Stickler DJ, Lear JC, Morris NS, Macleod SM, Downer A, Cadd DH, et al. Observations on the adherence of *Proteus mirabilis* onto polymer surfaces. *J Appl Microbiol* 2006;**100**:1028–33.
56. Matsukawa M, Kunishima Y, Takahashi S, Takeyama K, Tsukamoto T. Bacterial colonization on intraluminal surface of urethral catheter. *Urology* 2005;**65**:440–4.
57. Mobley HT. Virulence of Proteus mirabilis. In: Mobley HL, Warren JW, editors. *Urinary tract infections: Molecular pathogenesis and clinical management*. ASM Press; 1996. p. 245–70.
58. Sabbuba NA, Mahenthiralingam E, Stickler DJ. Molecular epidemiology of *Proteus mirabilis* infections of the catheterized urinary tract. *J Clin Microbiol* 2003;**41**:4961–5.
59. Sabbuba NA, Stickler DJ, Mahenthiralingam E, Painter DJ, Parkin J, Feneley RC. Genotyping demonstrates that the strains of *Proteus mirabilis* from bladder stones and catheter encrustations of patients undergoing long-term bladder catheterization are identical. *J Urol* 2004;**171**:1925–8.
60. Mathur S, Sabbuba NA, Suller MT, Stickler DJ, Feneley RC. Genotyping of urinary and fecal *Proteus mirabilis* isolates from individuals with long-term urinary catheters. *Eur J Clin Microbiol Infect Dis* 2005;**24**:643–4.
61. Rocha SP, Pelayo JS, Elias WP. Fimbriae of uropathogenic *Proteus mirabilis*. *FEMS Immunol Med Microbiol* 2007;**51**:1–7.
62. Jones BD, Mobley HL. Genetic and biochemical diversity of ureases of *Proteus, Providencia,* and *Morganella* species isolated from urinary tract infection. *Infect Immun* 1987;**55**:2198–203.
63. Jones BV, Mahenthiralingam E, Sabbaba NA, Stickler DJ. Role of swarming in the formation of crystalline *Proteus mirabilis* biofilms on urinary catheters. *J Med Microbiol* 2005;**54**:807–13.
64. Jones BV, Young R, Mahenthiralingam E, Stickler DJ. Ultrastructure of *Proteus mirabilis* swarmer cell rafts and role of swarming in catheter-associated urinary tract infection. *Infect Immun* 2004;**72**:3941–50.
65. Allison C, Lai HC, Hughes C. Co-ordinate expression of virulence genes during swarm-cell differentiation and population migration of *Proteus mirabilis*. *Mol Microbiol* 1992;**6**:1583–91.

66. Falkinham III JO, Hoffman PS. Unique developmental characteristics of the swarm and short cells of *Proteus vulgaris* and *Proteus mirabilis*. *J Bacteriol* 1984;**158**:1037–40.
67. Sabbuba N, Hughes G, Stickler DJ. The migration of *Proteus mirabilis* and other urinary tract pathogens over Foley catheters. *BJU Int* 2002;**89**:55–60.
68. Raz R, Schiller D, Nicolle LE. Chronic indwelling catheter replacement before antimicrobial therapy for symptomatic urinary tract infection. *J Urol* 2000;**164**:1254–8.
69. Tenney JH, Warren JW. Bacteriuria in women with long term catheters: paired comparison of indwelling and replacement catheters. *J Infect Dis* 1988;**157**:199–202.
70. Orr PH, Nicolle LE, Duckworth H, Brunka J, Kennedy J, Murray D, et al. Febrile urinary infection in the institutionalized elderly. *Am J Med* 1996;**100**:71–7.
71. Loeb M, Bentley DW, Bradley S, Crossley K, Gantz N, Garibaldi R, et al. Development of minimum criteria for the initiation of antibiotics in residents of long-term-care facilities: results of a consensus conference. *Infect Control Hosp Epidemiol* 2001;**22**:120–4.
72. Pratt RJ, Pellowe CM, Wilson JA, Loveday HP, Harper PJ, Jones SRLJ, et al. epic2: National evidence-based guidelines for preventing healthcare-associated infections in NHS hospitals in England. *J Hosp Infect* 2007;**65**(suppl. 1):S1–64.
73. Gould CV, Umscheid CA, Agarwal RK, Kuntz G, Pegues DA. Healthcare Infection Control Practices Advisory Committee. Guideline for prevention of catheter-associated urinary tract infections, 2009. *Infect Control Hosp Epidemiol* 2010;**31**:326–91.
74. Lo E, Nicolle L, Classen D, Arias KM, Podgorny K, Anderson DJ, et al. Strategies to prevent catheter-associated urinary tract infections in acute care hospitals. *Infect Control Hosp Epidemiol* 2008;**29** (suppl. 1):S41–50.
75. Saint S, Kaufman SR, Rogers MAM, Baker PD, Ossenkop K, Lipsky BA. Condom versus indwelling urinary catheters: a randomized trial. *J Am Geriatr Soc* 2006;**54**:1055–61.
76. National Institute on Disability and Rehabilitation Consensus Statement The prevention and management of urinary tract infections among people with spinal cord injuries. *J Am Paraplegia Soc* 1992;**15**:194–204.
77. Li L, Wen J, Wang L, Li YP, Li Y. Is routine indwelling catheterisation of the bladder for caesarean section necessary? A systematic review. *BJOG* 2011;**118**:400–9.
78. Zaouter C, Wuethrich P, Miccoli M, Carli F. Early removal of urinary catheter leads to greater post-void residuals in patients with thoracic epidural. *Acta Anaesthesiol Scand* 2012;**56**:1020–5.
79. Meddings J, Rogers AM, May M, Saint S. Systematic review and meta-analysis: reminder systems to reduce catheter-associated urinary tract infections and urinary catheter use in hospitalized patients. *Clin Infect Dis* 2010;**51**:550–60.
80. Schumm K, Lam TB. Types of urethral catheters for management of short-term voiding problems in hospitalized adults. *Cochrane Database of Systematic Reviews* 2008; (Issue 2) Art. No. CD004013.
81. Crnich CJ, Drinka PJ. Does the composition of urinary catheter influence clinical outcomes and the results of research studies. *Infect Control Hosp Epidemiol* 2007;**28**:102–3.
82. Srinivasan A, Karchmer T, Richards A, Song X, Perl T. A prospective trial of a novel, silicone-based, silver-coated Foley catheter for the prevention of nosocomial urinary tract infections. *Infect Control Hosp Epidemiol* 2006;**27**:38–43.
83. Pickard R, Lam T, MacLennan G, Start K, Kilonzo M, McPherson G, et al. (2013). Reducing symptomatic urinary tract infection in hospitalised adults requiring short-term catheterisation: a multicentre, randomised controlled trial. *Lancet* 2012;**380**:1927–35.
84. Stensballe J, Tvede M, Looms D, Lippert FK, Dahl B, Tønnesen E, et al. Infection risk with nitrofurazone-impregnated urinary catheters in trauma patients: a randomized trial. *Ann Intern Med* 2007;**147**:285–93.
85. Lee SJ, Kim SW, Cho YH, Shin WS, Lee SE, Kim CS, et al. A comparative multicentre study on the incidence of catheter-associated urinary tract infection between nitrofurazone-coated and silicone catheters. *Int J Antimicrob Agents* 2004;**24**(suppl. 1):S65–9.
86. Shimoni Z, Rodrig J, Kamma N, Froom P. Will more restrictive indications decrease rates of urinary catheterisation? An historical comparative study. *BMJ Open* 2012;**22**:e000473.

87. Fakih MG, Watson SR, Greene MT, Kennedy EH, Olmsted RN, Krein SL, et al. Reducing inappropriate urinary catheter use: a statewide effort. *Arch Intern Med* 2012;**172**:255–60.
88. Dyc NG, Pena ME, Shemes SP, Rey JE, Szpunar SM, Fakih MG. The effect of resident peer-to-peer education on compliance with urinary catheter placement indications in the emergency department. *Postgrad Med J* 2011;**87**:814–8.
89. Marigliano A, Barbadoro P, Pennacchietti L, D'Errico MM, Prospero E. Active training and surveillance: 2 good friends to reduce urinary catheterization rate. *Am J Infect Control* 2012;**40**:692–5.
90. Titsworth WL, Hester J, Correia T, Reed R, Williams M, Guin P, et al. Reduction of catheter-associated urinary tract infections among patients in a neurological intensive care unit: a single institution's success. *J Neurosurg* 2012;**116**:911–20.
91. Saint S, Kowalski CP, Forman J, Damschroder L, Hofer TP, Kaufman SR, et al. A multicenter qualitative study on preventing hospital-acquired urinary tract infection in US hospitals. *Infect Control Hosp Epidemiol* 2008;**29**:333–41.
92. Saint S, Olmsted RN, Fakih MG, Kowalski CP, Watson SR, Sales AE, et al. Translating health care-associated urinary tract infection prevention research into practice via the bladder bundle. *Jt Comm J Qual Patient Saf* 2009;**35**:449–55.
93. Burns AC, Petersen NJ, Garza A, Arya M, Patterson JE, Naik AD, et al. Accuracy of a urinary catheter surveillance protocol. *Am J Infect Control* 2012;**40**:55–8.
94. Wright MO, Kharasch M, Beaumont JL, Peterson LR, Robicsek A. Reporting catheter-associated urinary tract infections: denominator matters. *Infect Control Hosp Epidemiol* 2011;**32**:635–40.
95. Fakih MG, Greene MT, Kennedy EH, Meddings JA, Krein SL, Olmsted RN, et al. Introducing a population-based outcome measure to evaluate the effect of interventions to reduce catheter-associated urinary tract infection. *Am J Infect Control* 2012;**40**:359–64.

Presence and Control of *Legionella pneumophila* and *Pseudomonas aeruginosa* Biofilms in Hospital Water Systems

Ginny Moore and Jimmy Walker

INTRODUCTION

Government legislation states that water supplied for drinking, washing and food preparation must be 'wholesome' at the time and point of supply, meaning it must not contain any micro-organism, parasite or substance at a concentration 'which would constitute a potential danger to human health'.[1] Public water supplies are closely controlled and monitored and water companies must demonstrate that the treatment processes they use effectively achieve, maintain and/or restore the wholesomeness of their water.

Most health-related water-quality issues are the result of microbial contaminants. Contamination of water by faecal material containing pathogenic organisms (e.g., *Escherichia coli* O157, *Shigella* spp. and *Cryptosporidium*) has caused numerous disease outbreaks.[2–6] The disinfection of water has played a major role in reducing the incidence of waterborne disease[7]; in the United Kingdom all public drinking water is disinfected before being supplied, usually by the addition of a chemical oxidant such as chlorine or chlorine dioxide. Demonstrating the absence of coliforms, more specifically *E. coli*, in water indicates that the system is neither contaminated with faecal organisms nor vulnerable to faecal contamination.[1,8]

The susceptibility of the coliform bacteria to chlorine has been demonstrated[9–11]; however, maintaining a free residual chlorine concentration within a distribution network does not completely inhibit the growth or regrowth of micro-organisms.[12,13] In addition, the absence of traditional microbial indicators of water quality (coliforms) may not reflect the disinfection response of other more opportunistic pathogens.[11,14]

Water and water systems harbour a diversity of micro-organisms.[15,16] Microbial resistance to chlorine and other disinfectants differs depending on the type of organism and its physiological state. Resistance is enhanced in the presence of organic and inorganic molecules which protect micro-organisms from the action of biocides by

Biofilms in Infection Prevention and Control
Doi: http://dx.doi.org/10.1016/B978-0-12-397043-5.00017-7

physically shielding them from disinfection[17] and/or by reacting with the antimicrobial agent, reducing its bioavailability.[18] Such organic and inorganic molecules also provide a nutrient source for bacterial growth.[19–21]

In comparison to any bulk fluid, nutrient molecules are likely to be at a higher concentration on or in close proximity to a surface.[22] Organic and inorganic molecules rapidly adsorb to water distribution surfaces. Pipe linings may also leach nutrients.[23,24] The ability of bacterial cells to achieve a close association with such surfaces enables them to readily scavenge available nutrients. Once cells are attached and under favourable conditions, they can multiply, form biofilm micro-colonies and produce extracellular polysaccharides. Surface-associated bacteria are more resistant to disinfectants than corresponding planktonic cells, and biofilms growing at the water–pipe material interface facilitate the survival and growth of pathogens.[25,26]

In a water distribution network, the continuous flow of water through large diameter pipes limits biofilm formation. However, as the water passes into pipes of smaller diameter, longer residence times, resulting from variable flow rates and usage levels, allow microbial contaminants to accumulate.[27] Biofilms can harbour large numbers of micro-organisms.[28] Nonetheless, it is the bacteria that detach from the biofilm and leave the water system that pose the greatest risk in terms of waterborne infection.

To colonise new surfaces, individual cells must be able to disperse from a mature biofilm and reattach elsewhere.[29] Some organisms (e.g., *Pseudomonas aeruginosa*) produce enzymes[30] that cleave constituents of the extracellular polymeric matrix releasing cells from the surface.[22] Detachment is also caused by physical forces such as shearing—the continual removal of small portions of the biofilm via fluid dynamic forces—and abrasion caused by the collision of particles from the bulk fluid.[22] Detachment of biofilm aggregates can lead to high levels of local contamination in the water phase.[31] If these aggregates contain pathogens, the water leaving and/or the aerosols generated from local outlets represent a risk factor for onward transmission.

Many of the bacteria that make up the basic water distribution biofilm occur naturally in aquatic and soil environments. Although mostly harmless to healthy individuals, they can cause serious opportunistic infections, particularly in infants, the elderly and individuals with a weakened immune system. Consequently, hospitals and other healthcare facilities are high-risk environments. The susceptibility of the patients, coupled with the complexity of hospital water systems, mean biofilms can play an important role in waterborne nosocomial infection.

In the United Kingdom, nosocomial (healthcare-associated) infections cost the National Health Service £1 billion per annum, affect 8% of all hospitalised patients—23% in intensive care units (ICUs)—and are responsible for 5000 deaths annually.[32] The contribution that water supplies make to disease burden remains a subject of debate.[33,34] Nonetheless, a number of nosocomial outbreaks have been described in which a waterborne transmission route has been implicated (Table 17.1).

Table 17.1 Outbreaks Associated with Different Types of Waterborne Micro-organisms

Source/reservoir	Location	Organism	Likely mode of transmission	Details of outbreak/infection	Ref
Taps					
Hot water supply	Military treatment facility (Texas, USA)	*Mycobacterium simiae*	Ingestion of, or showering with, contaminated water	14 identified cases over a 15-month period; 1 death	146
Hot water supply	Operating theatre (Paris, France)	*Mycobacterium xenopi*	Use of contaminated water to rinse surgical instruments	58 infected patients (spinal infection) over a 10- year period	147
Tap aerator	Paediatric oncology unit (Freiburg, Germany)	*Acinetobacter junii*	Contamination of staff hands during hand washing	3 infected patients (bacteraemia) over a 3-month period	148
Tap faucet	Intensive care unit (Barcelona, Spain)	*Serratia marcescens*	Consumption of contaminated water when taking oral medication	19 cases over a 12-month period; 10 infections and 9 colonisations	149
Tap faucet	Haematology ward (Haifa, Israel)	*Stenotrophomonas maltophilia*		2 deaths (mucocutaneous and soft tissue infection)	150
Tap aerator	Neonatal intensive care unit (Nijmegen, Netherlands)	*Stenotrophomonas maltophilia*	Use of contaminated water to wash pre-term infants	5 identified cases over a 2-month period; 1 death	151
Tap water	Intensive care unit (California, USA)	*Legionella dumoffii*	Direct wound contact—use of contaminated water to bathe patients	3 identified cases; 2 deaths (sternal wound infection)	152
Tap faucet	Haematology ward (Lyon, France)	*Legionella pneumophila* SG 5	Aspiration of (cold) water when washing	1 death (multi-organ failure)	153
Tap water	Intensive care unit (New York, USA)	*Legionella pneumophila* SG 6	Aspiration of nasogastric tube solutions diluted with tap water	2 identified cases in same week	154
Tap faucet	Intensive care unit (Ulm, Germany)	*Pseudomonas aeruginosa*	Washing, tooth brushing, or cleaning of dental prostheses	5 infected patients during a 7-month surveillance period	155
Sink taps and traps	Intensive care unit (Clichy, France)	*Pseudomonas aeruginosa*	Use of contaminated water to rinse re-usable enteral feed containers	36 identified cases over a 16-month period; 27 infections, 9 colonisations, 1 death	156
Tap rosette	Neonatal unit (Belfast, Northern Ireland)	*Pseudomonas aeruginosa*	Use of contaminated water to wash babies during nappy change	15 identified cases over a 2-month period; 5 infections, 10 colonisations, 3 deaths	116

(*Continued*)

Table 17.1 Outbreaks Associated with Different Types of Waterborne Micro-organisms (Continued)

Source/reservoir	Location	Organism	Likely mode of transmission	Details of outbreak/infection	Ref
Showers					
Shower hose	Bone marrow transplant unit (Minnesota, USA)	*Mycobacterium mucogenicum*	Contamination of central venous catheters during bathing.	6 infected patients (bacteraemia) over a 3-month period	157
Shower head	Bone marrow transplant unit (Helsinki, Finland)	*Pseudomonas aeruginosa*	Contact with contaminated water during bathing	5 infected patients (bacteraemia) over a 3-month period	158
Hot water supply	General hospital (Värnamo, Sweden)	*Legionella pneumophila* SG 1	Inhalation of aerosols generated by shower nozzle	31 identified cases (28 patients and 3 staff) over a 2-month period; 3 deaths	159
Hot water supply	Community hospital (New York, USA)	*Legionella pneumophila* SG 1	Inhalation of aerosols generated by shower nozzle	7 identified cases (6 patients and 1 employee) over a 7-month period	160
Water supply	Transplant unit (Oxford, UK)	*Legionella pneumophila*	Inhalation of aerosols generated within cubicle of post-op shower bath	2 identified cases	161
Baths					
Re-circulating hot water	University hospital (Nagoya, Japan)	*Legionella pneumophila* SG 10	Inhalation of aerosols generated by all-day-running-hot-water bath	1 death (acute respiratory failure)	162
Water supply	Delivery room (Turin, Italy)	*Legionella pneumophila* SG 1	Aspiration of contaminated water during delivery (birthing pool)	1 case of pneumonia in a 7-day-old neonate	163
Swimming pool	Physiotherapy unit (Halifax, Canada)	*Pseudomonas aeruginosa*	Contact with contaminated water during hydrotherapy sessions	15 cases of *Pseudomonas* folliculitis(10 patients, 5 staff) over a 3- week period	164
Whirlpool bath	Haematology unit (Iowa, USA)	*Pseudomonas aeruginosa*	Immersion in contaminated water; water contaminated by pool drain	7 infected patients (bacteraemia) over a 14-month period; 4 deaths	165
Bath toys	Paediatric oncology ward (Melbourne, Australia)	*Pseudomonas aeruginosa*	Contact with contaminated water retained within bath toys	8 infected patients	166

Water bath used to thaw fresh frozen plasma	Neonatal intensive care unit (Brussels, Belgium)	*Pseudomonas aeruginosa*	Cross-contamination between water bath and staff hands	4 colonised or infected new-borns over a period of 1 week; 3 deaths	167
Water bath used to calibrate thermometers	Intensive care units (Winnipeg, USA)	*Pseudomonas (Burkholderia) cepacia*	Cross-contamination between ventilator thermometer and tubing; inhalation of aerosol	23 colonisations or infections over a 10-month period	168
Respiratory equipment					
Re-usable oxygen humidifier	Post-anaesthesia recovery (Valencia, Spain)	*Legionella pneumophila* SG 1	Inadequate cleaning/disinfection; inhalation of contaminated aerosols	3 infected patients over a 6- week period	169
Oxygenhumidifier	Cardiology; nephrology ward (Turin, Italy)	*Legionella pneumophila* SG 1	Use of contaminated water; inhalation of contaminated aerosols	5 cases of fatal pneumonia during a 12-month period	170
Nebulizer	General hospital (Quebec City, Canada)	*Legionella dumoffii*	Malfunction of water distillation system; inhalation of contaminated aerosols	5 infected patients over an 11-month period; 3 deaths	171
Jet nebulizer and room humidifier	University hospital (Illinois, USA)	*Legionella pneumophila*	Use of contaminated tap water to fill reservoir; inhalation of contaminated aerosols	5 identified cases over a 5-month period	172
Medication nebulizer	Community hospital (USA)	*Legionella pneumophila* SG 3	Use of contaminated tap water to wash nebulizer; inhalation of contaminated aerosols	13 identified cases over a 4- year period; 4 deaths	173
Room humidifier	Neonatal unit (Nicosia, Cyprus)	*Legionella pneumophila* SG 3	Use of contaminated tap water to fill reservoir; inhalation of contaminated [cold mist] aerosol	9 of 32 newborns delivered over a 17- day period acquired infection; 3 deaths	174
Ice					
Ice machine	Veteran medical ward (New York, USA)	*Mycobacterium fortuitum*	Ingestion of contaminated ice	30 colonisations over a 4-month period	175
Ice machine	Intensive care unit (New York, USA)	*Legionella pneumophila* SG 6	Microaspiration of contaminated ice or ice water	1 case of pneumonia	176

(*Continued*)

Table 17.1 Outbreaks Associated with Different Types of Waterborne Micro-organisms (Continued)

Source/reservoir	Location	Organism	Likely mode of transmission	Details of outbreak/infection	Ref
Ice machine	Transplant unit (Copenhagen, Denmark)	*Legionella pneumophila* SG 1	Microaspiration of contaminated ice received for moisturising the oral mucosa	ice machine implicated in two of three infections occurring over a 2- year period	177
Ice	Intensive care unit (Hawaii, USA)	*Ewingella americana*	Contamination of the ice bath used to cool syringes for cardiac output determinations	4 cases of bacteraemia over a 6-week period	178
Ice	Intensive care unit (USA)	*Flavobacterium* spp.	Contamination of catheters by syringes cooled in contaminated ice	14 cases of bacteraemia over a 5-month period	179
Water features					
Decorative fountain	Stem cell transplantation unit (Maryland, USA)	*Legionella pneumophila* SG 1	Stagnation of water during maintenance; inhalation of contaminated aerosol	2 neutropenic patients developedpneumonia 2 weeks after admission to unit	180
Decorative fountain	Hospital lobby (Wisconsin, USA)	*Legionella pneumophila* SG 1	Inadequate disinfection of decorative foam material; inhalation of contaminated aerosol	8 cases over a 4- week period. Patients were out-patients at time of acquisition	181
Cleaning equipment					
Cleaning solutions, mops and cloths	Haematology-oncology unit (Bonn, Germany)	*Pseudomonas aeruginosa*	Contamination of high-contact surfaces during cleaning with contaminated cleaning materials	6 infections over a 5- week period; 2 deaths	182
Dishwasher	Neonatal intensive care unit (Madrid, Spain	*Pseudomonas aeruginosa*	Contamination of feeding bottles during washing/preparation	12 infected and 18 colonised neonates over a 3-month period; 2 deaths	183
Mineral water	Neonatal intensive care unit (La Réunion, France)	*Pseudomonas aeruginosa*	Use of contaminated mineral water, (contaminated at source,) to prepare milk	19 identified cases	184

End uses of water in hospitals are numerous and varied. Infection can occur by ingestion of contaminated water; inhalation of contaminated aerosols; or through contact with skin, mucous membranes, eyes and ears (see Table 17.1). Although a number of different organisms have been reported to cause infection (e.g., nontuberculous mycobacteria and *Stenotrophomonas maltophilia*), two organisms of particular concern, especially in terms of hospital water supplies and biofilm control, are *P. aeruginosa* and *Legionella pneumophila*.

PSEUDOMONAS AERUGINOSA

Pseudomonas spp. is a common micro-organism that has the ability to survive and adapt to nutrient-rich or poor conditions resulting in its colonisation and survival in a wide range of internal and external environments.[35] *P. aeruginosa* is able to adapt to different environments partly because of the high percentage of transcriptional regulators that allow the cells to adapt rapidly to changing environmental conditions.[36] Particular strains, such as *P. aeruginosa*, are part of the endogenous flora of hospital patients (2.6–24%) and are able to act as opportunistic pathogens. This is because immunocompromised patients are vulnerable to infections as a result of a lowered immunity.[37] As *P. aeruginosa* is such a ubiquitous micro-organism, it has become a challenge within a range of healthcare sectors and outbreaks caused by *P. aeruginosa* have been reported in a variety of settings (e.g., adult ICUs, neonatal ICUs, medical wards, haematology units and burns units) where acquisition rates have varied between 6 and 32%.[38–46]

Pseudomonas spp. Biofilm Formation

Cystic fibrosis (CF) patients often carry *P. aeruginosa* as part of their lung flora, and it may be persistently present as part of the consortia recovered from the lung despite aggressive antimicrobial therapy.[47,48] The moist lung with its plentiful supply of nutrients is an ideal environment for the growth of this adaptable micro-organism because persistence will be, in part, due to the ability of these bacteria to form surface-associated biofilm communities enmeshed in an extracellular matrix. The CF lung is initially colonised with nonmucoid *P. aeruginosa* strains, but with time mucoid variants emerge as a result of the overproduction of alginate. These variants become the predominant lung pathogen conferring a selective advantage for *P. aeruginosa* in the CF airway.[49]

The biofilm matrix is a poorly defined mixture of protein, polysaccharide and DNA[50,51] but has been visualised as a highly organised and coordinated assembly of both polysaccharide and DNA components.[52] Two potential polysaccharide biosynthetic loci, *psl* (polysaccharide synthesis locus) and *pel* (polysaccharide encoding locus) of *P. aeruginosa* have been identified as loci that are important in biofilm initiation and formation in nonmucoid *P. aeruginosa* strains[53–55]; their expression has been shown to be elevated in variants isolated from aging *P. aeruginosa* PAO1 biofilms.[56]

Locus *psl* has been shown to play a prominent role during biofilm development by promoting cell-surface and intercellular interactions.[52] Cystic fibrosis patients who are positive for *P. aeruginosa* may disseminate droplets and aerosols that come in contact with a wide range of surfaces; these include those in the patient's immediate environment (i.e., bed cabinet, bed rails, TV screen and controller) as well as further afield—the handwashing station and their own nebulisers and masks.[57] Patients, staff and visitors will come into contact with these surfaces thus providing potential transmission routes.

Another area where *P. aeruginosa* are of concern is in infected wounds, including diabetic foot ulcers, pressure ulcers and venous leg ulcers in burns units, where polymicrobial infections are a major cause of morbidity in burn patients.[58] Colonisation, which precedes infection, is followed by biofilm formation and this process has been shown to start almost immediately after admission (46.5 % within 24 hours) with 90% of patients being colonised within 7 days.[59]

In the biofilm mode of growth, *P. aeruginosa* expresses two types of quorum-sensing (QS) population density-dependent systems: LasI-LasR and RhlI-RhlR. Both QS systems contribute to the pathology of cutaneous wound infections[60,61] and LasI-LasR and RhlI-RhlR have been shown to regulate the expression of virulence factors, such as exoenzyme S (ExoS) and exotoxin A (ToxA), which can further induce apoptosis in macrophages and neutrophils.[62,63]

Because the treatment of burn patients can often involve regular showering or washing of wound areas to debride the skin, micro-organisms are dispersed into the environment either on skin cells or in droplets of water that can then contaminate environmental surfaces. The larger particles (>10 microns) drop out of suspension and contaminate surfaces within a 2 to 3 m range.[64] In a wet room environment where the debridement would take place, it is difficult to control the dispersal and presence of micro-organisms, such as *P. aeruginosa*, which will grow on a wide range of moist surfaces as biofilm. This process often leads to contaminated wet cubical areas, handwashing basins, shower hoses and drains; thus a range of microbial biofilms result in the contamination of other patients and healthcare professionals. As a consequence of this mechanism of dispersal, *P. aeruginosa* is provided with a vehicle with which to colonise a variety of surfaces. Where moisture is present, biofilm growth persists, potentially leading to an ongoing dissemination process as patients arrive and depart.[65]

Neonatal units are where some of the most vulnerable patients are in augmented care units. Babies requiring neonatal surgery or neonatal cardiology services, and infants that need therapeutic cooling, are transferred to a neonatal intensive care unit (NICU). All neonates of less than 27 weeks gestation and those of longer gestation who require ongoing intensive care are centralised in the NICU. Many of the babies may not have fully developed antibodies and consequently have poor immune systems and as a result are susceptible to infections.[66–72]

Water has become a recognised route of transfer for *P. aeruginosa* while selective antibiotic pressures also increase acquisition rates in ICUs.[73] A number of studies have shown the ubiquity of *P. aeruginosa* in water samples. In one study *P. aeruginosa* was found in 11.4% of 484 tap water samples taken from patients' rooms and isolated from 38 patients.[74] Although 11 (52.4%) of the 21 taps were contaminated with a patient strain, 7 of the *P. aeruginosa* strains were isolated from the hands of healthcare workers (HCWs); these had the same genotype as that from the last patient they had touched in six cases and in the seventh with the last tap water sample used. More than half of *P. aeruginosa* carriage in patients in that study was acquired via tap water or cross-transmission.

LEGIONELLA PNEUMOPHILA

Legionellae are Gram-negative bacteria found naturally in freshwater environments (e.g., rivers, lakes and reservoirs). As such, they may also inhabit the municipal water supply and be found in cooling towers[75] and industrial, commercial and domestic hot and cold water systems.[76–79] Legionellae are normally present in the natural environment in low numbers. However, human-made water systems, particularly those that are poorly designed, installed or maintained, can provide favourable growth conditions and legionellae numbers can increase rapidly. A colonised water system that is not appropriately managed can act as the source of major outbreaks of legionellosis.

Legionellosis, primarily Legionnaires' disease (a serious form of pneumonia) but also including non-pneumonic legionellosis and Pontiac Fever—a self-limiting flu-like illness—can result when droplets or aerosols containing legionella bacteria are inhaled and deposited in the lungs. Macrophages ingest the bacteria but, rather than being destroyed, legionellae create a protective vacuole and replicate within the cell. When the host cell is eventually killed, large numbers of bacteria are released, each capable of infecting other macrophages and perpetuating the infection.

In the United Kingdom, Legionnaires' disease is most commonly, but not exclusively, caused by *Legionella pneumophila*; it is a rare but serious disease. Between 2009 and 2011, there were 934 confirmed cases in England and Wales, 355 (38%) of which occurred while the affected person was travelling abroad.[80] It is likely, however, that many more cases are unreported and/or undiagnosed. It is estimated that Legionnaires' disease accounts for 3% (6000 cases) of community-acquired pneumonias that occur in the United Kingdom each year.[80] Community-acquired Legionnaires' disease has a fatality rate of 12%. This increases to 24% if the infection is acquired while in hospital.

Patient exposure to *Legionella pneumophila* in the hospital can occur while showering, bathing or drinking (water or ice) or through contact with contaminated medical equipment (e.g., medication nebulisers) rinsed with tap water (see Table 17.1). To minimise the risk of exposure, measures should be introduced to prevent proliferation of the organism, build up of biofilm in the water system and exposure to aerosols.[24]

The majority of legionellae are biofilm associated.[81] A range of plumbing materials have been shown to support biofilm development and the persistence of *L. pneumophila*.[24,82] However, there is limited knowledge about the mediators of *Legionella* biofilm formation.[83,84] A glycosaminoglycan-binding adhesin (*Lcl*) has been described and is thought to play an important role in initial surface attachment and cell–cell interactions.[83] Although secreted by *L. pneumophila*, this adhesin has not been detected in other *Legionella* species. Since *Lcl* is also thought to facilitate attachment to human lung epithelial cells,[85] this may be why human legionellosis is primarily associated with *L. pneumophila*.

Although capable of surface attachment, *L. pneumophila*, under dynamic flow conditions, does not form a robust biofilm[86]; however, the organism can be rapidly incorporated into pre-established biofilms.[24,82,87] In addition in the natural environment, the attachment and/or detachment of *Legionella* spp. may be influenced by bacteria likely to be present in water systems. Biofilms formed by *Flavobacterium* spp,[84] *Mycobacterium chelonae*,[88] *Acinetobacter lwoffii*[89] and *Pseudomonas putida*[90] have all been shown to be conducive to the attachment and persistence of *L. pneumophila*.

In contrast, a monospecies biofilm, formed by *P. fluorescens* or *P. aeruginosa*, can have an antagonistic effect perhaps by producing inhibitory substances—that is, bacteriocin (antibacterial)-like substances—or QS molecules that suppress *L. pneumophila* growth and/or biofilm formation.[89,91] Legionellae are capable of necrotrophic feeding[92] and the complex nutrients available within a biofilm may be sufficient to support their growth. However, many studies have suggested that although *L. pneumophila* will survive and persist within a biofilm, for it to multiply, protozoa must also be present.[93,94] Free-living protozoa are widely dispersed within water distribution systems. They have the capacity to establish and stimulate biofilm formation by adhering to surfaces and secreting metabolic substances.[95] They also serve as vehicles for the multiplication and dispersal of bacteria. *Legionella* is an endoparasite of free-living protozoa such as *Hartmannella vermiformis*[28,96] and *Acanthamoeba* spp.[94,97,98]

The ability of *L. pneumophila* to invade amoebae in the same way as it infects human macrophages allows the organism to multiply in a nutrient-rich, environmentally buffered compartment,[99] even in unfavourable conditions. Rupture of the host cell results in large numbers of legionellae being released into the water phase, free to be ingested by non-infected amoebae, incorporated into an existing biofilm or dispersed in aerosols.

Intracellular growth reportedly enhances the infectivity of *L. pneumophila*.[94] This and the issues described previously highlight the potential problem in controlling *L. pneumophila*. Should disinfection target *Legionella*, all bacteria, biofilms or protozoa?[99]

LEGISLATION AND GUIDANCE FOR CONTROL OF MICRO-ORGANISMS

The majority of hospital water supplies will be provided by a third party who has a duty to supply water that is wholesome and free of microbial pathogens. The Health and Safety Executive's (HSE) Approved Code of Practice (ACOP) and guidance document—*Legionnaires' Disease: The Control of Legionella Bacteria in Water Systems* (L8)[100]—provides practical advice on how to comply with UK Health and Safety law with respect to the control of *Legionella*; that is, specifically, the Health and Safety at Work Act 1974 and the Control of Substances Hazardous to Health Regulations 1999 (COSHH).*

To comply with their legal duties, employers and those responsible for the control of premises should:

- Identify and assess sources of risk
- Prepare a scheme for preventing or controlling the risk
- Implement, manage and monitor precautions
- Keep records of the precautions
- Appoint a person to be managerially responsible

Since legionellae are commonly found in almost all natural water sources, it is perhaps unrealistic to expect those managing the risk of Legionnaires' disease to completely eradicate the organism from water. Rather, efforts should focus on minimising the proliferation of *Legionella* spp. within the distribution system.

As well as complying with the HSE's ACOP, the design, installation and operational management of hot and cold water systems in any National Health Service (NHS) premises should also comply with recommendations outlined by the Department of Health in its Health Technical Memorandum—*The Control of Legionella, Hygiene, 'Safe' Hot Water, Cold Water and Drinking Water Systems* (HTM 04-01). However, following the death of four babies in Northern Ireland, it was clear that there was insufficient guidance for the sampling and detection of *Pseudomonas aeruginosa*. As a result in 2013, the Department of Health published an addendum to HTM 04-01 which contained specific guidance on *P. aeruginosa* and water quality in augmented care.

The addendum introduced the concept of a multidisciplinary, risk-management approach to the microbiological safety of water which was first described in *Legionella* control documentation from the World Health Organisation in 2007.[101] The HTM 04-01 Addendum on *P. aeruginosa* guidance recommends that a multidisciplinary Water Safety Group—comprised of the Director of Infection Prevention and Control (DIPC), consultant microbiologists, members of both the infection control and estates/facilities teams, together with senior nurses from relevant augmented care units—commission and develop a 'Water Safety Plan'. The plan should establish good practice for local water distribution and supply by identifying potential microbiological hazards caused by *P. aeruginosa* and/or other opportunistic pathogens and consider the practical aspects of appropriate infection control.

*Note: As of September 2013, the ACOP and associated guidance are out for consultation and are expected to be published in early 2014.

Colonisation of Hospital Water Systems

Plumbing systems in buildings are generally more contaminated than the main water supply. Reasons include a reduced disinfectant residual in the building plumbing network and regions of pipework (e.g., dead legs and blind ends) that facilitate colonisation by allowing water to stagnate. Hospital water systems are particularly complex and the risk of infection is increased during periods of construction and refurbishment. Intermittent water use and/or changes in flow rate can lead to periods of stagnation, and intermittent pressure differentials can result in intrusion of external contaminants and/or descalement, all of which can lead to an increased concentration of potential pathogens in the water.[102,103]

Tap handles, handwashing basins and waste outlets can become contaminated with bacteria during use.[104–108] The incoming water may be contaminated or bacteria may be transferred from a patient to the sink via HCWs' hands either during handwashing or when discarding clinical waste (e.g., water used for bathing patients). If these bacteria include potential pathogens, the residues that develop on sink surfaces represent a risk factor for onward transmission.

Automation allows handwashing to be non-touch and sensor taps have been introduced into many hospitals, both for water conservation and in the belief that avoiding contact with hands will reduce cross contamination. Evidence for microbially contaminated taps is limited and, although the authors have not been able to find any evidence for hospital-acquired infections linked to tap surfaces, one must consider that tap handles are touched before performing hand hygiene on many occasions during the day.[104,105]

Sensor taps have been implicated, however, as a source of *P. aeruginosa*.[43,109–111] The complex internal configuration of sensor mixer taps, together with the wide range of materials used in their manufacture, may lead to points of concentrated bacteria growth.[112,113] Thermostatic mixer valves (TM**Vs**) mix the hot and cold water supplies to produce a temperature-controlled flow at approximately 41°C. They were initially introduced to prevent scalding particularly during full-body immersion of high-risk patients, including young children, the elderly and the mentally and physically challenged.[100,114]

Indeed, there are concerns about the risk of patients, particularly the elderly and children, burning themselves on the body of mixer taps incorporating TMV3 thermostatic mixing valves.[115] However, TMVs generally are now used at all outlets. Although tap outlets are supplied with water at a safe and comfortable temperature, once the hands are no longer near the infra-red sensor, flow rate ceases and water stagnates within the faucet at a temperature conducive to bacterial growth.[109] Colonisation of the outlet may be facilitated by the high surface area to volume ratio associated with particular types of flow straightener.[116]

Control of Biofilms in Healthcare Water Systems

There are a number of ways in which the presence of micro-organisms in a water system can be controlled and Water Safety Plans should incorporate the technical requirements that relate to the planning and avoidance of water flow stagnation, temperature control and disinfection.[27] There is no 'magic bullet' and the control of biofilms is complex and may take considerable time to achieve despite using a range of procedures.[117]

Cleaning Regimes

Based on the national standards (National Patient Safety Agency, 2010), UK hospitals can institute their own cleaning and decontamination policies. The Revised NHS Cleaning Manual[2009] provides step-by-step instructions on cleaning methods for each specific area (e.g., handwashing basins). Disinfectant use is advised and up-to-date advice on detergent types is provided. New guidance was issued by Northern Ireland in 2012, specifying the use of multiple cloths for the cleaning of handwashing basins and the removal of biofilm to reduce the potential for retrograde contamination.[118]

The cleanliness of clinical handwashing basins should be considered a critical control point, and it is therefore important to ensure that the cleaning of them, and the taps, is undertaken in a way that does not allow cross-contamination from a bacterial source.[106–108] For example, there is a microbial contamination risk for tap outlets if the same cloth is used to clean the bowl of the basin, or surrounding area, before the tap.[119]

Thermal Control and Disinfection

Water temperatures greater than 60°C are considered inhibitory for *Legionella* spp. and other non-sporulating bacteria.[120] Effective thermal disinfection can be carried out by raising the temperature of the hot water system and flushing all outlets, faucets and showerheads for at least five minutes.[121] If temperature is used as the means of controlling *Legionella* spp., the hot water circulating loop should be designed to give a return temperature to the calorifier of 50°C or above, with 55°C at the supply to the draw-off point farthest away in the circulating system.[121]

For thermal disinfection, the calorifier/heater temperature must be sufficiently high to ensure that the temperature in all parts of the circulating system, and at the return connection, does not fall below 60°C. Each tap or draw-off point should then be run sequentially from the nearest point to the outlet fartherest away for a period of at least five minutes at full temperature.[121] However, thermal disinfection is a temporary measure and the water system will become recolonised within weeks of the recirculating water returning to baseline temperature.[122]

Water Outlet Removal and Replacement

Where a water outlet has tested positive for *Legionella* spp. or *P. aeruginosa*, there is a high likelihood that a biofilm will have formed on a number of the components. The

Figure 17.1 ***Stainless steel water tap outlet.*** This type of outlet can be easily removed from the wall mounting for cleaning, decontamination or sterilisation.

removal and dismantling of the tap allows access for direct manual cleaning and further decontamination techniques such as descaling and disinfection.[121,123] In addition, TMVs should be serviced every six months in accordance with guidelines.[121]

In central sterile services departments, automated washer disinfectors, commonly used to clean surgical instruments, may provide effective microbial decontamination of tap components because of the high temperatures achieved (~80°C) and can deliver a validated process with each tap being individually bagged and labelled. Heat-resistant components allow for a more effective decontamination process. However, some tap components cannot withstand the high temperatures of washer disinfector processing or steam sterilisation. Tap manufacturers should be consulted for advice in selecting a particular disinfection and decontamination process and the Water Safety Group should advise, based on microbiological monitoring where appropriate, when taps should be replaced.

A number of tap manufacturers have addressed this issue by manufacturing removable tap outlets that can be autoclaved (Figure 17.1) or have disposable spout components (Figure 17.2). Traditionally, taps were produced via a process known as rough sand casting resulting in the internal lumens being very rough. Other tap manufacturers have now produced smooth bore outlets that replace the flow straightener (Figure 17.3) or designed the inner surface of the tap to be smooth to prevent biofilm formation. As yet there is no published scientific evidence that has independently evaluated the impact of the smooth surfaces on biofilm colonisation or control.

Chemical Disinfection

Water is usually delivered to consumer buildings with a low concentration of active chlorine disinfectant to ensure that it is fit for drinking. Oxidising chemicals, such as

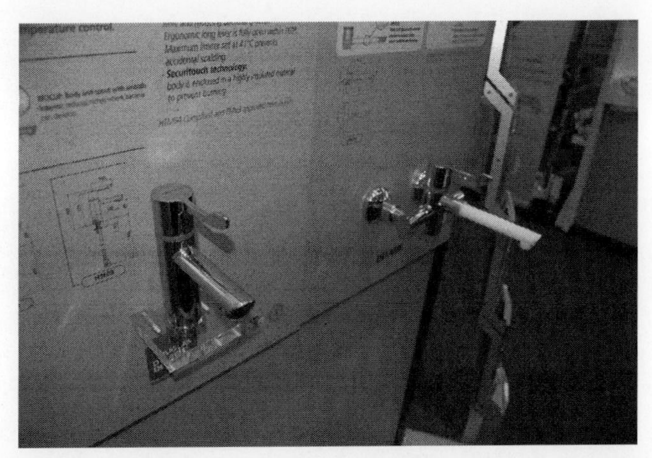

Figure 17.2 ***Two different types of tap.*** The tap on the left has a smooth internal bore and can be dismounted and decontaminated, while the one on the right has a disposable removable spout.

Figure 17.3 ***Smooth bore terminal flow straightener.*** This type of straightener is screwed into the tap outlet and is lined with alloys to reduce microbial contamination and biofilm growth.

chlorine, are the most commonly used biocides for controlling biofilms in water systems. All water systems have a chlorine demand as a result of the soluble oxidisable demand, and the greater that demand the less that will be available for biofilm control.[124]

Chlorine will irreversibly oxidise protein and other organic constituents resulting in disruption of the cell membrane that will lead to the loss of cell permeability and normal enzyme activity and hydrolysis of organic constituents and, subsequently, cell death. The use of chlorine also can be affected by pH and is most effective at pH 6 to 6.5. It is generally accepted that chlorine is less effective against biofilms than planktonic cells.

As buildings are constructed and commissioned, there is an advantage to ensuring that the building does not become contaminated with micro-organisms that would lead to the build-up of microbial biofilms composed of *L. pneumophila* and *P. aeruginosa*; disinfection can assist with that process. Previous publications indicate that once a building's water system has become contaminated, eradication can become very difficult to achieve.[125,126] The majority of hospitals use tanks to store the incoming water to ensure that there is a resilient supply in the event of delivery problems. These tanks should have tight-fitting lids to prevent the ingress of organic matter and should be kept clean; they should be designed to have a turnover that will prevent stagnation and biofilm formation.[121]

It may be appropriate (e.g., during an outbreak) to provide supplementary disinfection, such as hyperchlorination, to cleanse the system of microbiological contamination. This involves injection of a chlorine-based disinfectant at a level that achieves 50 mg/l free chlorine in the tank and ensures that every outlet is flushed until there is a smell of chlorine. Exposure periods would typically be concentration/time-dependent (e.g., 50 mg/l for one hour or longer) periods for lower concentrations and then be drained and flushed.[100,127]

Monochloramine

Monochloramine, which is used in low concentrations as a secondary disinfectant in water systems, is much more stable and does not dissipate as rapidly as free chlorine. Traditionally, ammonia is added to water first followed by chlorine gas to form mononchloramine which is thought to react specifically with nucleic acids but not sugars.[128] Monochloramine is considered more effective than free chlorine in controlling *Legionella* spp. in biofilms in large distribution systems.[129]

Chlorine Dioxide

Chlorine dioxide is commonly used as a method of water system disinfection and has been used for biofilms and *Legionella* control over a wide pH range.[117,130–132] It is a powerful oxidant that disrupts protein synthesis, making it an efficient antimicrobial. However, penetration of biofilms can be challenging and chlorine dioxide can take months to achieve microbial control, can corrode pipework and readily decomposes, particularly in hot water, though higher concentrations can be used in the hot water supply.[130,133,134]

Concentrations of 0.5 mg/l are effective against planktonic and sessile *Legionella* in hot water systems. However, The Drinking Water Inspectorate advises a maximum of 0.5 mg/l chlorine dioxide in drinking water and concentrations need to be closely monitored.[100]

Ionisation

Silver-copper ionisation of hospital water systems has been used in the control of *Legionella* spp. either alone or in combination with other disinfectants, yet eradication is often difficult to achieve.[135–139] *In vitro* ionisation has also been shown to be effective at eradicating *P. aeruginosa*. However, more recently during a European review period that ended in September 2011, no manufacturer provided sufficient evidence to support copper for use as a biocide in water systems. Consequently, the European Commission announced that as of February 1, 2013, marketing and use of elemental copper as a biocide will no longer be allowed under the Biocidal Products Directive.[140] The Health and Safety Executive has since submitted an 'essential use derogation' to the European Commission to allow for the continued use of copper in *Legionella* control systems within the United Kingdom.

Ultraviolet Irradiation

Ultraviolet (UV) irradiation has been used successfully as a biocide for a broad spectrum of planktonic bacteria in water processing, food and air disinfection and has been used for *Legionella* eradication.[141,142] However, the intensity of UV irradiation and the amount of time that the micro-organisms are exposed to radiation has a bearing on efficacy.

Exposing incoming water to UV irradiation to reduce the planktonic bioburden of the peripheral water system and locating a UV lamp closer to the point-of-use (POU) may be beneficial. Showers and taps have been designed and fitted with UV lamps located immediately before the outlet for microbial control and are currently being installed in a number of hospitals.

UV disinfection leaves no residual chemical agent in the water. In taps it cannot prevent the recontamination of water and subsequent colonisation of flow straighteners or the external surface of the tap body, which may become contaminated through exogenous sources or subsequent contamination of waste traps.[106,143]

POINT-OF-USE WATER FILTRATION

In outbreaks where water is implicated as the source, filters can be effective in providing *P. aeruginosa* free water. A long-term pre- and post-POU filtration study carried out in an ICU, where all taps tested positive for *P. aeruginosa*, concluded that the use of POU filtration reduced the risk of patient colonisation by 85% and a reduction in invasive infections of 56%.[144] While filters are often considered to be expensive, studies have shown that POU filtering can be cost effective when compared to the cost of an outbreak.[145]

Despite this, filters are considered a temporary solution to be used during outbreaks or where engineering solutions fail to provide *P. aeruginosa* free water[119]; so, the origin of the contamination must be investigated and addressed. Continual use may result in biofilm formation within the filter outlet itself, as a consequence of hand-to-filter contact.

There are some practical considerations that need to be taken into account when using filters, including the following:

- Does the tap outlet have a connection that will enable a filter to be fitted?
- Once fitted, is there sufficient flow through the filter to enable effective hand-washing?
- Is there sufficient space between the body of the filter and the handwashing basin for the hands to be inserted without touching and contaminating the filter body?

Small basins are particularly prone to the latter consideration because of the reduced space under the tap once the filter is fitted.

Point-of-use filters should be considered primarily as a temporary measure until a permanent safe engineering solution is developed, although long-term use of such filters may be required in some cases.[123]

CONCLUSION

Biofilms are a major challenge in water systems at healthcare premises. Water systems are not sterile and micro-organisms and pathogens will readily form a biofilm on a range of different materials in them. While the use of a biocide can often be successful in a laboratory trial, their use in a water system is more difficult to achieve because of the complex nature of the built environment. As a consequence, control of biofilms has to be focussed and specifically based on risk assessments to prevent the occurrence of infections in healthcare setting. For micro-organisms, such as *L. pneumophila*, this would involve recognising that the highest risk in a healthcare environment would be devices that disperse aerosols such as showers. However, taps and other uses of water must also be considered.

The presence of *Pseudomonas aeruginosa* in water in an augmented care unit and particularly NICUs should be assessed and control measures put in place accordingly. Addressing contamination as a result of *L. pneumophila* and *P. aeruginosa* will also prevent contamination by a range of other waterborne Gram-negative pathogens such as *S. maltophilia*.

Biofilms cannot be eradicated, but it is important that appropriate control measures be put in place to protect vulnerable patients in healthcare environments. A wide range of control measures can be used but those responsible for their implementation must be aware of the limitations of the methods and that they should be reviewed regularly to ensure that they are still appropriate and effective.

Acknowledgment The views expressed are those of the authors and not necessarily those of the Public Health England. The authors would like to thank Mr. Allan Bennett for his constructive comments on the manuscript.

REFERENCES

1. Anonymous. *The Water Supply (Water Quality) Regulations 2000*; <www.legislation.gov.uk/uksi/2000/3184/contents/made>; 2000.
2. Chalmers RM, Aird H, Bolton FJ. Waterborne *Escherichia coli* O157. *Symp Ser* 2000;**29**:124S–132SS.
3. Karanis P, Kourenti C, Smith H. Waterborne transmission of protozoan parasites. A worldwide review of outbreaks and lessons learnt. *J Water Health* 2007;**5**:1–38.
4. Ewald PW. Waterborne transmission and the evolution of virulence among gastrointestinal bacteria. *Epidemiol Infect* 1991;**106**(1):83–119.
5. Galal-Gorchev H. Chlorine in water disinfection. *Pure Appl Chem* 1996;**68**:1731–5.
6. Furtado C, Adak GK, Stuart JM, Wall PG, Evans HS, Casemore DP. Outbreaks of waterborne infectious intestinal disease in England and Wales, 1992-5. *Epidemiol Infect* 1998;**121**(1):109–19.
7. Cabral JP. Water microbiology. Bacterial pathogens and water. *Int J Environ Res Public Health* 2010;**7**(10):3657–703.
8. Edberg S, Rice E, Karlin R, Allen M. *Escherichia coli*: the best biological drinking water indicator for public health protection. *J Appl Microbiol* 2000;**88**(S1):106S–116SS.
9. Rice EW, Clark RM, Johnson CH. Chlorine inactivation of *Escherichia coli* O157:H7. *Emerg Infect Dis* 1999;**5**(3):461–3.
10. Berman D, Rice EW, Hoff JC. Inactivation of particle-associated coliforms by chlorine and monochloramine. *Appl Environ Microbiol* 1988;**54**(2):507–12.
11. Kuchta JM, States SJ, McNamara AM, Wadowsky RM, Yee RB. Susceptibility of *Legionella pneumophila* to chlorine in tap water. *Appl Environ Microbiol* 1983;**46**(5):1134–9.
12. LeChevallier MW, Cawthon CD, Lee RG. Factors promoting survival of bacteria in chlorinated water supplies. *Appl Environ Microbiol* 1988;**54**(3):649–54.
13. Payment P. Poor efficacy of residual chlorine disinfectant in drinking water to inactivate waterborne pathogens in distribution systems. *Can J Microbiol* 1999;**45**(8):709–15.
14. Sobsey MD. Inactivation of health-related microorganisms in water by disinfection processes. *Water Sci Technol* 1989;**21**(3):179–95.
15. Emtiazi F, Schwartz T, Marten SM, Krolla-Sidenstein P, Obst U. Investigation of natural biofilms formed during the production of drinking water from surface water embankment filtration. *Water Res* 2004;**38**(5):1197–206.
16. Hoefel D, Monis P, Grooby W, Andrews S, Saint C. Profiling bacterial survival through a water treatment process and subsequent distribution system. *J Appl Microbiol* 2005;**99**(1):175–86.
17. Ridgway HF, Olson BH. Chlorine resistance patterns of bacteria from two drinking water distribution systems. *Appl Env Micro* 1982;**44**(4):972–87.
18. Lu W, Kiéné L, Lévi Y. Chlorine demand of biofilms in water distribution systems. *Water Res* 1999;**33**(3):827–35.
19. LeChevallier MW, Schulz W, Lee RG. Bacterial nutrients in drinking water. *App Environ Microbiol* 1991;**57**(3):857–62.
20. Simoes LC, Azevedo N, Pacheco A, Keevil CW, Vieira MJ. Drinking water biofilm assessment of total and culturable bacteria under different operating conditions. *Biofouling* 2006;**22**(1-2):91–9.
21. Chu C, Lu C, Lee C. Effects of inorganic nutrients on the regrowth of heterotrophic bacteria in drinking water distribution systems. *J Environ Manage* 2005;**74**(3):255–63.
22. Donlan RM. Biofilms: microbial life on surfaces. *Emerg Infect Dis* 2002;**8**(9):881–90.
23. Morton SC, Zhang Y, Edwards MA. Implications of nutrient release from iron metal for microbial regrowth in water distribution systems. *Water Res* 2005;**39**(13):2883–92.
24. Rogers J, Dowsett A, Dennis P, Lee J, Keevil C. Influence of plumbing materials on biofilm formation and growth of *Legionella pneumophila* in potable water systems. *Appl Env Micro* 1994;**60**(6):1842–51.
25. Hassett DJ, Elkins JG, Ma JF, McDermott TR. *Pseudomonas aeruginosa* biofilm sensitivity to biocides: use of hydrogen peroxide as model antimicrobial agent for examining resistance mechanisms. *Methods Enzymol* 1999;**310**:599–608.

26. Walker JT, Mackerness CW, Rogers J, Keevil CW. Biofilm—a haven for waterborne pathogens. In: Lappin-Scott HM, Costerton JW, editors. *Microbial biofilms*. Cambridge University Press; 1995. p. 196–204.
27. Exner M, Kramer A, Lajoie L, Gebel J, Engelhart S, Hartemann P. Prevention and control of health care-associated waterborne infections in health care facilities. *Am J Infect Control* 2005;**33**(5 Suppl 1): S26–40.
28. Kuiper MW, Wullings BA, Akkermans AD, Beumer RR, Van Der Kooij D. Intracellular proliferation of *Legionella pneumophila* in *Hartmannella vermiformis* in aquatic biofilms grown on plasticized polyvinyl chloride. *Appl Env Micro* 2004;**70**(11):6826–33.
29. Eginton PJ, Gibson H, Holah J, Handley PS, Gilbert P. Quantification of the ease of removal of bacteria from surfaces. *J Ind Microbiol* 1995;**15**(4):305–10.
30. Boyd A, Chakrabarty AM. Role of alginate lyase in cell detachment of *Pseudomonas aeruginosa*. *Appl Env Micro* 1994;**60**(7):2355–9.
31. Wingender J, Flemming HC. Biofilms in drinking water and their role as reservoir for pathogens. *Int J Hyg Environ Health* 2011;**214**(6):417–23.
32. Smyth ET, McIlvenny G, Enstone JE, Emmerson AM, Humphreys H, Fitzpatrick F, et al. Four country healthcare associated infection prevalence survey 2006: overview of the results. *J Hosp Infect* 2008; **69**(3):230–48.
33. Anaissie EJ, Penzak SR, Dignani MC. The hospital water supply as a source of nosocomial infections: a plea for action. *Arch Intern Med* 2002;**162**(13):1483–92.
34. Hunter PR. National disease burden due to waterborne transmission of nosocomial pathogens is substantially overestimated. *Arch Intern Med* 2003;**163**(16):1974; author reply, 5.
35. Mena KD, Gerba CP. Risk assessment of *Pseudomonas aeruginosa* in water. *Rev Environ Contam Toxicol* 2009;**201**:71–115.
36. Stover CK, Pham XQ, Erwin AL, Mizoguchi SD, Warrener P, Hickey MJ, et al. Complete genome sequence of *Pseudomonas aeruginosa* PAO1, an opportunistic pathogen. *Nature* 2000;**406** (6799):959–64.
37. Morrison Jr. AJ, Wenzel RP. Epidemiology of infections due to *Pseudomonas aeruginosa*. *Rev Infect Dis* 1984;**6**(Suppl. 3):S627–42.
38. Jumaa P, Chattopadhyay B. Outbreak of gentamicin, ciprofloxacin-resistant *Pseudomonas aeruginosa* in an intensive care unit, traced to contaminated quivers. *J Hosp Infect* 1994;**28**(3):209–18.
39. Becks VE, Lorenzoni NM. *Pseudomonas aeruginosa* outbreak in a neonatal intensive care unit: a possible link to contaminated hand lotion. *Am J Infect Control* 1995;**23**(6):396–8.
40. Cobben NAM, Drent M, Jonkers M, Wouters EFM, Vaneechouette M, Stobberingh EE. Outbreak of severe *Pseudomonas aeruginosa* respiratory infections due to contaminated nebulizers. *J Hosp Infect* 1996;**33**:63–70.
41. Richet H, Escande MC, Marie JP, Zittoun R, Lagrange PH. Epidemic *Pseudomonas aeruginosa* serotype O16 bacteremia in hematology–oncology patients. *J Clin Microbiol* 1989;**27**(9):1992–6.
42. Kolmos HJ, Thuesen B, Nielsen SV, Lohmann M, Kristoffersen K, Rosdahl VT. Outbreak of infection in a burns unit due to *Pseudomonas aeruginosa* originating from contaminated tubing used for irrigation of patients. *J Hosp Infect* 1993;**24**(1):11–21.
43. Durojaiye OC, Carbarns N, Murray S, Majumdar S. Outbreak of multidrug-resistant *Pseudomonas aeruginosa* in an intensive care unit. *J Hosp Infect* 2011;**78**(2):154–5.
44. Thuong M, Arvaniti K, Ruimy R, de la Salmoniere P, Scanvic-Hameg A, Lucet JC, et al. Epidemiology of *Pseudomonas aeruginosa* and risk factors for carriage acquisition in an intensive care unit. *J Hosp Infect* 2003;**53**(4):274–82.
45. Bonten MJ, Bergmans DC, Speijer H, Stobberingh EE. Characteristics of polyclonal endemicity of *Pseudomonas aeruginosa* colonization in intensive care units. Implications for infection control. *Am J Respir Crit Care Med* 1999;**160**(4):1212–9.
46. Kelsey M. Pseudomonas in augmented care: Should we worry? *J Antimicrob Chemother* 2013
47. Williams HD, Davies JC. Basic science for the chest physician: *Pseudomonas aeruginosa* and the cystic fibrosis airway. *Thorax* 2012;**67**(5):465–7.
48. Rudkjobing VB, Thomsen TR, Alhede M, Kragh KN, Nielsen PH, Johansen UR, et al. The microorganisms in chronically infected end-stage and non-end-stage cystic fibrosis patients. *FEMS Immunol Med Microbiol* 2012;**65**(2):236–44.

49. Govan JR, Deretic V. Microbial pathogenesis in cystic fibrosis: mucoid *Pseudomonas aeruginosa* and *Burkholderia cepacia*. *Microbiol Rev* 1996;**60**(3):539–74.
50. Ryder C, Byrd M, Wozniak DJ. Role of polysaccharides in *Pseudomonas aeruginosa* biofilm development. *Curr Opin Microbiol* 2007;**10**(6):644–8.
51. Ma L, Wang S, Wang D, Parsek MR, Wozniak DJ. The roles of biofilm matrix polysaccharide *Psl* in mucoid *Pseudomonas aeruginosa* biofilms. *FEMS Immunol Med Microbiol* 2012;**65**(2):377–80.
52. Ma L, Jackson KD, Landry RM, Parsek MR, Wozniak DJ. Analysis of *Pseudomonas aeruginosa* conditional psl variants reveals roles for the *psl* polysaccharide in adhesion and maintaining biofilm structure postattachment. *J Bacteriol* 2006;**188**(23):8213–21.
53. Friedman L, Kolter R. Genes involved in matrix formation in *Pseudomonas aeruginosa* PA14 biofilms. *Mol Microbiol* 2004;**51**(3):675–90.
54. Friedman L, Kolter R. Two genetic loci produce distinct carbohydrate-rich structural components of the *Pseudomonas aeruginosa* biofilm matrix. *J Bacteriol* 2004;**186**(14):4457–65.
55. Franklin MJ, Nivens DE, Weadge JT, Howell PL. Biosynthesis of the *Pseudomonas aeruginosa* extracellular polysaccharides, alginate, Pel, and Psl. *Front Microbiol* 2011;**2**:167.
56. Kirisits MJ, Prost L, Starkey M, Parsek MR. Characterization of colony morphology variants isolated from *Pseudomonas aeruginosa* biofilms. *Appl Environ Microbiol* 2005;**71**(8):4809–21.
57. Pitchford KC, Corey M, Highsmith AK, Perlman R, Bannatyne R, Gold R, et al. *Pseudomonas* species contamination of cystic fibrosis patients' home inhalation equipment. *J Pediatr* 1987;**111**(2):212–6.
58. Coetzee E, Rode H, Kahn D. *Pseudomonas aeruginosa* burn wound infection in a dedicated paediatric burns unit. *S Afr J Surg* 2013;**51**(2):50–3.
59. Taneja N, Chari P, Singh M, Singh G, Biswal M, Sharma M. Evolution of bacterial flora in burn wounds: key role of environmental disinfection in control of infection. *Int J Burns Trauma* 2013;**3**(2):102–7.
60. Zhao G, Hochwalt PC, Usui ML, Underwood RA, Singh PK, James GA, et al. Delayed wound healing in diabetic (db/db) mice with *Pseudomonas aeruginosa* biofilm challenge: a model for the study of chronic wounds. *Wound Repair Regen* 2010;**18**(5):467–77.
61. Nakagami G, Morohoshi T, Ikeda T, Ohta Y, Sagara H, Huang L, et al. Contribution of quorum sensing to the virulence of *Pseudomonas aeruginosa* in pressure ulcer infection in rats. *Wound Repair Regen* 2011;**19**(2):214–22.
62. Storey DG, Ujack EE, Rabin HR, Mitchell I. *Pseudomonas aeruginosa* lasR transcription correlates with the transcription of lasA, lasB, and toxA in chronic lung infections associated with cystic fibrosis. *Infect Immun* 1998;**66**(6):2521–8.
63. Rumbaugh KP, Griswold JA, Iglewski BH, Hamood AN. Contribution of quorum sensing to the virulence of *Pseudomonas aeruginosa* in burn wound infections. *Infect Immun* 1999;**67**(11):5854–62.
64. Lowbury EJ, Babb JR, Ford PM. Protective isolation in a burns unit: the use of plastic isolators and air curtains. *J Hyg* 1971;**69**(4):529–46.
65. Chetchotisakd P, Phelps CL, Hartstein AI. Assessment of bacterial cross-transmission as a cause of infections in patients in intensive care units. *Clin Infect Dis* 1994;**18**(6):929–37.
66. Thoni A, Mussner K, Ploner F. Water birthing: retrospective review of 2625 water births. Contamination of birth pool water and risk of microbial cross-infection. *Minerva Ginecol* 2010;**62**(3):203–11.
67. Sanchez-Carrillo C, Padilla B, Marin M, Rivera M, Cercenado E, Vigil D, et al. Contaminated feeding bottles: the source of an outbreak of *Pseudomonas aeruginosa infections* in a neonatal intensive care unit. *Am J Infect Control* 2009;**37**(2):150–4.
68. Thoni A, Zech N, Moroder L, Ploner F. Water contamination and infection rate after water births. *Gynakol Geburtshilfliche Rundsch* 2007;**47**(1):33–8.
69. Gerardin P, Farny K, Simac C, Laurent AF, Grandbastien B, Robillard PY. *Pseudomonas aeruginosa* infections in a neonatal care unit at Reunion Island. *Arch Pediatr* 2006;**13**(12):1500–6.
70. Zabel LT, Heeg P, Goelz R. Surveillance of *Pseudomonas aeruginosa*-isolates in a neonatal intensive care unit over a one-year period. *Int J Hyg Environ Health* 2004;**207**(3):259–66.
71. Orrett FA. Fatal multi-resistant *Pseudomonas aeruginosa* septicemia outbreak in a neonatal intensive care unit in Trinidad. *Ethiop Med J* 2000;**38**(2):85–91.

72. Muyldermans G, de Smet F, Pierard D, Steenssens L, Stevens D, Bougatef A, et al. Neonatal infections with *Pseudomonas aeruginosa* associated with a water-bath used to thaw fresh frozen plasma. *J Hosp Infect* 1998;**39**(4):309–14.
73. Berthelot P, Chord F, Mallaval F, Grattard F, Brajon D, Pozzetto B. Magnetic valves as a source of faucet contamination with *Pseudomonas aeruginosa*? *Intensive Care Med* 2006;**32**(8):1271.
74. Rogues AM, Boulestreau H, Lasheras A, Boyer A, Gruson D, Merle C, et al. Contribution of tap water to patient colonisation with *Pseudomonas aeruginosa* in a medical intensive care unit. *J Hosp Infect* 2007;**67**(1):72–8.
75. Türetgen I, Sungur EI, Cotuk A. Enumeration of *Legionella pneumophila* in cooling tower water systems. *Environ Monit Assess* 2005;**100**(1–3):53–8.
76. Leoni E, De Luca G, Legnani PP, Sacchetti R, Stampi S, Zanetti F. *Legionella* waterline colonization: detection of *Legionella* species in domestic, hotel and hospital hot water systems. *J Appl Microbiol* 2005;**98**(2):373–9.
77. Wadowsky RM, Yee RB, Mezmar L, Wing EJ, Dowling JN. Hot water systems as sources of *Legionella pneumophila* in hospital and nonhospital plumbing fixtures. *Appl Environ Microbiol* 1982; **43**(5):1104–10.
78. Arnow PM, Weil D, Para MF. Prevalence and significance of *Legionella pneumophila* contamination of residential hot-tap water systems. *J Infect Dis* 1985;**152**(1):145–51.
79. Stout JE, Yu VL, Muraca P. Isolation of *Legionella pneumophila* from the cold water of hospital ice machines: implications for origin and transmission of the organism. *Infect Control* 1985;**6**(4):141–6.
80. Naik FC, Zhao H, Harrison TG, Phin N. Legionnaires' disease in England and Wales, 2011. Health Protection Report; 2012.
81. Fields BS, Benson RF, Besser RE. *Legionella* and Legionnaires' disease: 25 years of investigation. *Clin Microbiol Rev* 2002;**15**(3):506–26.
82. Rogers J, Dowsett A, Dennis P, Lee J, Keevil C. Influence of temperature and plumbing material selection on biofilm formation and growth of *Legionella pneumophila* in a model potable water system containing complex microbial flora. *Appl Env Micro* 1994;**60**(6):1585–92.
83. Mallegol J, Duncan C, Prashar A, So J, Low DE, Terebeznik M, et al. Essential roles and regulation of the *Legionella pneumophila* collagen-like adhesin during biofilm formation. *PloS One* 2012; **7**(9):e46462.
84. Stewart CR, Muthye V, Cianciotto NP. *Legionella pneumophila* persists within biofilms formed by *Klebsiella pneumoniae, Flavobacterium* sp., and *Pseudomonas fluorescens* under dynamic flow conditions. *PloS One* 2012;**7**(11):e50560.
85. Duncan C, Prashar A, So J, Tang P, Low DE, Terebiznik M, et al. *Lcl* of *Legionella pneumophila* is an immunogenic *GAG* binding adhesin that promotes interactions with lung epithelial cells and plays a crucial role in biofilm formation. *Infect Immun* 2011;**79**(6):2168–81.
86. Mampel J, Spirig T, Weber SS, Haagensen JA, Molin S, Hilbi H. Planktonic replication is essential for biofilm formation by *Legionella pneumophila* in a complex medium under static and dynamic flow conditions. *Appl Env Micro* 2006;**72**(4):2885–95.
87. Declerck P. Biofilms: the environmental playground of *Legionella pneumophila. Environ Microbiol* 2010; **12**(3):557–66.
88. Gião MS, Azevedo NF, Wilks SA, Vieira MJ, Keevil CW. Interaction of *Legionella pneumophila* and *Helicobacter pylori* with bacterial species isolated from drinking water biofilms. *BMC Microbiol* 2011; **11**(1):57.
89. Guerrieri E, Bondi M, Sabia C, de Niederhausern S, Borella P, Messi P. Effect of bacterial interference on biofilm development by *Legionella pneumophila. Curr Microbiol* 2008;**57**(6):532–6.
90. Vervaeren H, Temmerman R, Devos L, Boon N, Verstraete W. Introduction of a boost of *Legionella pneumophila* into a stagnant-water model by heat treatment. *FEMS Microbiol Ecol* 2006;**58**(3):583–92.
91. Kimura S, Tateda K, Ishii Y, Horikawa M, Miyairi S, Gotoh N, et al. *Pseudomonas aeruginosa Las* quorum sensing autoinducer suppresses growth and biofilm production in *Legionella* species. *Microbiology* 2009;**155**(6):1934–9.
92. Temmerman R, Vervaeren H, Noseda B, Boon N, Verstraete W. Necrotrophic growth of *Legionella pneumophila. Appl Env Micro* 2006;**72**(6):4323–8.

93. Murga R, Forster TS, Brown E, Pruckler JM, Fields BS, Donlan RM. Role of biofilms in the survival of *Legionella pneumophila* in a model potable-water system. *Microbiol* 2001;**147**(Pt 11):3121–6.
94. Declerck P, Behets J, Margineanu A, van Hoef V, De Keersmaecker B, Ollevier F. Replication of *Legionella pneumophila* in biofilms of water distribution pipes. *Microbiol Res* 2009;**164**(6):593–603.
95. Hoffmann R, Michel R. Distribution of free-living amoebae (FLA) during preparation and supply of drinking water. *Int J Hyg Environ Health* 2001;**203**(3):215–9.
96. Wadowsky RM, Wilson TM, Kapp NJ, West AJ, Kuchta JM, States SJ, et al. Multiplication of *Legionella* spp. in tap water containing *Hartmannella vermiformis*. *Appl Environ Microbiol* 1991;**57**(7):1950–5.
97. Kilvington S, Price J. Survival of *Legionella pneumophila* within cysts of *Acanthamoeba polyphaga* following chlorine exposure. *J Appl Bacteriol* 1990;**68**(5):519–25.
98. Holden EP, Winkler H, Wood D, Leinbach E. Intracellular growth of *Legionella pneumophila* within *Acanthamoeba castellanii* Neff. *Infect Immun* 1984;**45**(1):18–24.
99. Taylor M, Ross K, Bentham R. *Legionella*, protozoa, and biofilms: interactions within complex microbial systems. *Microbial Ecol* 2009;**58**(3):538–47.
100. Anonymous. *Legionnaires' disease. The control of Legionella bacteria in water systems. Approved Code of Practice and Guidance*; <www.hse.gov.uk/pubns/books/l8.htm>; 2000.
101. WHO. *Legionella and the prevention of legionellosis*; <www.who.int/water_sanitation_health/gdwqrevision/legionella/en/index.html>; 2007.
102. Mermel LA, Josephson SL, Giorgio CH, Dempsey J, Parenteau S. Association of Legionnaires' disease with construction: contamination of potable water? *Infect Control Hosp Epidemiol* 1995; **16**(2):76–81.
103. Nygård K, Wahl E, Krogh T, Tveit OA, Bøhleng E, Tverdal A, et al. Breaks and maintenance work in the water distribution systems and gastrointestinal illness: a cohort study. *Int J Epidemiol* 2007;**36**(4):873–80.
104. Casey AL, Adams D, Karpanen TJ, Lambert PA, Cookson BD, Nightingale P, et al. Role of copper in reducing hospital environment contamination. *J Hosp Infect* 2010;**74**(1):72–7.
105. Griffith CJ, Cooper RA, Gilmore J, Davies C, Lewis M. An evaluation of hospital cleaning regimes and standards. *J Hosp Infect* 2000;**45**(1):19–28.
106. Vergara-Lopez S, Dominguez MC, Conejo MC, Pascual A, Rodriguez-Bano J. Wastewater drainage system as an occult reservoir in a protracted clonal outbreak due to metallo-beta-lactamase-producing *Klebsiella oxytoca*. *Clin Microbiol Infect* 2013;**19**(11):E490–8.
107. Breathnach AS, Cubbon MD, Karunaharan RN, Pope CF, Planche TD. Multidrug-resistant *Pseudomonas aeruginosa* outbreaks in two hospitals: association with contaminated hospital waste-water systems. *J Hosp Infect* 2012;**82**(1):19–24.
108. Hota S, Hirji Z, Stockton K, Lemieux C, Dedier H, Wolfaardt G, et al. Outbreak of multidrug-resistant *Pseudomonas aeruginosa* colonization and infection secondary to imperfect intensive care unit room design. *Infect Control Hosp Epidemiol* 2009;**30**(1):25–33.
109. Halabi M, Wiesholzer-Pittl M, Schoberl J, Mittermayer H. Non-touch fittings in hospitals: a possible source of *Pseudomonas aeruginosa* and *Legionella* spp. *J Hosp Infect* 2001;**49**(2):117–21.
110. Merrer J, Girou E, Ducellier D, Clavreul N, Cizeau F, Legrand P, et al. Should electronic faucets be used in intensive care and hematology units? *Intensive Care Med* 2005;**31**(12):1715–8.
111. Yapicioglu H, Gokmen TG, Yildizdas D, Koksal F, Ozlu F, Kale-Cekinmez E, et al. *Pseudomonas aeruginosa* infections due to electronic faucets in a neonatal intensive care unit. *J Paediatr Child Health* 2012;**48**(5):430–4.
112. Sydnor ER, Bova G, Gimburg A, Cosgrove SE, Perl TM, Maragakis LL. Electronic-eye faucets: *Legionella* species contamination in healthcare settings. *Infect Control Hosp Epidemiol* 2012; **33**(3):235–40.
113. Waines PL, Moate R, Moody AJ, Allen M, Bradley G. The effect of material choice on biofilm formation in a model warm water distribution system. *Biofouling* 2011;**27**(10):1161–74.
114. Murray JP. A study of the prevention of hot tapwater burns. *Burns Incl Therm Inj* 1988;**14**(3):185–93.
115. Anonymous. Reducing risk with better tap design. *Health Estate* 2009;**63**(5):31–6.
116. Troop P. *RQIA Report: RQIA Independent Review of Pseudomonas Interim Report*; <www.rqia.org.uk/publications/rqia_reviews.cfm>; 2012.

117. Marchesi I, Marchegiano P, Bargellini A, Cencetti S, Frezza G, Miselli M, et al. Effectiveness of different methods to control legionella in the water supply: ten-year experience in an Italian university hospital. *J Hosp Infect* 2011;**77**(1):47–51.
118. Anonymous. *Policy for the provision and management of cleaning services*; <www.dhsspsni.gov.uk/sub-611-2012-policy-cleaning-services.pdf>; 2012.
119. Anonymous. *Water sources and potential Pseudomonas aeruginosa contamination of taps and water systems: Advice for augmented care units*. Department of Health; <www.dh.gov.uk/health/2013/03/pseudomonas-addendum/>; 2013.
120. Dennis PJ, Green D, Jones BP. A note on the temperature tolerance of *Legionella*. *J Appl Bacteriol* 1984;**56**(2):349–50.
121. Anonymous. *Water systems Health Technical Memorandum 04-01: The control of Legionella, hygiene, "safe" hot water, cold water and drinking water systems. Part B*; <http://publications.spaceforhealth.nhs.uk/stream.php?id=11s4oAn3ss8Q26588t56590=pnn2o9s6r46q>; 2006.
122. Lin YS, Stout JE, Yu VL, Vidic RD. Disinfection of water distribution systems for *Legionella*. *Semin Respir Infect* 1998;**13**(2):147–59.
123. Anonymous. *Health Technical Memorandum 04-01 Addendum: Pseudomonas aeruginosa—advice for augmented care units*; Department of Health; <www.gov.uk/government/publications/addendum-to-guidance-for-healthcare-providers-on-managing-pseudomonas-published>; 2013.
124. Characklis WG, Marshall KE. *Biofilms*. John Wiley & Sons Inc; 1990.
125. Garcia MT, Baladron B, Gil V, Tarancon ML, Vilasau A, Ibanez A, et al. Persistence of chlorine-sensitive *Legionella pneumophila* in hyperchlorinated installations. *J Appl Microbiol* 2008;**105**(3):837–47.
126. Perola O, Kauppinen J, Kusnetsov J, Karkkainen UM, Luck PC, Katila ML. Persistent *Legionella pneumophila* colonization of a hospital water supply: efficacy of control methods and a molecular epidemiological analysis. *APMIS* 2005;**113**(1):45–53.
127. Muraca PW, Yu VL, Goetz A. Disinfection of water distribution systems for *Legionella*: a review of application procedures and methodologies. *Infect Control Hosp Epidemiol* 1990;**11**(2):79–88.
128. Jacangelo JG, Olivieri VP. Aspects of the mode of action of monochloramine. In: Jolley RL, Bull RJ, Davis WP, Katz S, Roberts MH, Jacobs VA, editors. *Water chlorination, chemistry, environmental impact and health effects*. Lewis Publishers; 1985.
129. Moore MR, Pryor M, Fields B, Lucas C, Phelan M, Besser RE. Introduction of monochloramine into a municipal water system: impact on colonization of buildings by *Legionella* spp. *Appl Environ Microbiol* 2006;**72**(1):378–83.
130. Pavey NL, Roper M. *Chlorine dioxide water treatment—for hot and cold water services*. Oakdale Printing Co; 1998.
131. Walker JT, Mackerness CW, Mallon D, Makin T, Williets T, Keevil CW. Control of *Legionella pneumophila* in a hospital water system by chlorine dioxide. *J Ind Microbiol* 1995;**15**(4):384–90.
132. Loret JF, Robert S, Thomas V, Cooper AJ, McCoy WF, Levi Y. Comparison of disinfectants for biofilm, protozoa and *Legionella* control. *J Water Health* 2005;**3**(4):423–33.
133. Zhang Z, Stout JE, Yu VL, Vidic R. Effect of pipe corrosion scales on chlorine dioxide consumption in drinking water distribution systems. *Water Res* 2008;**42**(1-2):129–36.
134. Lin YE, Stout JE, Yu VL. Controlling *Legionella* in hospital drinking water: an evidence-based review of disinfection methods. *Infect Control Hosp Epidemiol* 2011;**32**(2):166–73.
135. Chen YS, Lin YE, Liu YC, Huang WK, Shih HY, Wann SR, et al. Efficacy of point-of-entry copper--silver ionisation system in eradicating *Legionella pneumophila* in a tropical tertiary care hospital: implications for hospitals contaminated with *Legionella* in both hot and cold water. *J Hosp Infect* 2008;**68**(2):152–8.
136. Cachafeiro SP, Naveira IM, Garcia IG. Is copper-silver ionisation safe and effective in controlling *Legionella*? *J Hosp Infect* 2007;**67**(3):209–16.
137. Modol J, Sabria M, Reynaga E, Pedro-Botet ML, Sopena N, Tudela P, et al. Hospital-acquired legionnaires' disease in a university hospital: impact of the copper-silver ionization system. *Clin Infect Dis* 2007;**44**(2):263–5.
138. Blanc DS, Carrara P, Zanetti G, Francioli P. Water disinfection with ozone, copper and silver ions, and temperature increase to control *Legionella*: Seven years of experience in a university teaching hospital. *J Hosp Infect* 2005;**60**(1):69–72.

139. Stout JE, Yu VL. Experiences of the first 16 hospitals using copper-silver ionization for *Legionella* control: implications for the evaluation of other disinfection modalities. *Infect Control Hosp Epidemiol* 2003;**24**(8):563–8.
140. Anonymous. *Copper ionisation systems*; <www.hse.gov.uk/legionnaires/faqs.htm#silver-copper-systems>; 2013.
141. Triassi M, Di Popolo A, Ribera D'Alcala G, Albanese Z, Cuccurullo S, Montegrosso S, et al. Clinical and environmental distribution of *Legionella pneumophila* in a university hospital in Italy: efficacy of ultraviolet disinfection. *J Hosp Infect* 2006;**62**(4):494–501.
142. Hall KK, Giannetta ET, Getchell-White SI, Durbin LJ, Farr BM. Ultraviolet light disinfection of hospital water for preventing nosocomial *Legionella* infection: a 13-year follow-up. *Infect Control Hosp Epidemiol* 2003;**24**(8):580–3.
143. Kim BR, Anderson JE, Mueller SA, Gaines WA, Kendall AM. Literature review—efficacy of various disinfectants against *Legionella* in water systems. *Water Res* 2002;**36**(18):4433–44.
144. Trautmann M, Halder S, Hoegel J, Royer H, Haller M. Point-of-use water filtration reduces endemic *Pseudomonas aeruginosa* infections on a surgical intensive care unit. *Am J Infect Control* 2008; **36**(6):421–9.
145. Holmes C, Cervia JS, Ortolano GA, Canonica FP. Preventive efficacy and cost-effectiveness of point-of-use water filtration in a subacute care unit. *Am J Infect Control* 2010;**38**(1):69–71.
146. Conger NG, O'Connell RJ, Laurel VL, Olivier KN, Graviss EA, Williams-Bouyer N, et al. Mycobacterium simiae outbreak associated with a hospital water supply. *Infect Control Hosp Epidemiol* 2004;**25**:1050–5.
147. Astagneau P, Desplaces N, Vincent V, Chicheportiche V, Botherel A-H, Maugat S, et al. *Mycobacterium xenopi* spinal infections after discovertebral surgery: investigation and screening of a large outbreak. *Lancet* 2001;**358**:747–51.
148. Kappstein I, Grundmann H, Hauer T, Niemeyer C. Aerators as a reservoir of Acinetobacter junii: an outbreak of bacteraemia in paediatric oncology patients. *J Hosp Infect* 2000;**44**:27–30.
149. Horcajada JP, Martínez JA, Alcón A, Marco F, De Lazzari E, de Matos A, et al. Acquisition of multidrug-resistant *Serratia marcescens* by critically ill patients who consumed tap water during receipt of oral medication. *Infect Control Hosp Epidemiol* 2006;27:774–7.
150. Sakhnini E, Weissmann A, Oren I. Fulminant *Stenotrophomonas maltophilia* soft tissue infection in immunocompromised patients: an outbreak transmitted via tap water. *Am J Med Sci* 2002;**323**:269–72.
151. Verweij PE, Meis JF, Christmann V, Van der Bor M, Melchers WJ, Hilderink BG, et al. Nosocomial outbreak of colonization and infection with *Stenotrophomonas maltophilia* in preterm infants associated with contaminated tap water. *Epidemiol Infect* 1998;**120**:251–6.
152. Lowry PW, Blankenship RJ, Gridley W, Troup NJ, Tompkins LS. A cluster of *Legionella* sternal-wound infections due to postoperative topical exposure to contaminated tap water. *N Engl J Med* 1991;**324**:109–13.
153. Brûlet A, Nicolle MC, Giard M, Nicolini FE, Michallet M, Jarraud S, et al. Fatal nosocomial *Legionella pneumophila* infection due to exposure to contaminated water from a washbasin in a hematology unit. *Infect Control Hosp Epidemiol* 2008;**29**:1091–3.
154. Venezia RA, Agresta MD, Hanley EM, Urquhart K, Schoonmaker D. Nosocomial legionellosis associated with aspiration of nasogastric feedings diluted in tap water. *Infect Control Hosp Epidemiol* 1994;**15**:529–33.
155. Trautmann M, Michalsky T, Wiedeck H, Radosavljevic V, Ruhnke M. Tap water colonization with *Pseudomonas aeruginosa* in a surgical intensive care unit (ICU) and relation to Pseudomonas infections of ICU Patients. *Infect Control Hosp Epidemiol* 2001;**22**:49–52.
156. Bert F, Maubect E, Bruneau B, Berry P, Lambert-Zechovsky N. Multi-resistant *Pseudomonus aeruginosa* outbreak associated with contaminated tap water in a neurosurgery intensive care unit. *J Hosp Infect* 1998;39:53–62. In addition, see RQIA. Independent review of incidents of *Pseudomonas aeruginosa* infection in neonatal units in Northern Ireland. Belfast: The Regulation and Quality Improvement Authority; 2012.
157. Kline S, Cameron S, Streifel A, Yakrus MA, Kairis F, Peacock K, et al. An outbreak of bacteremias associated with *Mycobacterium mucogenicum* in a hospital water supply. *Infect Control Hosp Epidemiol* 2004;**25**:1042–9.

158. Lyytikäinen O, Golovanova V, Kolho E, Ruutu P, Sivonen A, Tiittanen L, et al. Outbreak caused by tobramycin-resistant *Pseudomonas aeruginosa* in a bone marrow transplantation unit. *Scand J Infect Dis* 2001;**33**:445–9.
159. Darelid J, Bengtsson L, Gästrin B, Hallander H, Löfgren S, Malmvall BE, et al. An outbreak of Legionnaires' disease in a Swedish hospital. *Scand J Infect Dis* 1994;**26**:417–25.
160. Hanrahan JP, Morse DL, Scharf VB, Debbie JG, Schmid GP, McKinney RM, et al. A community hospital outbreak of *legionellosis*. Transmission by potable hot water. *Am J Epidemiol* 1987; **125**:639–49.
161. Tobin JO'H, Dunnill MS, French M, Morris PJ, Beare J, Fisher-Hoch S, et al. Legionnaires' disease in a transplant unit: isolation of the causative agent from shower baths. *Lancet* 1980;**316**:118–21.
162. Torii K, Iinuma Y, Ichikawa M, Kato K, Koide M, Baba H, et al. A case of nosocomial *Legionella pneumophila* pneumonia. *Jpn J Infect Dis* 2003;**56**:101–2.
163. Franzin L, Scolfaro C, Cabodi D, Valera M, Tovo PA. *Legionella pneumophila* pneumonia in a newborn after water birth: a new mode of transmission. *Clin Infect Dis* 2001;**33**:e103–4.
164. Schlech WF, Simonsen N, Sumarah R, Martin RS. Nosocomial outbreak of *Pseudomonas aeruginosa folliculitis* associated with a physiotherapy pool. *Can Med Assoc J* 1986;**134**:909–13.
165. Berrouane YF, McNutt L-A, Buschelman BJ, Rhomberg PR, Sanford MD, Hollis RJ, et al. Outbreak of severe *Pseudomonas aeruginosa* infections caused by a contaminated drain in a whirlpool bathtub. *Clin Infect Dis* 2000;**31**:1331–7.
166. Buttery JP, Alabaster SJ, Heine RG, Scott SM, Crutchfield RA, Garland SM. Multiresistant *Pseudomonas aeruginosa* outbreak in a pediatric oncology ward related to bath toys. *Pediatr Infect Dis J* 1998;**17**:509–13.
167. Muyldermans G, De Smet F, Pierard D, Steenssens L, Stevens D, Bougatef A, et al. Neonatal infections with *Pseudomonas aeruginosa* associated with a water-bath used to thaw fresh frozen plasma. *J Hosp Infect* 1998;**39**:309–14.
168. Conly JM, Klass L, Larson L, Kennedy J, Low DE, Harding GK. *Pseudomonas cepacia* colonization and infection in intensive care units. *Can Med Assoc J* 1986;**134**:363–6.
169. Bou R, Ramos P. Outbreak of nosocomial Legionnaires' disease caused by a contaminated oxygen humidifier. *J Hosp Infect* 2009;**71**:381–3.
170. Moiraghi A, Castellani Pastorist M, Barral C, Carle F, Sciacovelli A, Passarinol G, et al. Nosocomial *legionkllosis* associated with use of oxygen bubble humidifiers and underwater chest drains. *J Hosp Infect* 1987;**10**:47–50.
171. Joly JR, Déry P, Gauvreau L, Coté L, Trépanier C. Legionnaires' disease caused by *Legionella dumoffii* in distilled water. *Can Med Assoc J* 1986;**135**:1274–7.
172. Arnow PM, Chou T, Weil D, Shapiro EN, Kretzschmar C. Nosocomial Legionnaires' disease caused by aerosolized tap water from respiratory devices. *J Infect Dis* 1982;**146**:460–7.
173. Mastro TD, Fields BS, Breiman RF, Campbell J, Plikaytis BD, Spika JS. Nosocomial Legionnaires' disease and use of medication nebulizers. *J Infect Dis* 1991;**163**:667–71.
174. Yiallouros PK, Papadouri T, Karaoli C, Papamichael E, Zeniou M, Pieridou-Begatzouni D, et al. First outbreak of nosocomial *Legionella* infection in term neonates caused by a cold mist ultrasonic humidifier. *Clin Infect Dis* 2013 doi: 10.1093/cid/cit176.
175. Laussucq S, Baltch AL, Smith RP, Smithwick RW, Davis BJ, Desjardin EK, et al. Nosocomial *Mycobacterium fortuitum* colonization from a contaminated ice machine. *Am Rev Respir Dis* 1988;**138**:891–4.
176. Graman PS, Quinlan GA, Rank JA. Nosocomial legionellosis traced to a contaminated ice machine. *Infect Control Hosp Epidemiol* 1997;**18**:637–40.
177. Bangsborg JM, Uldum S, Jensen JS, Bruun BG. Nosocomial legionellosis in three heart-lung transplant patients: case reports and environmental observations. *Eur J Clin Microbiol Infect Dis* 1995;**14**:99–104.
178. Pien FD. Nosocomial *Ewingella americana* bacteremia in an intensive care unit. *Arch Intern Med* 1986;**146**:111–2.

179. Stamm WE, Colella JJ, Anderson RL, Dixon RE. Indwelling arterial catheters as a source of nosocomial bacteremia. An outbreak caused by *Flavobacterium* species. *N Engl J Med* 1975; **292**:1099–102.
180. Palmore TM, Stock F, White M, Bordner M, Michelin A, Bennett JE, et al. A cluster of nosocomial Legionnaire's disease linked to a contaminated hospital decorative water fountain. *Infect Control Hosp Epidemiol* 2009;**30**:764–8.
181. Haupt TE, Heffernan RT, Kazmierczak JJ, Nehls-Lowe H, Rheineck B, Powell C, et al. An outbreak of Legionnaires' disease associated with a decorative water wall fountain in a hospital. *Infect Control Hosp Epidemiol* 2012;**33**:185–91.
182. Engelhart S, Krizek L, Glasmachery A, Fischnaller E, Markleinz G, Exner M. *Pseudomonas aeruginosa* outbreak in a haematology-oncology unit associated with contaminated surface cleaning equipment. *J Hosp Infect* 2002;**52**:93–8.
183. Sánchez-Carrillo C, Padilla B, Marín M, Rivera M, Cercenado E, Vigil D, et al. Contaminated feeding bottles: the source of an outbreak of *Pseudomonas aeruginosa* infections in a neonatal intensive care unit. *Am J Infect Control* 2009;**37**:150–4.
184. Naze F, Jouen E, Randriamahazo RT, Simac C, Laurent P, Blériot A, et al. *Pseudomonas aeruginosa* outbreak linked to mineral water bottles in a neonatal intensive care unit: fast typing by use of high-resolution melting analysis of a variable-number tandem-repeat locus. *J Clin Microbiol* 2010;**48**:3146–52.

CHAPTER EIGHTEEN

Wound Infection and Biofilms

Sara McCarty, Eleri M. Jones, Simon Finnegan, Emma Woods, Christine A. Cochrane and Steven L. Percival

INTRODUCTION

The cost of treating chronic wounds in the Unites States alone has been estimated at $7 billion per annum, and wound types are documented to be increasing at a rate of 10% per year. Non-healing wounds constitute a significant problem to the medical profession, and it is now considered that biofilms play a critical role in preventing a chronic wound from healing. A *wound* is described as being 'the loss of continuity of epithelium with or without the loss of underlying connective tissue'.[1] Wound infections develop as a result of a complex association of factors. If the protective nature of the skin is breached, an inflammatory response is initiated presenting clinically in the form of redness, pain, raised temperature and swelling.[2] These events occur in order to restore homeostasis at the compromised site.

Infection in a wound depends on many variables ranging from pre-, intra- and post-operative care to the patient's pre-existing medical conditions to name but a few; this makes predicting the likelihood of which wound will become infected very difficult. In a 2002 survey by the Nosocomial Infection National Surveillance Service (NINSS), it was reported that the incidence of hospital-acquired infections in surgical wounds was 10%, with a cost to the National Health Service of £1 billion per annum.[3] It has also been estimated that patients with a surgical site infection require an additional 6.5 days in hospital thus doubling the stay's cost.[4]

The impact of micro-organisms on wound healing is to date poorly understood, but evidence of bacteria within chronic wounds are now documented to be responsible for causing a delay in healing rates.[5,6] If this statement is true, reducing a wound's bioburden should help reduce the factors that impede healing. This can be achieved with the use of topical antimicrobials which may be helpful in overcoming the deliriious effects of bacteria. It is well documented that chronic wounds are polymicrobial; however, it is only recently that microbiologists have begun to address and recognise the significance and importance of an interacting biofilm community and its relationship to the disease process.

Biofilms in Infection Prevention and Control
Doi: http://dx.doi.org/10.1016/B978-0-12-397043-5.00018-9

It would appear that the wound per se is capable of supporting a biofilm. For a biofilm to develop in a wound, a number of stages are required—namely, microbial attachment, proliferation, quorum sensing (QS), exopolymer production and dispersal/detachment. Biofilms are well documented as being recalcitrant and able to withstand treatment with antimicrobial therapy. The existence of biofilms in an acute partial-thickness wound[7] and in chronic human wounds[8] has been documented and thus major advances are being made in this area.[5,6]

Although the wound-healing process is affected by both intrinsic and extrinsic factors, it has to be acknowledged that the actual presence of bacteria does not constitute wound infection. This is because microbial colonisation of a wound is a 'normal' event, with both acute and chronic wounds being inhabited by different populations of micro-organisms. The survival and proliferation of bacteria within wounds depends on the efficiency of the host's immune system and the availability of the necessary chemical and physical factors used to reduce the bacteria's presence.

In wound microbiology the clinician has to distinguish between different levels of bioburden evident in a chronic wound (i.e., contamination, colonisation or infection). Visually this is not possible and requires microbiological sampling in conjunction with clinical observations of the wounds' state and condition. Contamination of a wound refers to the microbial presence without multiplication, while colonisation is the adhered microbes present on and within the wound bed. In contrast, critical colonisation refers to microbial multiplication in a wound but without a host reaction and obvious clinical signs of infection. Critical colonisation also refers to the inability of the wound to maintain a balance between the increasing bioburden and an effective immune system—the wound has become compromised but is not yet demonstrating overt clinical signs of infection other than not healing. A wound is infected when multiplication of micro-organisms occurs resulting in a host reaction. At this stage therefore, it would be advantageous to assist in the re-establishment of the microbial balance or homeostasis in a wound. This can often be achieved with the use of antimicrobials that help reduce the wounds microbial load.

It is the intention of this review to address the role that biofilms may play in wound healing and highlight methods that can be used to prevent and treat them.

BIOLOGY OF WOUND HEALING

The skin is the human body's most important line of defence, protecting tissue from external factors such as noxious chemicals and micro-organisms. Once there is a breach in this defence, the body's main goal is to reestablish this barrier as quickly as possible to limit both the loss of tissue and fluids (e.g., blood) and the invasion of the delicate body tissues by particulates and micro-organisms. The wound-healing response is a cascade of events that occur as a consequence of a trauma that leads to the physical

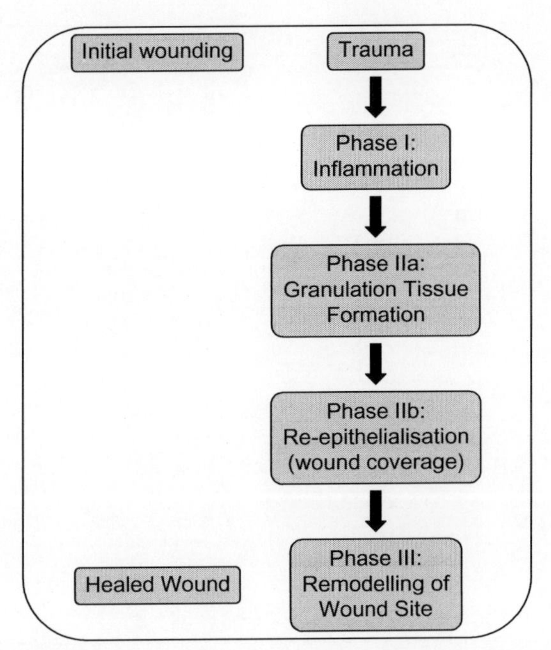

Figure 18.1 ***Simplified depiction of the stages of wound healing in which complex biochemical and cellular processes are involved.***

breakdown of the skin barrier. This healing response has been described as a series of overlapping phases (Figure 18.1), with the appearance and resolution of each phase being orchestrated by numerous signalling mechanisms that ultimately lead to a healed wound.

One of the earliest—and most important—phases in the healing process is the inflammatory response. Inflammatory cells (e.g., neutrophils and macrophages) are involved in the removal of tissue contamination and damaged skin components from the wound site in preparation for the formation of the granulation tissue—a tissue that is essential if the wound is to heal properly. These inflammatory cells also provide numerous signalling molecules that promote the influx of other cells that are responsible for the production of new tissue components such as blood vessels and connective tissue. In addition, through bacterial killing mechanisms, inflammatory cells are central in preventing the establishment of excessive numbers of micro-organisms within an open wound that could lead to infection.

Wounds that do not heal in a timely manner can be an indication of an underlying systemic problem of an individual who presents with a non-healing wound. Numerous underlying disease processes can play a major role in how effective the wound-healing response can be to an injury. Chronic venous insufficiency and diabetes are two examples of underlying disease processes that can lead to the formation of debilitating non-healing (chronic) wounds, commonly termed 'ulcers'.

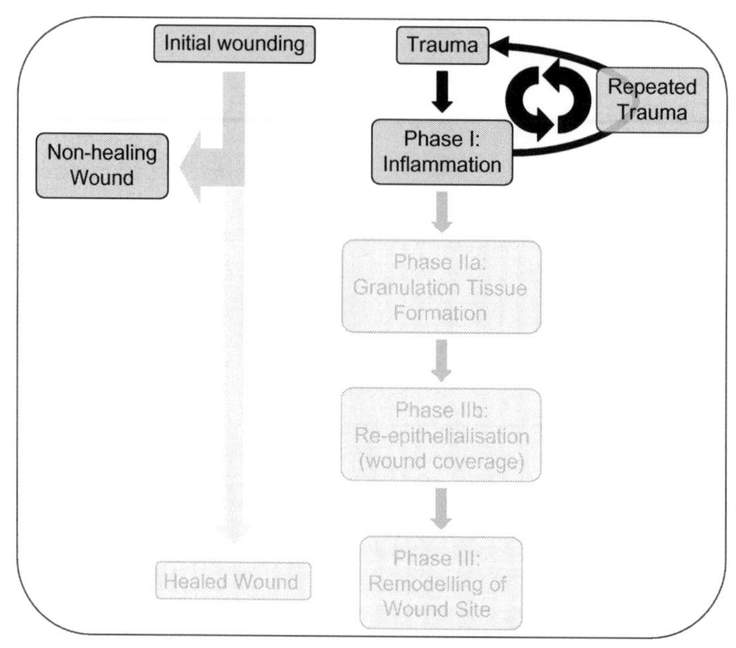

Figure 18.2 ***Wound healing and its association with repeated trauma causing continuing insult to the affected tissue.***

Recent studies have suggested that inflammation—the key phase in the wound-healing response (see Figure 18.1)—may play a major role in the formation and persistence of ulcers. If uncontrolled, the repeated stimulation of a localised inflammatory response at a wound site, particularly in a lower limb, can lead to the formation of a non-healing wound (Figure 18.2).

The signals that ensure that inflammatory cells progress through the wound optimally in a normally healing wound, appear to be disrupted in a non-healing wound which has important consequences for healing. As stated earlier, inflammatory cells play a major role in removing tissue debris from newly formed wound sites. The cells release protein-degrading enzymes (proteinases) that are responsible for breaking down the protein components of the debris, aiding in debris removal. Proteinases are very damaging enzymes and, therefore, they are tightly regulated by (1) being produced only where and when they are required and (2) being controlled by specific proteinase inhibitors. Because of the nature of the underlying aetiology of the diseases where ulceration is seen, there is an elevated and uncontrolled production of these proteinases that leads to tissue breakdown or persistence of the wound.

One of the problems with non-healing wounds is the susceptibility of them to adhesion and colonisation by micro-organisms. Bacterial colonisation of the beds of

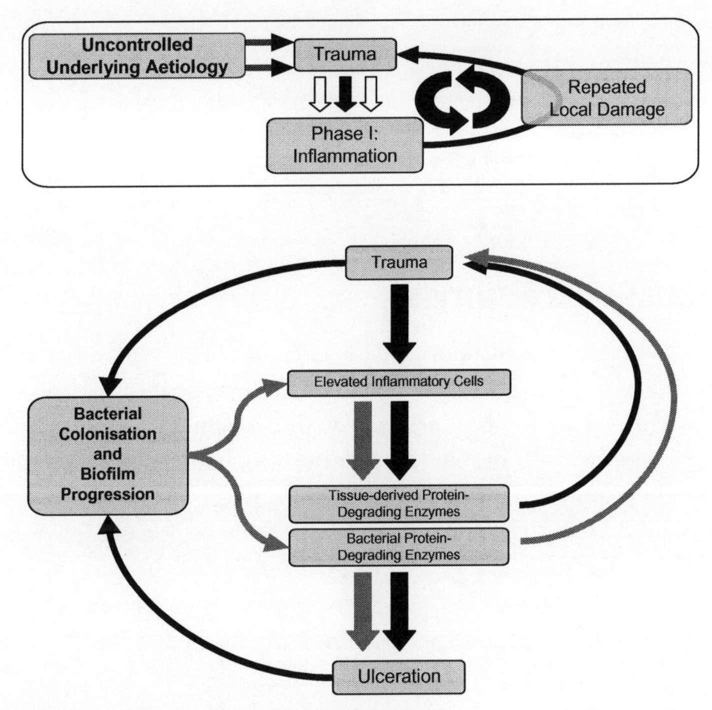

Figure 18.3 ***Bacterial involvement in the healing process.***

non-healing wounds allows for the establishment of populations of micro-organisms at the wound site.

The exact nature of the bacterial populations is of great medical interest, but one aspect of the biology of a micro-organism that is likely to be a major force in the maintenance of non-healing wounds is the production of microbial-specific protein-degrading enzymes (proteinases). Bacterial cells produce their own cocktail of proteinases that are useful for their basic physiological processes. However, bacterially derived enzymes together with the human-derived enzymes also contribute to the total proteinase levels within a contaminated wound site. In ulcers, where there is a sustained contamination of the wound site by micro-organisms, these proteinases themselves stimulate a further inflammatory response within the tissue, leading to—among other things—further proteinase production (Figure 18.3).

Therefore, the level of bacterial contamination of a wound may have profound effects on the pathology of non-healing wounds. In addition, if the bacterial population within the wound site can establish an environment that can help protect the community of micro-organisms from its environment (e.g., by the formation of a biofilm), then removal of this aspect of the pathology, which has both an indirect and direct affect on the persistence of the wound, becomes an important aspect of the patient's non-healing wound care and treatment.

Key Points

- Wound healing progresses through a series of overlapping stages in which a variety of cellular and matrix components act together to regain the integrity of damaged tissue.
- Recent studies have suggested that inflammation is the key phase in the wound-healing response.

MICROBIAL COMMUNITIES

The process of wound healing can be interrupted by a number of factors that include contamination with a foreign object or micro-organisms.[9] Inhibiting factors must be removed in order for healing of a wound to proceed normally.[10] When the inhibiting agent is a microbial biofilm, its identification and removal is fundamental for successful wound healing.[9]

Clinical biofilms have been identified macroscopically in non-healing human wounds as slough, which is often thick and yellow, or as a slimy and shiny layer on the surface of the wound bed.[11] Such observations can be confirmed using direct or indirect microscopic visualisation. Direct methods involve viewing the intact biofilm using confocal microscopy,[12] atomic force microscopy[13] or scanning electron microscopy (SEM).[14–16] SEM is a useful tool because it allows for a high resolution and a high magnification of the biofilm landscape.[17] Indirect identification involves manual removal of the biofilm by sharp debridement or by sonication, followed by subsequent procedures that quantify the removed biofilm.

In previous experiments, direct examination of wound tissue has demonstrated that biofilm-residing bacteria can be visualised as clusters of cells[18] encased by a visible layer of an extracellular polymeric substance (EPS).[14] Microscopic examination of wound exudates also can be used to confirm the presence of a bacterial biofilm.[19] The microscopic distribution of bacteria throughout wound tissue is suggestive of their phenotypic state. Evenly spaced bacteria indicate non-biofilm-forming isolates, while clusters of single species and multi-species bacteria support the biofilm diagnosis.[20] The significance of bacterial clustering has been previously reported in human wounds[15] and in equine wounds.[21]

A chronic wound environment may consist of devitalised sloughy tissue and a mixed community of aerobic and obligate anaerobic bacteria.[22] Anaerobes are found to be extremely prevalent in wounds[23] and work in synergy with aerobic bacteria. In fact combinations of anaerobic and aerobic bacteria have been shown to produce levels of sepsis or disease that could not be induced by individual species[24] in such bacterial species as *Prevotella melaninogenicus*, *Porphyromonas asaccharolytica* and *Peptostreptococcus micros*[25]; *Porphyromonas asaccharolytica* and *Klebsiella pneumoniae*[22,26]; *Escherichia coli* and

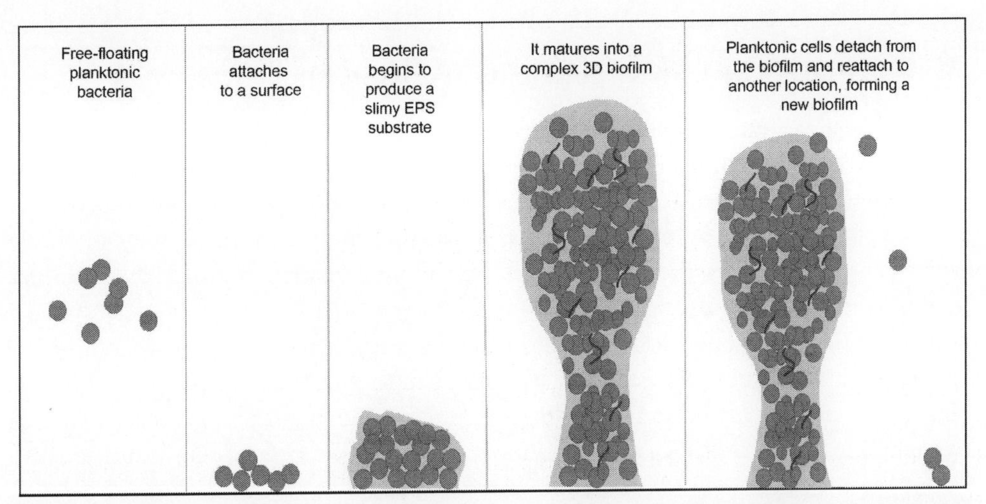

Figure 18.4 ***Diagram summarising the steps in the attachment and development of a bacterial biofilm on a surface.***

Bacteriodes fragilis[22,27]; and *Staphylococcus aureus*.[22] Because chronic wounds constitute a polymicrobial environment, treatment regimes need to be targeted at all micro-organisms not individual species and genera of bacteria, when a wound is critically colonised[5,6] and infected.[27]

Wound healing and infection are influenced by the relationship between the ability of bacteria to create a stable, prosperous community within a wound environment and the ability of the host to control the degree of bacterial colonisation. Since bacteria are able to rapidly form their own protective biofilm, the ability of the host to control these organisms is likely to decrease as the biofilm community develops. Within a stable, climax biofilm community, interactions between aerobic and anaerobic bacteria are likely to increase their net pathogenic effect, thus enhancing their potential to cause infection and delay healing.

Communities of bacteria have a direct effect on wound healing through the production of destructive enzymes and toxins, which indirectly promote a chronic inflammatory state (see Figure 18.4). A prolonged exposure to chronic wound bacteria leads to an ongoing inflammatory response; consequentially, numerous lytic enzymes could have a detrimental affect on the cellular processes involved in wound healing. Bacteria (e.g., *Pseudomonas aeruginosa*) release proteinases, which are known to affect growth factors and many other tissue proteins that are necessary for the wound-healing process.[28,29] Along with an increase in the bacterial bioburden of a wound comes the production of exudate. Exudate is known to degrade growth factors which will affect cell proliferation and wound healing.[30]

THE POTENTIAL SIGNIFICANCE OF BIOFILMS IN WOUNDS

Bacteria within a biofilm are reported to be less susceptible to the immune defence system when compared to planktonic bacteria. Consequently, a biofilm-associated wound infection may persevere and progress from an acute to a chronic infection. It has been shown that if antibodies are present at the biofilm interface the polymeric matrix generally renders them ineffective. The biofilm matrix is also able to inhibit chemotaxis and degranulation by polymorphonucleocytes (PMNs) and macrophages and depress the lymphoproliferative response of monocytes to polyclonal activators.

The persistent nature of the biofilm also is known to cause tissue damage. PMNs have been shown to be ineffective at engulfing bacteria in biofilms causing them to release large amounts of pro-inflammatory enzymes and cytokines. Consequently, this leads to chronic inflammation and destruction of nearby tissues.

While a moist wound environment is acknowledged to be able to support the development of bacterial biofilms, at present there is very little clinical evidence to substantiate this claim. Serralta and colleagues provided some superficial evidence that biofilms may form in wounds, suggesting their role in significantly enhancing inflammation and infection and ultimately delaying healing.[7]

Exogenous and endogenous sources of micro-organisms, in particular the nose, skin, mouth and gut, are involved in the advancement of biofilms and subsequent community interactions in wounds. It is likely that the early contaminants on a wound's surface are skin flora which adhere to the wound, proliferate and synthesize EPS. Although the host would initiate a normal immune response to try and establish and maintain some form of 'homoeostasis', the biofilm would remain firmly attached to the host's tissue. As a biofilm develops on a wound surface, a climax community would develop, implying stable associations and integrations of function between microbial populations in the wound bed.

At this stage the micro-organisms, while interfering with the wound-healing process, may not necessarily induce any clinical signs of infection. It may be fitting then to consider the use of a broad-spectrum topical antimicrobial agent for biofilm control. Without some form of control of microbial progression, a transition from an early 'healthy' biofilm to a 'pathogenic' wound biofilm may develop and ultimately lead to clinical infection. If the net pathogenic effect of the biofilm community exceeds the host's immune response, wound healing is likely to be compromised.

Treatment of Wounds

Once a wound has been clinically diagnosed as infected, the main method of treatment is to reduce the bacterial bioburden. This is generally achieved by using an array of antimicrobial compounds containing silver or iodine.[5,6,31] Routine usage of topical antibiotics is not justifiable for colonised or indeed infected wounds[32] because of the prevalence and possible enhancement of antibiotic-resistant bacteria.

Biofilms and Wounds

Biofilms are commonly associated with chronic wounds and are far more ubiquitous than thought previously. Biofilms are microbial accretions, which form readily on solid surfaces (either biological or non-biological), including the skin, and are surrounded by a self-produced extracellular matrix of hydrated EPS.[33] Indeed, the dermal wound provides micro-organisms with an ideal environment in which to flourish and the biofilm mode of growth offers protection from external environmental stress and inhibitors.[34] The biofilm offers a range of benefits to the residing micro-organisms, including greater protection from the host immune response, resistance from antimicrobial interventions and easier gene transfer between micro-organisms; this can lead to the sharing of beneficial characteristics such as increased virulence.[34,35]

Key Points

- A biofilm is defined as a community of micro-organisms irreversibly attached to a surface and encased in an EPS, with increased resistance to host cellular and chemical responses.[36]
- EPS consists of polysaccharides, proteins and nucleic acids making up around 80% of a biofilm.
- Biofilm bacteria are reported to be less susceptible to the immune defence system.

BIOFILM FORMATION

The development of a biofilm community occurs in several phases (refer to Figure 18.4), forming within a few hours, and is influenced by its environmental surroundings. Once in a biofilm state, the bacteria have great resistance to antimicrobials and prove to be difficult to remove in a clinical setting. Further advantages to being in the form of a biofilm are an increased metabolic efficiency, substrate accessibility, enhanced resistance to environmental stress and antimicrobials and an overall increased ability to cause infection and disease.[37]

Initial attachment: The first step in becoming a biofilm is the initial attachment of the cell-surface of free-floating planktonic micro-organisms to a surface; it can be reversible.

Growth and proliferation: Bacterial cells begin to divide and proliferate forming a micro-colony. As they colonise, the surface micro-organisms secrete and encase themselves with an extracellular polymeric substance.

Maturation: Once established in a biofilm environment, the micro-organisms secrete QS molecules, changing the biofilm phenotype related to growth and gene transcription.

Detachment: Planktonic bacteria detach from the mature biofilm, which are then able to disperse and develop another biofilm in a new location.

It is now understood that many chronic infections are the result of the biofilm mode of microbial growth which may delay the healing process. The National Institutes of Health estimates that pathogenic biofilms are involved in as many as 80% of all human infectious diseases. Therefore, the presence of a biofilm in a wound environment is expected to play a role in the delay of wound healing and in the development of chronic wounds (Figure 18.5). The first association of biofilms with wounds was when bacterial colonisation was discovered on sutures and staples removed from a healed surgical incision wound site.[39]

It is a known fact that micro-organisms contribute to wound infections; however, major controversy still exists as to the exact mechanism and their significance in non-healing wounds. One suggestion is that the density of the micro-organisms is the key factor in determining whether the wound heals or not. Another argument is that it is not the density, but the presence and interaction of specific micro-organisms within the wound, that delays healing. Others argue that micro-organisms have no role in the delay of wound healing; in addition:

- Bendy et al. first described the clinical significance of microbial density within wound healing. It was reported that healing of decubitus ulcers could only progress when the bacterial load was less than 106 CFU/ml (colony forming units/ml) of wound fluid.[40]
- It has been widely published that the majority of wounds are polymicrobial with the presence of both aerobes and anaerobes found in non-healing wounds.

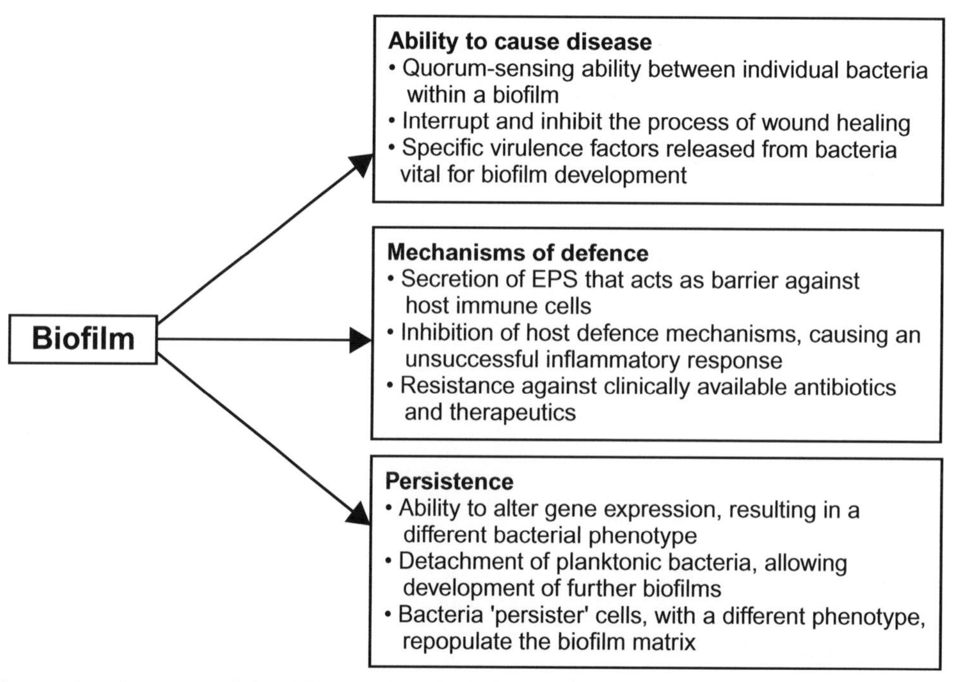

Figure 18.5 ***Summary of the different characteristics of a bacterial biofilm.*** *Source: Adapted from Seth et al.*[38]

- Evidence shows that multiple micro-organisms delay wound healing, and if four or more species are present, the healing outcome is likely to be very poor.

There is also controversy as to how much bacterial load affects wound healing. The first value described to clinically delay healing was colonisation of 106 CFU/ml.[40] Further studies investigating this argue that even smaller amounts of bacteria can cause a delay in healing. Work by Robson and Heggers, which has continued for more than 30 years, demonstrates that infection within acute and chronic wounds is present when bacterial load is 105 CFU/g of tissue.[41] Then, in more recent studies, it has been demonstrated by Breidenbach and Trager that the critical level of bacterial loading required to cause an infection is much lower at a value of 104 CFU/g of tissue.[42]

The microbial diversity of biofilms found in wounds is very high with a variety of species isolated from various wounds. One study showed that an average of 5.4 bacterial species were isolated from one chronic wound but the specific species varied between each wound.[43]

Biofilms in Acute and Surgical Wounds

Wounds are categorised simply as either acute or chronic. Acute wounds are caused by external damage to the skin and may include surgical wounds, minor cuts and abrasions, burns or more traumatic wounds such as lacerations or gunshot injuries. Acute wounds are expected to heal within a timely manner even though specific treatment for each may vary.

With maximum blood perfusion, surgical wounds heal quickly through delivery of oxygen, nutrients and immune cells to the site of injury, allowing minimal opportunity for colonisation of micro-organisms. Biofilms are much less likely to be observed in acute wounds compared to chronic wounds. Therefore, they do not have such a detrimental affect in the process of wound healing and acute wounds can successfully heal within a few weeks as expected.

Key Points

- Acute wounds are usually tissue injuries that heal completely and in the usual expected manner, within the expected time frame; typically, this is before 8 weeks have passed.
- Biofilms are less prominent in acute wounds.
- Biofilm formation may be a factor in the non-healing observed in chronic wounds.

Biofilms in Burn Wounds

A major complication with burn wounds is microbial infection, around 75% of associated burn deaths are as a result of developing sepsis related to the presence of a microbial or biofilm infection.[44] The surface of a wound in both partial-thickness and full-thickness burns provides a favourable environment for attachment of micro-organisms and development of biofilms. It is a protein-rich environment with avascular necrotic tissue providing bacteria with all the nutrients it requires to form a biofilm.

Kennedy and colleagues also found evidence of biofilm formation in burn wound biopsies using light and electron microscopy. Microbiological cultures revealed that these biopsies contained a range of micro-organisms, including both Gram-positive and Gram-negative bacteria.[45] Samples taken from ulcerated areas showed widespread presence of neutrophils and evidence of microbial invasion. The presence of carbohydrates also was found to be associated with those areas with microbial presence, indicating the presence of EPSs and the development of biofilms.[45]

Furthermore, burn wound specimens taken as early as 7 days post-injury were shown to be positive for microbial invasion.[45] Gram-positive bacteria located deep within the sweat glands and hair follicles that survive the initial burn injury will have colonised the wound surface within 48 hours, unless antimicrobial agents have been used along with primary treatments. Within 7 days burn wounds will be colonised with other micro-organisms (see Table 18.1). Studies have recently confirmed the implications of biofilms in burn wounds. Experimental studies of partial-thickness burn wounds in an animal model showed the development of biofilms within 48 to 72 hours. Human strains of *P. aeruginosa* isolated from burn wounds can develop into mature biofilms within 10 hours *in vitro*.[46]

Biofilms in Chronic Wounds

A chronic wound is defined as one that does not progress through the usual phases of healing and therefore fails to heal in a timely manner. The majority of clinicians consider a wound to be chronic when it has not healed within six weeks and shows signs of prolonged inflammation, a defective wound matrix and failure of re-epithelialisation. Chronic wounds, unlike acute wounds, in some of the worst cases can persist for years. Chronic wounds affect up to 2% of the UK population and are a significant burden to the patient in addition to the escalation of healthcare costs. Chronic wounds are divided into the following major categories:

- Venous leg ulcers
- Diabetic foot ulcers
- Pressure ulcers

Table 18.1 The Most Prevalent Micro-organisms Isolated from Burn Wounds

Aerobes	Anaerobes
P. aeuroginsa	*Peptostreptococcus* spp.
S. aureus	*Bacteroides* spp.
Klebsiella spp.	*Propionibacterium acnes*
Eneterococcus spp.	
Candida spp.	

Direct evidence of the presence of bacterial biofilms within chronic wound tissue has been presented in both humans and animals. The 2008 James et al. study found that 60% of chronic wound specimens from human subjects showed evidence of biofilm formation via molecular analyses compared with only 6% of acute wound specimens. Furthermore, molecular analyses of chronic wound specimens revealed polymicrobial communities.[15] The chronic wounds took more than 3 months to heal effectively whereas all the acute wounds had fully healed within 3 weeks. This proves that not only are biofilms present in chronic wounds but it also suggests that they play a significant role in impairment of the healing process.

Westgate and colleagues also presented evidence of bacterial biofilms in equine chronic wounds; stained tissue samples showed evidence of biofilms within 8 out of 13. Furthermore, those bacterial isolates obtained from chronic and acute wounds showed significantly higher biofilm-forming potential (BFP) than those isolates obtained from uninjured skin ($P<0.05$).[10] With the BFP, typically, being described as the average biofilm concentration within a given area, after an allotted time.

Experimental models of chronic infected wounds have also provided evidence of biofilm formation. Schaber et al. demonstrated that *Pseudomonas aeruginosa* is capable of forming biofilms within 8 hours of infection in thermally injured mice.[47] The authors noted that SEM images revealed matrix-like structures and/or 'bacterial flocs' in association with the *P. aeruginosa* aggregates (Figure 18.6), which provides evidence of biofilm formation.[47]

Porcine partial-thickness wounds inoculated with *Staphylococcus aureus* in an experimental wound model showed evidence of biofilm formation 48 hours after inoculation and occlusion.[14] Furthermore, this study revealed that the biofilm-like communities formed in these wounds demonstrated increased antimicrobial resistance compared with their planktonic phenotype.[14] Interestingly, polymorphonuclear cells were observed on the surface of the biofilm, but not within the biofilm. This suggests that neutrophils were unable to penetrate the EPS biofilm matrix, thus offering the bacteria residing within the biofilm a survival advantage over their planktonic counterparts.[14,35]

Serralta et al. used a porcine wound model to study formation of biofilms; partial-thickness wounds were created on three pigs and were challenged with *P.aeruginosa* then covered. Wounds were washed vigorously after 72 hours and wound cultures were stained with Congo red dye to look for EPS.[7] A thick EPS was detected surrounding bacteria, indicating a biofilm.[7]

Biofilms have been shown to impair cutaneous wound healing in a number of ways. One way in which bacteria can cause infection is through the evasion of the host's immune defence system; biofilms offer residing cells protection from such attack. Phagocytic cells are less effective at destroying bacteria within a biofilm.[48] The biofilm provides protection from antibodies complement and the cells of the immune system.[48] There are a number of ways in which the biofilm mode of growth provides

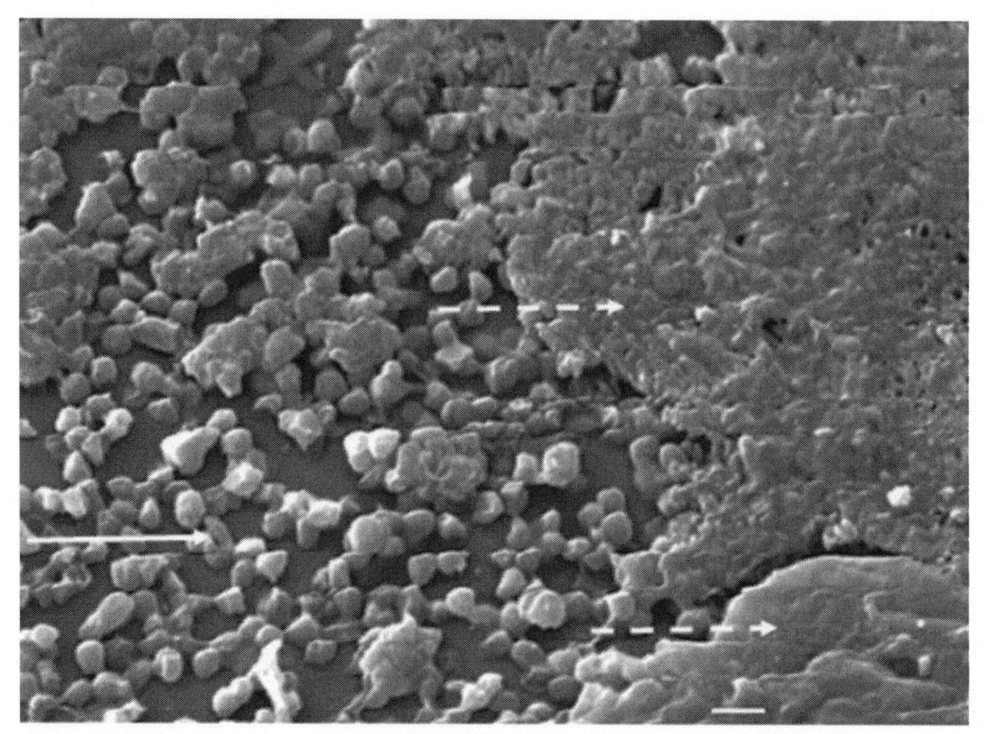

Figure 18.6 ***Scanning electron micrograph of a*** **S. aureus** ***reference strain ATCC 25923.*** This image shows evidence of a slime layer. Bacteria were attached to the surface (*straight arrows*). EPS material covering bacteria (*dashed arrows*) (magnification = × 6000; bar marker = 1 micron).

a defence mechanism from the host's immune system. The biofilm matrix has been shown to interfere with phagocytic activity[49] and the stimulation of the complement system also has been shown to be affected by the presence of a biofilm.[50] Leid and colleagues also demonstrated that leucoytes penetrating the biofilm are unable to engulf the residing bacteria in an *in vitro S.aureus* biofilm model.[51]

Antibiotic resistance is a characteristic noted in biofilm infections. This resistance appears to be a consequence of the close cell–cell contact within the biofilm, allowing for the effective transfer of plasmids which can encode antimicrobial resistance.[33] Some studies have suggested that the biofilm prevents the penetration of the antimicrobial agent to the cells within the biofilm matrix.[52,53] The slow microbial growth observed within a biofilm has also been correlated with a reduction in the efficacy of antimicrobial treatment.[54]

Biofilms have been shown to impair wound healing by delaying re-epithelialisation, as demonstrated by the Schierle team.[19] Cutaneous splinted mouse wounds were inoculated with *S. aureus* and *S. epidermidis* and biofilms formed after approximately three days post-inoculation. Quantification of the epithelial gap present in the wounds revealed that re-epithelialisation was disrupted, with the epithelial gaps significantly larger in the wounds colonised by bacterial biofilms compared to uninfected control wounds.[19] Furthermore, with the use of both exogenous and endogenous disruption of biofilm signalling, the authors demonstrated that the inhibition of re-epithelialisation was a result of biofilm formation.[19]

Unfortunately, there is no simple method to quickly identify the presence of a biofilm within a clinical setting. The identification of bacterial aggregates combined with an infection in a wound can be made and the most suitable course of treatment selected. Even though research into new treatments for cutaneous wound biofilm infection is still in its infancy, a number of treatment modalities have been postulated, including the use of antimicrobial solutions and wound dressings.

Key Points

- The three catergories of chronic wounds are: venous leg ulcers, diabetic foot ulcers and pressure ulcers.
- If a wound has not healed after 12 weeks, it is classified as chronic.
- Identification of biofilms in chronic wounds is difficult.

Debridement

Physical intervention in the form of debridement is one of the most effective methods to reduce the physical presence of bacterial biofilms in a wound. Debridement of necrotic tissue and wound exudate strategies range from the use of simple dressings to provide a moist wound milieu conducive to accelerated and progressive healing, to the more invasive method of surgical removal of the necrotic tissue. Necrotic tissue provides a favourable environment for microbial growth; therefore, its removal is vital to reducing bacterial load within a wound. Removal of the contaminated tissue exposes the healthier tissue and aids wound healing.

Antimicrobials

Various antimicrobial agents have different modes of action, but their overall effects on the bacteria are similar. They either exert a microbiostatic effect by stopping or reducing cell division and inhibiting growth of the bacteria, or a microbicidal effect by directly killing the micro-organisms. The most common antimicrobials used in wound care are honey, silver, iodine and related iodophores—that is, povidone iodine (PI) and cadexomer iodine (CI). These are often incorporated into dressings for the killing of bacteria and to facilitate wound healing.

Antimicrobial Effect on Biofilms

Percival and others assessed the efficacy of a silver wound dressing on biofilms grown *in vitro*. Silver-containing wound dressing samples were applied to 24-hour biofilms composed of *P. aeruginosa*, *Enterobacter cloacae*, *S. aureus* or a mixed bacterial biofilm; then bacterial viability was evaluated.[55] Visualisation of the biofilms following treatment with the silver dressing samples also were conducted using scanning confocal microscopy. Results revealed that following a 48-hour treatment period, total bacterial kill was achieved in both the tested mono- and poly-microbial biofilms.[55]

An earlier study also demonstrated that silver ions can act to destabilise the biofilm matrix by reducing the intermolecular forces within the EPS.[13] Hill and colleagues developed an *in vitro* model of chronic wound biofilms to test the efficacy of silver-containing and iodine-containing wound dressings.[56] The authors of this study found that the iodine-based dressing completely disrupted fully established 7-day-old mixed bacterial biofilms; whereas, the silver-containing wound dressings had no effect on 7-day-old biofilms.[56]

Light microscopy and SEM has been used previously to identify the susceptibility of mixed species biofilms to a number of antibiotic commercial products, including iodine dressings. Results showed that a 1% PI solution was capable of biofilm disruption and was superior to alternative silver-based dressings; however, this effect was minimal.[56]

In contrast, the PI-containing dressings (Inadine® and Iodoflex®) demonstrated complete and efficient destruction of bacteria in a constant depth film fermentor (CDFF)-generated biofilm model. In comparison, an iodine-impregnated dressing (Betadine® cream/gauze) showed no significant reduction in *P. aeruginosa* cell counts and only a slight reduction of the *S. aureus* bacteria counts. This study used a sophisticated *in vitro* biofilm model which attempted to closely mimic chronic wound biofilms.[56]

Brown and colleagues[57] undertook a study to investigate the relationship between the glycocalyx matrix secreted by biofilm cells and the resistance of *P. aeruginosa* biofilms to PI. Transmission electron microscopy (TEM) showed that, while the amount of glycocalyx material per cell was not significantly different between biofilm and planktonic cultures, the distribution of glycocalyx around bacterial cells differed between cells residing in the biofilm and planktonic states. Addition of alginic acid to the planktonic cells resulted in a slight increase in resistance to PI. Interestingly, it was concluded that PI resistance of bacteria in the biofilm state is because of the protective layering of cells within this glycocalyx matrix, increasing the time needed to detach cells required for contact with iodine in the inner biofilm layer.[57]

A strategy aimed at addressing the problems associated with biofilm infections in wounds has been termed 'biofilm-based wound care' (BBWC), which incorporates aggressive wound debridement alongside a number of anti-biofilm interventions such as QS inhibitors and the use of substances that interfere with the interactions within the EPS and metabolism or block bacterial attachment.[35] The use of such anti-biofilm agents in chronic wound management is thought to be more beneficial than the use of antimicrobials because they are less cytotoxic.

The following are the potential anti-biofilm agents:

- Lactoferrin
- Xylitol
- Gallium
- Dispersin B
- Ethylenediaminetetraacetic acid (EDTA)
- Acetyl salicylic acid
- Honey

Honey, at a low concentration of 0.5% (v/v) has been shown by Lee and colleagues to significantly reduced biofilm formation in *E. coli.*[58] Interestingly, transcriptome analyses showed that honey significantly repressed QS and virulence genes prevalent in biofilm associated bacteria.

The effectiveness of such interventions, however, requires further investigation. Currently not one single strategy has proved to be effective in suppressing and removing the entire biofilm from a wound. Clinically, multiple and simultaneous strategies must be used to successfully manage and prevent a biofilm infection.

Key Points

- Antimicrobial agents either exert a microbiostatic effect by stopping or reducing cell division or a microbicidal effect by killing micro-organisms.
- Both silver and iodine are to a degree effective on some biofilms.
- A multi-factorial approach is still the only way to address biofilms.

CONCLUSION

Open wounds constitute an environment which is moist and one that is conducive to biofilm formation. Evidence of fibronectin, found particularly on the surface of venous leg ulcers, are known to be an adhesive site for bacteria, keratinocytes and fibroblasts. By aiding bacterial attachment, biofilm formation in a wound environment will commence. As the biofilm develops, it is probable that simply the presence of the biofilm and the extracellular components released from the sessile bacteria will have the capacity to damage the surrounding tissue.

In addition, because the phagocytic cells released from the host are not successful at fully penetrating the biofilm evident in the wound bed, the released enzymes during the phagocytic process will lead to extensive tissue insult.[59] In conjunction with this, if there is synergy between the human proteases released during wound healing and bacterial proteases further tissue breakdown would be inevitable and wound healing will be hindered.

Non-healing and infected wounds are microbiologically diverse. This diversity has a number of advantages to micro-organisms because it is likely to help microbial interactions and enhance their survival. However, these interactions are likely to have detrimental effects on the host. Wound infections due to these colonising bacteria are diagnosed on the basis of clinical signs.[60] The objective of infection control and, in particular, wound care is to minimise the risk of the transmission of micro-organisms and to prevent patients from acquiring an infection. However, despite an array of infection-control procedures, this often is not achievable. Failing this, to help reduce the risk of the wound becoming infected, the microbial load in the wound bed should be significantly reduced to allow the host immune system to prevail.

By reducing the number of micro-organisms, their related interactions and pathogenicity also can be reduced. Through the correct use of currently available antimicrobials, the bioburden can be reduced significantly, and therefore aid in wound healing. The development of new treatment interventions aimed at biofilm infections in chronic cutaneous wounds, a focal point for future research, provides much promise.

REFERENCES

1. Wu L, Brucker M, Gruskin E, Roth SI, Mustoe TA. Differential effects of PDGF-BB in accelerating wound healing on aged versus young animals: the impact of tissue hypoxia. *Plast Reconstr Surg* 1997;**99**:815.
2. Calvin M. Cutaneous wound repair. *Wounds* 1998;**1**:12–32.
3. NINSS *Surveillance of surgical site infection in English hospitals: a national surveillance and quality improvement programme.* Public Health Laboratory Service; 2002.
4. Plowman RE. The socioeconomic burden of hospital acquired infection. *Euro Surveill* 2000;**5**:49–50.
5. Percival SL, Bowler PG. Biofilms and their potential role in wound healing. *Wounds* 2004;**16**(7):234–40.
6. Percival SL, Bowler PG. Bacterial interactions in wounds and potential effects on wound healing. *World Wide Wounds* 2004; July.
7. Serralta VW, Harrison-Balestra C, Cazzaniga AL, Davis SC, Mertz PM. Lifestyles of bacteria in wounds: presence of biofilms. *Wounds* 2001;**13**(1):29–34.
8. Bello YM, Falabella AF, Cazzaniga AL, Harrison-Balestra C, Mertz PM. Are biofilm present in human chronic wounds? Presented at the symposium on advanced wound care and medical research forum on wound repair, Las Vegas, April 30–May 3; 2001.
9. Edwards R, Harding KG. Bacteria and wound healing. *Actual Pharm Biol Cl* 2004;**17**:91–6.
10. Westgate SJ, Percival SL, Knottenbelt DC, Clegg PD, Cochrane CA. Microbiology of equine wounds and evidence of bacterial biofilms. *Vet Microbiol* 2011;**150**(1–2):152–9.
11. Rondas AALM, Schols JMGA, Stobberingh EE, Price PE. Definition of infection in chronic wounds by Dutch nursing home physicians. *Int Wound J* 2009;**6**:267–74.
12. Akiyama H, Oono T, Saito M, Iwatsuki K. Assessment of cadexomer iodine against *Staphylococcus aureus* biofilm *in vivo* and *in vitro* using confocal laser scanning microscopy. *J Dermatol* 2004;**31**:529–34.
13. Chaw KC, Manimaran M, Tay FEH. Role of silver ions in destabilization of intermolecular adhesion forces measured by atomic force microscopy in *Staphylococcus epidermidis* biofilms. *Antimicrob Agents Chemother* 2005;**49**(12):4853–9.
14. Davis SC, Ricotti C, Cazzaniga A, Welsh E, Eaglstein W, Mertz PM. Microscopic and physiologic evidence for biofilm-associated wound colonization *in vivo*. *Wound Repair Regen* 2008;**16**:23–9.
15. James GA, Swogger E, Wolcott R, DeLancey Pulcini E, Secor P, Sestrich J, et al. Biofilms in chronic wounds. *Wound Repair Regen* 2008;16:37–44.
16. Sun Y, Dowd SE, Smith E, Rhoads DD, Wolcott RD. *In vitro* multispecies Lubbock chronic wound biofilm model. *Wound Repair Regen* 2008;**16**:805–13.

17. Schaudinn C, Carr G, Gorur A, Jaramillo D, Costerton JW, Webster P. Imaging of endodontic biofilms by combined microscopy (FISH/cLSM-SEM). *J Microsc* 2009;**235**:124–7.
18. Lewandowski L, Stoodley AP, Roe F. Internal mass transport in heterogeneous biofilms: recent advances. *Corrosion,* paper no. 222. NACE; 1995.
19. Schierle CF, De La Garza M, Mustoe TA, Galiano RD. Staphylococcal biofilms impair wound healing by delaying reepithelialization in a murine cutaneous wound model. *Wound Repair Regen* 2009;**17**:354–9.
20. Fazli M, Bjarnsholt T, Kirketerp-Moller K, Jorgensen B, Andersen AS, Krogfelt KA, et al. Nonrandom distribution of *Pseudomonas aeruginosa* and *Staphylococcus aureus* in chronic wounds. *J Clin Microbiol* 2009;**47**:4084–9.
21. Cochrane CA, Freeman K, Woods E, Welsby S, Percival SL. Biofilm evidence and the microbial diversity of horse wounds. *Can J Microbiol* 2009;**55**:197–202.
22. Rhoads DD, Cox SB, Rees EJ, Sun Y, Wolcott RD. Clinical identification of bacteria in human chronic wound infections: culturing vs. 16S ribosomal DNA sequencing. *BMC Infect Dis* 2012;**12**:321.
23. Wolcott RD, Gontcharova V, Sun Y, Zischakau A, Dowd SE. Bacterial diversity in surgical site infections: not just aerobic cocci any more. *J Wound Care* 2009;**18**(8):317–23.
24. Brook I. Role of encapsulated anaerobic bacteria in synergistic infections. *CRC Crit Rev Microbiol* 1987;**14**:171–93.
25. Sundqvist GK, Eckerbom MI, Larsson AP, Sjogren UT. Capacity of anaerobic bacteria from necrotic dental pulps to induce purulent infections. *Infect Immun* 1979;**25**:685–93.
26. Mayrand D, McBride BC. Exological relationships of bacteria involved in a simple, mixed anaerobic infection. *Infect Immun* 1980;**27**:44–50.
27. Hansson C, Hoborn J, Moller A, Swanbeck G. The microbial flora in venous leg ulcers without clinical signs of infection. Repeated culture using a validated standardised microbiological technique. *Acta Derm Venereol* 1995;**75**:24–30.
28. Steed DL, Donohoe D, Webster MW, Lindsley L. Effect of extensive debridement and treatment on the healing of diabetic foot ulcers. Diabetic Ulcer Study Group. *J Am Coll Surg* 1996;**183**:61–4.
29. Travis J, Potempa J, Maeda H. Are bacterial proteinases pathogenic factors? *Trends Microbiol* 1995;**3**: 405–7.
30. Falanga V, Grinnell F, Gilchrest B, Maddox YT, Moshell A. Workshop on the pathogenesis of chronic wounds. *J Invest Dermatol* 1994;**102**:125–7.
31. Tonks A, Cooper RA, Price AJ, Molan PC, Jones KP. Stimulation of TNF-alpha release in monocytes by honey. *Cytokine* 2001;**14**:240–2.
32. Bergman A, Yanai J, Weiss J, Bell D, David MP. Acceleration of wound healing by topical application of honey. An animal model. *Am J Surg* 1983;**145**:374–6.
33. Davis SC, Martinez L, Kirsner R. The diabetic foot: the importance of biofilms and wound bed preparation. *Curr Diab Rep* 2006;**6**(6):439–45.
34. Percival S, Bowler P. Understanding the effects of bacterial communities and biofilms on wound healing. *World Wide Wounds* 2004; June.
35. Thomson CH. Biofilms: do they affect wound healing? *Int Wound J* 2011;**8**:63–7.
36. Percival SL, Hill KE, Williams DW, Hooper SJ, Thomas DW, Costerton JW. Biofilms: possible strategies for suppression in chronic wounds. *Nurs Stand* 2012;**23**(32):64–72.
37. Percival S, Hill KE. A review of the scientific evidence for biofilms in wounds. *Wound Repair Regen* 2012;**20**:647–57.
38. Seth AK, Geringer MR. *In vivo* modelling of biofilm-infected wounds: a review. *J Surg Res* 2012; **178**:330–8.
39. Gristina AG, Price JL. Bacterial colonisation of percutaneous sutures. *Surgery* 1985;**98**(1):12–19.
40. Bendy RH, Nuccio PA. Relationship of quantitative wound bacterial counts to healing of decubiti: effect of topical gentamicin. *Antimicrob Agents Chemother* 1964;**4**:147–55.
41. Robson MC, Heggers JP. Delayed wound closures based on bacterial counts. *J Surg Oncol* 1970; **2**:379–83.
42. Breidenbach WC, Trager S. Quantitative culture technique and infection in complex wounds of the extremities closed with free flaps. *Plast Reconstr Surg* 1995;**95**:860–5.
43. Thomsen T, Aasholm M. The bacteriology of chronic venous leg ulcer examined by culture-independent molecular methods. *Wound Repair Regen* 2010;**18**:38–49.

44. Church D, Elsayed S. Burn wound infections. *Clin Microbiol Rev* 2006;**19**(2):403.
45. Kennedy P, Brammah S, Wills E. Burns, biofilm and a new appraisal of burn wound sepsis. *Burns* 2010;**36**:49–56.
46. Harrison-Balestra C, Cazzaniga AL. A wound-isolated *Pseudomonas aeruginosa* grows a biofilm in vitro within 10 hours and is visualized by light microscopy. *Dermatol Surg* 2003;**29**:631–5.
47. Schaber JA, Triffo WJ, Suh SJ, Oliver JW, Hastert MC, Griswold JA, et al. *Pseudomonas aeruginosa* forms biofilms in acute infection independent of cell-to-cell signalling. *Infect Immun* 2007;**75**(8):3715–21.
48. Kharazmi A. Mechanisms involved in the evasion of the host defence by *Pseudomonas aeruginosa. Immunol Lett* 1991;**30**(2):201–5.
49. Shiau AL, Wu CL. The inhibitory effect of *Staphylococcus epidermidis* slime on the phagocytosis of murine peritoneal macrophages is interferon-independent. *Microbiol Immunol* 1998;**42**(1):33–40.
50. Jensen ET, Kharazmi A, Garred P, Kronborg G, Fomsgaard A, Mollnes TE, et al. Complement activation by *Pseudomonas aeruginosa* biofilms. *Microb Pathog* 1993;**15**(5):377–88.
51. Leid JG, Shirtliff ME, Costerton JW, Stoodley P. Human leukocytes adhere to, penetrate, and respond to *Staphylococcus aureus* biofilms. *Infect Immun* 2002;**70**(11):6339–45.
52. De Beer D, Srinivasan R, Stewart PS. Direct measurement of chlorine penetration into biofilms during disinfection. *Appl Environ Microbiol* 1994;**60**(12):4339–44.
53. Suci PA, Mittelman MW, Yu FP, Geesey GG. Investigation of ciprofloxacin penetration into *Pseudomonas aeruginosa* biofilms. *Antimicrob Agents Chemother* 1994;**38**(9):2125–33.
54. Tuomanen E, Cozens R, Tosch W, Zak O, Tomasz A. The rate of killing of *Escherichia coli* by β-lactam antibiotics is strictly proportional to the rate of bacterial growth. *J Gen Microbiol* 1986; **132**(5):1297–304.
55. Percival SL, Bowler P, Woods EJ. Assessing the effect of an antimicrobial wound dressing on biofilms. *Wound Repair Regen* 2008;**16**:52–7.
56. Hill KE, Malic S, McKee R, Rennison T, Harding KG, Williams DW, et al. An *in vitro* model of chronic wound biofilms to test wound dressings and assess antimicrobial susceptibilities. *J Antimicrob Chemother* 2010;**65**(6):1195–206.
57. Brown ML, Aldrich HC, Gauthier JJ. Relationship between glycocalyx and povidone-iodine resistance in *Pseudomonas aeruginosa* (ATCC 27853) biofilms. *Appl Environ Microbiol* 1995;**61**(1):187–93.
58. Lee J, Park J, Kim J, Neupane GP, Cho MH, Lee C, et al. Low concentrations of honey reduce biofilm formation, quorum sensing, and virulence in *Escherichia coli* O157:H7. *Biofouling* 2011;**27**(10):1095–104.
59. Holden-Lund C. Effects of relaxation with guided imagery on surgical stress and wound healing. *Res Nurs Health* 1988;**11**:235–44.
60. Cutting KF, White RJ, Mahoney P, Harding K. Clinical identification of wound infection: a Delphi approach. *Identifying criteria for wound infection*. London: EWMA Position Document, MEP; 2005.
61. Trengove NJ, Bielefeldt-Ohmann H, Stacey MC. Mitogenic activity and cytokine levels in non-healing and healing chronic leg ulcers. *Wound Repair Regen* 2000;**8**:10.
62. Ng CS, Wan S, Yim AP. Pulmonary ischaemic are perfusion injury: role of apoptosis. *Eur Respir J* 2005;**25**:356.
63. Trengove NJ, Stacey MC, Macauley S, et al. Analysis of the acute and chronic wound environments: the role of proteases and their inhibitors. *Wound Repair Regen* 1999;**7**:442.
64. Rotstein OD, Kao J. The spectrum of *Escherichia coli—bacteroides fragilis* pathogenic synergy in an intraabdominal infection model. *Can J Microbiol* 1988;**34**:352–7.

INDEX

Note: Page numbers followed by "*f*", "*t*", and "*b*" refers to figures, tables, and boxes, respectively.

B

C

D

I

R

S

T

U

V

W

X

Z

Printed and bound by CPI Group (UK) Ltd, Croydon, CR0 4YY
11/06/2025
01899189-0014